New Trends in Statistical Physics of Complex Systems

New Trends in Statistical Physics of Complex Systems

Special Issue Editor

Antonio Maria Scarfone

MDPI • Basel • Beijing • Wuhan • Barcelona • Belgrade

Special Issue Editor
Antonio Maria Scarfone
Istituto dei Sistemi Complessi–CNR
c/o Politecnico di Torino
Italy

Editorial Office
MDPI
St. Alban-Anlage 66
4052 Basel, Switzerland

This is a reprint of articles from the Special Issue published online in the open access journal *Entropy* (ISSN 1099-4300) from 2017 to 2018 (available at: https://www.mdpi.com/journal/entropy/special_issues/Statistical_Complex_Systems)

For citation purposes, cite each article independently as indicated on the article page online and as indicated below:

LastName, A.A.; LastName, B.B.; LastName, C.C. Article Title. *Journal Name* **Year**, *Article Number*, Page Range.

ISBN 978-3-03897-469-7 (Pbk)
ISBN 978-3-03897-470-3 (PDF)

Contents

About the Special Issue Editor

Antonio Maria Scarfone is a theoretical physics researcher working mainly in the field of statistical mechanics. He was graduated from the University of Torino in 1996 and took a Ph.D. in Physics at Politecnico of Torino in 2000. He has been a postdoctoral researcher at INFM (2000–2002) and got a research fellowship at the University of Cagliari in 2003. Since 2004, he has carried out research activities at CNR-ISC based in Politecnico of Torino. Currently, he has written approximately 100 scientific papers on international journals on the topics of statistical mechanics, kinetic theory, geometry information, non-linear classical and quantum dynamics, and non-commutative algebras. He is Editor-in-Chief of the Section *Statistical Physics* of *Entropy*, a member of the editorial board of *Advances in Mathematical Physics*, and belongs to the advisory panel of *Journal of Physics A*. Finally, he is one of the organizers of the SigmaPhi international conference series that has been held every three years starting from 2005.

entropy

MDPI

Editorial

New Trends in Statistical Physics of Complex Systems

Antonio M. Scarfone

Istituto dei Sistemi Complessi, Consiglio Nazionale delle Ricerche (ISC-CNR), c/o DISAT, Politecnico di Torino, Corso Duca degli Abruzzi 24, I-10129 Torino, Italy; antonio.scarfone@polito.it

Received: 23 November 2018; Accepted: 25 November 2018; Published: 27 November 2018

Keywords: generalized statistical mechanics; information theory; anomalous diffusion; stochastic processes; collective phenomena; disordered systems

---

A challenging frontier in physics concerns the study of complex and disordered systems. They are characterized by an elevated level of interconnection and interaction between the parts, with a richer global dynamic, which gives rise to the collective emerging behaviour of the entire system that are no longer recognized in the properties of the single individual entities. In this scenario, the methods of statistical physics, in understanding the properties of complex systems, have proved to be very promising.

In fact, statistical physics is so general that it still holds, in a much wider context, than that, on which the original theory was developed. Consequently, despite its largely recognized success in physics, considerable efforts have been made to extend the formalism of statistical physics, beyond its original application. In this way an increasing amount of physical and physical-like systems are now well-studied by using the standard tools of this theory.

Thus, in the last decades, we assisted in an intense research activity that has modified our comprehension of statistical physics, extending and renewing its applicability, considerably. Important developments, relating equilibrium and nonequilibrium statistical mechanics, kinetic theory, information theory, and others, have produced a new understanding of the properties of complex systems that requires, in many cases, the extension of the statistical physics theory, beyond its standard formalism.

In this volume, we have selected and invited a limited number of papers on several topical trends in the statistical physics of complex systems, based on the talks presented during the international conference SigmaPhi2017 held in Corfu, from the 10th until the 14th of July 2017 (http://sigmaphisrv.polito.it/), which gathered together, more than two hundred and fifty international scientists, working in all areas of physics and interdisciplinary applications in non-physical systems.

This special issue collects twelve regular research articles, which underwent the standard rigorous editorial process of the Entropy journal.

The first three articles [1–3] deal with non-extensive statistical mechanics and power-law distributions.

The paper "Equilibrium States in Two-Temperature Systems" by E.M.F. Curado and F.D. Nobre, proposed a non-linear (power-like) Fokker-Planck equation, describing a kinetic process governing the evolution toward the equilibrium of a system, whose underlying dynamics was characterized by two different diffusional mechanisms. The diffusion term in the kinetic equation is related to the sum of the two entropic forms, each of them associated to a given diffusion process. The corresponding diffusion coefficients introduce two distinct characteristic temperatures of the system, a situation that usually appears in nonequilibrium statistical mechanics. A physical application to type-II superconducting vortices was discussed as an illustrative example.

In the paper "Information-Length Scaling in a Generalized One-Dimensional Lloyd's Model" by J.A. Méndez-Bermúdez and R. Aguilar-Sánchez, a numerical study of the generalized one-dimensional Lloyd's model was presented. This model described a class of disordered systems characterized by a random variable whose density distribution function exhibits an asymptotically-slow decaying tail.

It was, in a sense, related to several disordered models that have been studied in the literature, like the bended random matrix model, the kicked-rotator model, the multiplex and multilayer random networks, and others, so that the result obtained in this paper could have a wide spectrum of applicability.

The paper "Oscillations in Multiparticle Production Processes" by G. Wilk and Z. Włodarczyk, dealt with the study of the power-law and quasi-power-law distributions, characterized by log-periodic oscillations that are ubiquitously observed in many different branches of science. Based on the nonextensive statistical mechanics, the authors presented a study of the available experimental data from the Large Hadron Collider (LHC) experiments, obtained from the multiparticle production processes, at high energies. The analysis concerned: (i) The log-periodic oscillations pattern decorating the power-like Tsallis distributions of the large transverse momenta spectra, and (ii) the oscillations of some coefficients in the recurrence relation defining the multiplicity of distributions.

The information theory is covered by the next three articles [4–6].

In the paper "Minimising the Kullback-Leibler Divergence for Model Selection in Distributed Nonlinear Systems" by O.M. Cliff, M. Prokopenko, and R. Fitch, the authors discussed, in the context of a nonlinear dynamical system, a possible decomposition of the Kullback-Leibler divergence, a fundamental quantity in information theory, in two information-measures, namely collective transfer entropy and stochastic interaction. This result has been derived by using rigorous methods, based on differential topology and, as highlighted in the paper, the approach proposed has potential applications beyond the study of a complex system theory, becoming relevant in a variety of contexts of artificial intelligence, like in machine learning.

The next paper "Conformal Flattening for Deformed Information Geometries on the Probability Simplex" by A. Ohara, dealt with the study of two important notions that play a relevant role in information geometry in characterizing statistical models with generalized exponential functions. The first one, the dual flatness, produced fruitful geometrical structures like the existence of canonical coordinate systems, a pair of conjugate potential functions and assures the existence of a canonical divergence. The second one, was concerned with the invariance of a geometric structure, a crucially valuable concept in developing mathematical statistics, which holds only when the statistical manifold has a special triplet given by the Riemannian metric and a pair of mutually dual-affine connections.

Always within the information geometry, the paper "The Volume of Two-Qubit States by Information Geometry" by M. Rexiti, D. Felice and S. Mancini, studied the volume of the set of two-qubits states with maximally disordered subsystems, by considering their phase space representation, in terms of probability distribution functions and by applying the classical Fisher information metric. The results obtained were compared with those derived by using two different versions of the quantum Fisher metric—the Helstrom and the Wigner–Yanase-like metrics. Although the absolute values of volumes of separable and entangled states differ in the two approaches, it was shown that their ratios are comparable, supporting the conclusion that classical Fisher information is able to capture the features of the volume of quantum states.

The last topic collects six articles and deals with collective phenomena, respectively, in condensed matter [7–10] and nuclear matter [11,12].

In the paper "Collective Motion of Repulsive Brownian Particles in Single-File Diffusion with and without Overtaking" by T. Ooshida, S. Goto and M. Otsuki, a system of repulsive Brownian particles confined in a (quasi-)one-dimensional channel, whose sub-diffusive behavior is known as the single-file diffusion, was studied. Collective dynamics was illustrated by the calculations of the two-particle displacement correlation of the system. It was shown both numerically and analytically that the overtaking processes only destroy the short-range correlations, leaving the long-range correlations nearly intact. Numerical solutions to the Langevin equation, with a large but finite interaction potential, were studied to clarify the effect of overtaking, while, when particles are allowed to overtake each other and, thereby, escape from the quasi-1D cage as a rare event, the effect of a non-zero overtaking rate on the displacement correlation was derived analytically.

In the paper "Strong- and Weak-Universal Critical Behaviour of a Mixed-Spin Ising Model with Triplet Interactions on the Union Jack (Centered Square) Lattice" by J. Strečka, the mixed spin-1/2 and spin-S Ising model, with a triplet interaction on the centered square lattice, was studied, by establishing a rigorous mapping correspondence with the symmetric eight-vertex model. This exact mapping equivalence was showed by employing two independent mechanisms—by exploiting the graph-theoretical formulation and by using the spin representation of the zero-field eight-vertex model. Then, it was shown that the critical exponents of the mixed spin-1/2 and spin-S Ising model, with a triplet interaction, fundamentally depend on the interaction anisotropy, as well as on the spin parity. This was in contrast to the universality conjecture, which states that the critical behavior of very different models may be characterized by the same set of critical exponents.

In the paper "Anomalous Statistics of Bose-Einstein Condensate in an Interacting Gas: An Effect of the Trap's Form and Boundary Conditions in the Thermodynamic Limit" by S. Tarasov, V. Kocharovsky, and V. Kocharovsky, the authors studied a mesoscopic system formed by a finite number of trapped particles in a three-dimensional rectangular box, with Dirichlet boundary conditions, in the mean-field Bogoliubov and Thomas-Fermi approximations. It was shown that this model yields the non-Gaussian condensate occupation statistics which is different from the non-Gaussian ideal-gas BEC statistics but shares with it, a similar dependence on the trap form and on the boundary conditions. Such a dependence does not vanish in the thermodynamic limit with the increase of the interparticle interaction, the number of trapped particles, and the volume of the system.

In the paper "Study on Bifurcation and Dual Solutions in Natural Convection in a Horizontal Annulus with Rotating Inner Cylinder Using Thermal Immersed Boundary-Lattice Boltzmann Method" by Y. Wei, Z. Wang, Y. Qian, and W. Guo, the mechanism of the rotation effect on bifurcation and on dual solutions, in a natural convection of a horizontal annulus, was studied, numerically. The mechanism of the rotation effect was quantified by linear speed of rotational inner cylinder and it was presented and analyzed by the streamlines and isotherms at different dimensionless linear speeds. The obtained results manifested the existence of three convection patterns which affect the heat transfer in different ways, where the linear speed determines the proportion of each convection.

The paper "Mathematical Realization of Entropy through Neutron Slowing Down" by B. Ganapol, D. Mostacci, and V. Molinari, deals with the study of a classical problem in the neutron transport theory. It was shown that the slowing down equation for elastic scattering of neutrons in an infinite homogeneous medium can be solved analytically. These solutions characterized the evolution of disorder associated with neutron–nucleus collisions. Starting from the monoenergetic neutrons configuration, that represents a complete ordering system, the subsequent scattering creates disorder by uniformly redistributing the neutron energy and the recoil energy transfer to field particles showing, in this way, an increasing entropy with increasing lethargy.

Finally, in the paper "Energy from Negentropy of Non-Cahotic Systems" by P. Quarati, A.M. Scarfone, and G. Kaniadakis the role of positive and negative contributions of entropy and free energy were explored, with their constraints, during a transition of a system, from a non-equilibrium to an equilibrium chaotic state. The amount of negentropy of a non-chaotic system is a source of energy that can be transferred to an internal or inserted subsystem. The subsystem increases its energy and can perform processes that otherwise would not happen like, for instance, the nuclear fusion of the inserted deuterons in a liquid metal matrix, among many others. A few evaluations concerning the non-ideal molecular gas, warm dense matter, and the nuclear matter were reported.

Acknowledgments: I express my thanks to the authors of the above contributions, to the journal Entropy, and MDPI for their support during this work.

Conflicts of Interest: The author declares no conflict of interest.

References

1. Curado, E.M.F.; Nobre, F.D. Equilibrium States in Two-Temperature Systems. *Entropy* **2018**, *20*, 183. [CrossRef]
2. Méndez-Bermúdez, J.A.; Aguilar-Sánchez, R. Information-Length Scaling in a Generalized One-Dimensional Lloyd's Model. *Entropy* **2018**, *20*, 300. [CrossRef]
3. Wilk, G.; Włodarczyk, Z. Oscillations in Multiparticle Production Processes. *Entropy* **2017**, *19*, 670. [CrossRef]
4. Cliff, O.M.; Prokopenko, M.; Fitch, R. Minimising the Kullback-Leibler Divergence for Model Selection in Distributed Nonlinear Systems. *Entropy* **2018**, *20*, 51. [CrossRef]
5. Ohara, A. Conformal Flattening for Deformed Information Geometries on the Probability Simplex. *Entropy* **2018**, *20*, 186. [CrossRef]
6. Rexiti, M.; Felice, D.; Mancini, S. The Volume of Two-Qubit States by Information Geometry. *Entropy* **2018**, *20*, 146. [CrossRef]
7. Ooshida, T.; Goto, S.; Otsuki, M. Collective Motion of Repulsive Brownian Particles in Single-File Diffusion with and without Overtaking. *Entropy* **2018**, *20*, 565. [CrossRef]
8. Strečka, J. Strong- and Weak-Universal Critical Behaviour of a Mixed-Spin Ising Model with Triplet Interactions on the Union Jack (Centered Square) Lattice. *Entropy* **2018**, *20*, 91. [CrossRef]
9. Tarasov, S.; Kocharovsky, V.; Kocharovsky, V. Anomalous Statistics of Bose-Einstein Condensate in an Interacting Gas: An Effect of the Trap's Form and Boundary Conditions in the Thermodynamic Limit. *Entropy* **2018**, *20*, 153. [CrossRef]
10. Wei, Y.; Wang, Z.; Qian, Y.; Guo, W. Study on Bifurcation and Dual Solutions in Natural Convection in a Horizontal Annulus with Rotating Inner Cylinder Using Thermal Immersed Boundary-Lattice Boltzmann Method. *Entropy* **2018**, *20*, 733. [CrossRef]
11. Ganapol, B.; Mostacci, D.; Molinari, V. A Mathematical Realization of Entropy through Neutron Slowing Down. *Entropy* **2018**, *20*, 233. [CrossRef]
12. Quarati, P.; Scarfone, A.M.; Kaniadakis, G. Energy from Negentropy of Non-Cahotic Systems. *Entropy* **2018**, *20*, 113. [CrossRef]

entropy

MDPI

Article

Equilibrium States in Two-Temperature Systems

Evaldo M. F. Curado * and Fernando D. Nobre

Centro Brasileiro de Pesquisas Físicas and National Institute of Science and Technology for Complex Systems, Rua Xavier Sigaud 150, Urca, Rio de Janeiro 22290-180, Brazil; fdnobre@cbpf.br

* Correspondence: evaldo@cbpf.br; Tel.: +55-(21)2141-7369

Received: 24 January 2018; Accepted: 24 February 2018; Published: 9 March 2018

Abstract: Systems characterized by more than one temperature usually appear in nonequilibrium statistical mechanics. In some cases, e.g., glasses, there is a temperature at which fast variables become thermalized, and another case associated with modes that evolve towards an equilibrium state in a very slow way. Recently, it was shown that a system of vortices interacting repulsively, considered as an appropriate model for type-II superconductors, presents an equilibrium state characterized by two temperatures. The main novelty concerns the fact that apart from the usual temperature T, related to fluctuations in particle velocities, an additional temperature θ was introduced, associated with fluctuations in particle positions. Since they present physically distinct characteristics, the system may reach an equilibrium state, characterized by finite and different values of these temperatures. In the application of type-II superconductors, it was shown that $\theta \gg T$, so that thermal effects could be neglected, leading to a consistent thermodynamic framework based solely on the temperature θ. In the present work, a more general situation, concerning a system characterized by two distinct temperatures θ_1 and θ_2, which may be of the same order of magnitude, is discussed. These temperatures appear as coefficients of different diffusion contributions of a nonlinear Fokker-Planck equation. An H-theorem is proven, relating such a Fokker-Planck equation to a sum of two entropic forms, each of them associated with a given diffusion term; as a consequence, the corresponding stationary state may be considered as an equilibrium state, characterized by two temperatures. One of the conditions for such a state to occur is that the different temperature parameters, θ_1 and θ_2, should be thermodynamically conjugated to distinct entropic forms, S_1 and S_2, respectively. A functional $\Lambda[P] \equiv \Lambda(S_1[P], S_2[P])$ is introduced, which presents properties characteristic of an entropic form; moreover, a thermodynamically conjugated temperature parameter $\gamma(\theta_1, \theta_2)$ can be consistently defined, so that an alternative physical description is proposed in terms of these pairs of variables. The physical consequences, and particularly, the fact that the equilibrium-state distribution, obtained from the Fokker-Planck equation, should coincide with the one from entropy extremization, are discussed.

Keywords: nonlinear Fokker-Planck equations; generalized entropies; nonextensive thermostatistics

1. Introduction

The linear Fokker-Planck equation (FPE) represents one of the most important equations of nonequilibrium statistical mechanics; it describes the time evolution of a probability density $P(\vec{x}, t)$ for finding a given particle at a position $\vec{x}$, at time t, diffusing under an external potential [1–4]. In the absence of external potential, the FPE reduces to the linear diffusion equation, usually associated with the description of the Brownian motion; a confining external potential yields the possibility of a stationary-state solution for a sufficiently long time. A particular interest in the literature is given to a harmonic confining potential, which leads to a Gaussian distribution as the stationary-state solution of the FPE [3,4].

It is very frequent nowadays, particularly within the realm of complex systems, to find dynamical behavior that falls out of the ambit of linear diffusion, usually called anomalous diffusion [5]. As typical examples, one may mention diffusion in media characterized by randomness, porosity, heterogeneity, as well as systems characterized by cooperative interactions among internal components, self-organization, and long-time memory. For dealing with these phenomena, one commonly uses a nonlinear (power-like) diffusion equation, known in the literature as porous-media equation [6]. Similarly to the linear case, by adding a confining potential one obtains a nonlinear Fokker-Planck equation (NLFPE) [7], as introduced in [8,9]. For a harmonic confining potential, this NLFPE presents a q-Gaussian distribution, typical of nonextensive statistical mechanics [10,11], as its stationary-state solution. In this way, the NLFPE introduced in [8,9] is associated with Tsallis entropy S_q [12] (where $q \in \Re$ is called entropic index), since the q-Gaussian solution coincides with the distribution that maximizes S_q.

Considering that statistical mechanics may be formulated by starting from a given statistical entropy [1,10], many entropic forms were introduced since the proposal of S_q, as attempts to generalize the standard Boltzmann-Gibbs (BG) formulation. Among those many, we may mention the entropic forms of [13–24]; a pedagogical and comprehensive classification of entropic forms is given in [21], whereas a discussion of how the volume of phase space defines its associated entropy may be found in [22]. Additionally, the connections of NLFPEs with nonadditive entropic forms were explored through generalized formulations of the H-theorem [7,25–47] and particularly, the NLFPE of [8,9] is also related to the entropy with S_q by an H-theorem.

Although one may pursue an analysis in arbitrary dimensions, by considering a probability density $P(x_1, x_2, \cdots, x_N, t)$, like those of [37,48–50], herein for simplicity, we will restrict ourselves to a one-dimensional space, described in terms of a probability density $P(x,t)$. In this case, a general NLFPE may be defined as [32,33,51]

$$\frac{\partial P(x,t)}{\partial t} = -\frac{\partial}{\partial x}\{A(x)\Psi[P(x,t)]\} + D\frac{\partial}{\partial x}\left\{\Omega[P(x,t)]\frac{\partial P(x,t)}{\partial x}\right\} , \qquad (1)$$

where D represents a diffusion coefficient with dimensions of energy divided by the viscosity coefficient, and the external force $A(x)$, with dimensions of force divided by the viscosity, is associated with a confining potential $\phi(x)$ $[A(x) = -d\phi(x)/dx]$. Herein, from now on, we will consider for simplicity the viscosity coefficient equal to one. The functionals $\Psi[P(x,t)]$ and $\Omega[P(x,t)]$ should satisfy certain mathematical requirements, e.g., positiveness and monotonicity with respect to $P(x,t)$ [32,33]; moreover, to ensure normalizability of $P(x,t)$ for all times one must impose the conditions,

$$P(x,t)|_{x\to\pm\infty} = 0\,; \quad \left.\frac{\partial P(x,t)}{\partial x}\right|_{x\to\pm\infty} = 0\,; \quad A(x)\Psi[P(x,t)]|_{x\to\pm\infty} = 0 \qquad (\forall t)\,. \qquad (2)$$

The NLFPE of Equation (1) recovers some well-known cases, as particular limits: (i) The linear FPE [1–4] for $\Psi[P(x,t)] = P(x,t)$ and $\Omega[P(x,t)] = 1$; (ii) The NLFPE introduced in [8,9], associated with nonextensive statistical mechanics, for $\Psi[P(x,t)] = P(x,t)$ and $\Omega[P(x,t)] = \mu[P(x,t)]^{\mu-1}$, where μ represents a real number, related to the entropic index through $\mu = 2 - q$. It should be mentioned that a large variety of NLFPEs, like the one related to nonextensive statistical mechanics, or in the general form of Equation (1), or even presenting nonhomogeneous diffusion coefficients in the nonlinear diffusion term, have been derived in the literature by generalizing standard procedures applied for the linear FPE [1–4], e.g., from approximations in the master equation [37,46,51–55], from a Langevin approach by considering a multiplicative noise [46,56–63], or form methods using ensembles and a projection approach [64].

Almost two decades ago, NLFPEs presenting more than one diffusive term appeared in the literature [51,52,65,66], and particularly, very general forms were derived by considering the continuum limit in a master equation with nonlinear transition probabilities [51,52]. A special interest was given to a concrete physical application, namely, a system of interacting vortices, currently used as a suitable model for type-II superconductors, which exhibited such a behavior [38,43,45,46,65]. This NLFPE, that appears as a particular case of the one derived in [51,52], presents two diffusive terms: (i) A linear contribution, obtained in the usual way, i.e., by applying an additive uncorrelated thermal noise in the system [2–4]; (ii) A nonlinear one, characterized by a power in the probability, like in the NLFPE of [8,9], which emerged from a coarse-graining approach in the vortex-vortex interactions. From these diffusive terms, two distinct temperatures were identified, respectively, the usual thermal temperature T, associated with the linear contribution, and an additional temperature θ (directly related to fluctuations in the vortex positions), defined from the diffusion coefficient of the nonlinear contribution. Moreover, it was shown that $\theta \gg T$ for typical type-II superconductors, so that thermal effects could be neglected, as an approximation [41]; based on this, a whole consistent thermodynamic framework was developed by considering the temperature θ and its conjugated entropy S_q, with $q = 2$ [41–44,67,68].

Motivated by these previous investigations, in the present work we focus on a NLFPE characterized by two diffusive terms, written in the general form [46]

$$\frac{\partial P(x,t)}{\partial t} = -\frac{\partial}{\partial x}\{A(x)\Psi[P(x,t)]\} + D_1\frac{\partial}{\partial x}\left\{\Omega_1[P(x,t)]\frac{\partial P(x,t)}{\partial x}\right\} + D_2\frac{\partial}{\partial x}\left\{\Omega_2[P(x,t)]\frac{\partial P(x,t)}{\partial x}\right\}, \quad (3)$$

where D_1 and D_2 represent diffusion coefficients, whereas the functionals ($\Psi[P(x,t)]$, $\Omega_1[P(x,t)]$, and $\Omega_2[P(x,t)]$) should satisfy similar mathematical requirements as mentioned above [32,33]. One sees easily that Equation (3) may be written in the form of Equation (1), provided that

$$\Omega[P(x,t)] = \frac{D_1}{D}\,\Omega_1[P(x,t)] + \frac{D_2}{D}\,\Omega_2[P(x,t)]\,. \quad (4)$$

Of course, the present approach can be generalized to cover situations with several (more than two) temperatures, by adding further nonlinear diffusive terms in (Equation (3)); this generalization is straightforward and will be addressed in future works.

In the analysis that follows, we discuss the physical aspects related to Equation (3), or equivalently, to Equations (1) and (4). In the next section we develop a generalized form of the H-theorem, from which the concepts of two temperatures, as well as their thermodynamically conjugated entropic forms appear, each of them associated with a given diffusion term. In Section 3 we work out the equilibrium solution of Equation (3), and we introduce a functional to be extremized, defined as a composition of the two entropic forms. We show that by choosing appropriately the Lagrange multipliers in this extremization procedure, one obtains an equation that coincides with the time-independent solution of Equation (3). In contrast to this functional, it is verified that the equilibrium solutions are not simple combinations of known equilibrium distributions, related to each diffusion contribution separately. In Section 4 we review briefly the physical application of type-II superconducting vortices, giving emphasis to its equilibrium distribution. Finally, in Section 5 we discuss the physical consequences of an equilibrium state characterized by two distinct temperatures and present our main conclusions, together with possible thermodynamic scenarios.

2. Generalized Forms of the H-Theorem

The H-theorem represents one of the most important results of nonequilibrium statistical mechanics, since it ensures that after a sufficiently long time, the associated system will reach an equilibrium state. In standard nonequilibrium statistical mechanics, it is usually proven by considering the BG entropy S_{BG}, and making use of an equation that describes the time evolution of the associated probability density, like the Boltzmann, or linear FPE (in the case of continuous probabilities), or the master equation (in the case of discrete probabilities) [1–4].

Recently, the H-theorem has been extended to generalized entropic forms by using NLFPEs [7,25–47]. In the case of a system under a confining external potential $\phi(x)$ [from which one obtains the external force appearing in Equation (1), or in Equation (3), $A(x) = -d\phi(x)/dx$], the H-theorem corresponds to a well-defined sign for the time derivative of the free-energy functional,

$$F[P] = U[P] - \gamma S[P]\,; \qquad U[P] = \int_{-\infty}^{\infty} dx\, \phi(x) P(x,t)\,, \tag{5}$$

where γ denotes a positive parameter with dimensions of temperature. Moreover, the entropy may be considered in the general form [32,33,35],

$$S[P] = k \int_{-\infty}^{\infty} dx\, g[P(x,t)]\,; \quad g(0) = g(1) = 0\,; \quad \frac{d^2 g}{dP^2} \leq 0\,, \tag{6}$$

where k represents a positive constant with entropy dimensions, that can be assumed as the Boltzmann constant, whereas the functional $g[P(x,t)]$ should be at least twice differentiable. Furthermore, the conditions that ensure normalizability of $P(x,t)$ for all times [cf. Equation (2)] are also used in the proof of the H-theorem. For completeness, we first prove the H-theorem for the BG entropy, making use of the linear FPE.

2.1. The H-Theorem from the Linear Fokker-Planck Equation

It is well-established that the BG entropy, defined following Equation (6) through $g[P(x,t)] = -P(x,t)\ln[P(x,t)]$, is directly related to the linear FPE, given as a particular case of Equation (1), with the functionals $\Psi[P(x,t)] = P(x,t)$ and $\Omega[P(x,t)] = 1$ [1–4]; moreover, in this case, one has the standard temperature in Equation (5), i.e., $\gamma = T$. Therefore, the time derivative of the free-energy functional becomes

$$\begin{aligned} \frac{dF}{dt} &= \frac{\partial}{\partial t}\left(\int_{-\infty}^{\infty} dx\, \phi(x)P(x,t) + kT\int_{-\infty}^{\infty} dx\, P(x,t)\ln[P(x,t)]\right) = \int_{-\infty}^{\infty} dx\, [\phi(x) + kT(\ln P + 1)]\,\frac{\partial P}{\partial t} \\ &= \int_{-\infty}^{\infty} dx\, [\phi(x) + kT\ln P]\left[-\frac{\partial[A(x)P(x,t)]}{\partial x} + D\frac{\partial^2 P(x,t)}{\partial x^2}\right]\,, \end{aligned} \tag{7}$$

where, in the second line we have used the particular case of Equation (1) for the time-derivative of the probability, and the normalization condition for all times, which implies on $\int_{-\infty}^{\infty} dx (\partial P/\partial t) = 0$. Then, we perform an integration by parts, use the conditions of Equation (2), and assume $D = kT$, to obtain

$$\frac{dF}{dt} = -\int_{-\infty}^{\infty} dx\, P(x,t)\left[-A(x) + \frac{kT}{P(x,t)}\frac{\partial P(x,t)}{\partial x}\right]^2 \leq 0\,, \tag{8}$$

Leading to a well-defined sign for the time-derivative of the free-energy functional. Since $U[P]$ and $S[P]$ are both finite for any P, this implies that for a finite value of γ, $F[P]$ is bounded from below and that the reached stationary state is stable.

2.2. The H-Theorem from Nonlinear Fokker-Planck Equations

Considering a procedure similar to the one presented above for the general forms of the free-energy functional in Equations (5) and (6), together with the NLFPE of Equation (1), the H-theorem may be proven (see, e.g., [32,33,35]), by considering $D = k\gamma$, and imposing the condition

$$-\frac{d^2 g[P]}{dP^2} = \frac{\Omega[P]}{\Psi[P]}\,, \tag{9}$$

Which relates the entropic form to a certain time evolution. Particular entropic forms and their associated NLFPEs were explored in [32], whereas families of NLFPEs (those characterized by the same ratio $\Omega[P]/\Psi[P]$) were studied in [35].

In what follows, we prove the H-theorem by making use of Equation (3); for that, we replace the free-energy functional of Equation (5) by

$$F[P] = U[P] - \theta_1 S_1[P] - \theta_2 S_2[P]\,; \qquad U[P] = \int_{-\infty}^{\infty} dx\, \phi(x) P(x,t)\,, \tag{10}$$

where θ_1 and θ_2 denote positive parameters with dimensions of temperature. Similarly to Equation (6), we now define

$$S_i[P] = k \int_{-\infty}^{\infty} dx\, g_i[P(x,t)]\,; \quad g_i(0) = g_i(1) = 0\,; \quad \frac{d^2 g_i}{dP^2} \leq 0\,; \quad (i = 1,2). \tag{11}$$

Hence, one obtains for the time derivative of the free-energy functional of Equation (10)

$$\begin{aligned}
\frac{dF}{dt} &= \frac{\partial}{\partial t}\left(\int_{-\infty}^{\infty} dx\, \phi(x)P(x,t) - k\theta_1 \int_{-\infty}^{\infty} dx\, g_1[P(x,t)] - k\theta_2 \int_{-\infty}^{\infty} dx\, g_2[P(x,t)]\right) \\
&= \int_{-\infty}^{\infty} dx \left[\phi(x) - k\theta_1 \frac{dg_1}{dP} - k\theta_2 \frac{dg_2}{dP}\right] \frac{\partial P}{\partial t} = \int_{-\infty}^{\infty} dx \left[\phi(x) - k\theta_1 \frac{dg_1}{dP} - k\theta_2 \frac{dg_2}{dP}\right] \\
&\times \left[-\frac{\partial}{\partial x}\{A(x)\Psi[P(x,t)]\} + D_1 \frac{\partial}{\partial x}\left\{\Omega_1[P(x,t)] \frac{\partial P(x,t)}{\partial x}\right\} + D_2 \frac{\partial}{\partial x}\left\{\Omega_2[P(x,t)] \frac{\partial P(x,t)}{\partial x}\right\}\right],
\end{aligned}$$

where in the last line we have substituted Equation (3) for the time derivative of the probability. Hence, carrying out an integration by parts and using the conditions of Equation (2), one obtains

$$\begin{aligned}
\frac{dF}{dt} &= - \int_{-\infty}^{\infty} dx\, \Psi[P(x,t)] \left[-A(x) + D_1 \frac{\Omega_1[P]}{\Psi[P]} \frac{\partial P(x,t)}{\partial x} + D_2 \frac{\Omega_2[P]}{\Psi[P]} \frac{\partial P(x,t)}{\partial x}\right] \\
&\times \left[-A(x) - k\theta_1 \frac{d^2 g_1}{dP^2} \frac{\partial P(x,t)}{\partial x} - k\theta_2 \frac{d^2 g_2}{dP^2} \frac{\partial P(x,t)}{\partial x}\right].
\end{aligned} \tag{12}$$

Remembering that the functional $\Psi[P(x,t)]$ was defined previously as a positive quantity, for a well-defined sign of the quantity above it is sufficient to impose the conditions,

$$D_1 = k\theta_1\,; \qquad D_2 = k\theta_2\,, \tag{13}$$

As well as

$$-\frac{d^2 g_1[P]}{dP^2} = \frac{\Omega_1[P]}{\Psi[P]}\,; \qquad -\frac{d^2 g_2[P]}{dP^2} = \frac{\Omega_2[P]}{\Psi[P]}\,, \tag{14}$$

Extending the condition of Equation (9) for two diffusion contributions. These conditions lead to the following generalized form for the H-theorem,

$$\frac{dF}{dt} = - \int_{-\infty}^{\infty} dx\, \Psi[P(x,t)] \left[-A(x) + D_1 \frac{\Omega_1[P]}{\Psi[P]} \frac{\partial P(x,t)}{\partial x} + D_2 \frac{\Omega_2[P]}{\Psi[P]} \frac{\partial P(x,t)}{\partial x}\right]^2 \leq 0\,. \tag{15}$$

Like before, since $F[P]$ [in Equation (10)] is finite for finite values of θ_1 and θ_2, the stationary solution satisfying $(dF/dt) = 0$ is a stable solution.

One should notice that substituting Equation (4) in Equation (9), and using the conditions of Equation (14), one gets

$$\frac{d^2 g}{dP^2} = \frac{D_1}{D} \frac{d^2 g_1[P]}{dP^2} + \frac{D_2}{D} \frac{d^2 g_2[P]}{dP^2}\,, \tag{16}$$

So that integrating twice with respect to $P(x,t)$, and using the conditions in Equations (6) and (11),

$$g[D_1, D_2, D, P(x,t)] = \frac{D_1}{D} g_1[P] + \frac{D_2}{D} g_2[P]\,. \tag{17}$$

Now, integrating with respect to the variable x, the following functional results

$$\Lambda[D_1,D_2,D,P(x,t)] = k\int_{-\infty}^{\infty} dx\; g[D_1,D_2,D,P(x,t)] = \frac{D_1}{D}S_1[P] + \frac{D_2}{D}S_2[P]\;. \tag{18}$$

We call attention to the fact that the functional $\Lambda[D_1, D_2, D, P(x,t)]$ above apparently depends on the probability, as well as on diffusion coefficients, a result that appears as a direct consequence of a NLFPE with more than one diffusion term. According to information theory, written in this form, this quantity should not be associated to an entropic form, since it violates one of its basic axioms, which states that an entropic form should depend only on the probability $P(x,t)$ [69]. Up to the moment, this functional may be understood as a linear combination of the two entropic forms $S_1[P]$ and $S_2[P]$, with coefficients (D_1/D) and (D_2/D), respectively; later on, based on thermodynamic arguments, we will argue that this functional may be interpreted also as an entropic functional depending only on the probabilities.

3. Equilibrium Distribution

In this section we work out the stationary-state (i.e., time-independent) solution of Equation (3), as well as the equilibrium distribution that results from an extremization procedure of the functional of Equation (18). As usual, the Lagrange parameters of this later approach will be defined appropriately so that these two results coincide; based on this, in the calculations that follow we refer to an equilibrium state, described by a distribution $P_{\rm eq}(x)$.

First, let us obtain the time-independent distribution of Equation (3); for this purpose, we rewrite it in the form of a continuity equation,

$$\frac{\partial P(x,t)}{\partial t} = -\frac{\partial J(x,t)}{\partial x}\;, \tag{19}$$

where the probability current density is given by,

$$J(x,t) = A(x)\Psi[P(x,t)] - D_1\Omega_1[P(x,t)]\frac{\partial P(x,t)}{\partial x} - D_2\Omega_2[P(x,t)]\frac{\partial P(x,t)}{\partial x}\;. \tag{20}$$

The solution $P_{\rm eq}(x)$ is obtained by setting $J_{\rm eq}(x) = 0$ (as required by conservation of probability [32]), so that

$$J_{\rm eq}(x) = A(x)\Psi[P_{\rm eq}(x)] - \{D_1\Omega_1[P_{\rm eq}(x)] + D_2\Omega_2[P_{\rm eq}(x)]\}\frac{dP_{\rm eq}}{dx} = 0\;, \tag{21}$$

Which may still be written in the form

$$A(x) = \left\{D_1\frac{\Omega_1[P_{\rm eq}(x)]}{\Psi[P_{\rm eq}(x)]} + D_2\frac{\Omega_2[P_{\rm eq}(x)]}{\Psi[P_{\rm eq}(x)]}\right\}\frac{dP_{\rm eq}}{dx}\;. \tag{22}$$

Integrating the equation above over x, and remembering that the external force was defined as $A(x) = -d\phi(x)/dx$, one gets,

$$\begin{aligned}\phi_0 - \phi(x) &= \int_{x_0}^{x} dx \left\{D_1\frac{\Omega_1[P_{\rm eq}(x)]}{\Psi[P_{\rm eq}(x)]} + D_2\frac{\Omega_2[P_{\rm eq}(x)]}{\Psi[P_{\rm eq}(x)]}\right\}\frac{dP_{\rm eq}}{dx} \\ &= \int_{P_{\rm eq}(x_0)}^{P_{\rm eq}(x)} \left\{D_1\frac{\Omega_1[P_{\rm eq}(x')]}{\Psi[P_{\rm eq}(x')]} + D_2\frac{\Omega_2[P_{\rm eq}(x')]}{\Psi[P_{\rm eq}(x')]}\right\} dP_{\rm eq}(x')\;,\end{aligned} \tag{23}$$

where $\phi_0 \equiv \phi(x_0)$. Now, one uses the relations in Equation (14), and carrying the integrations,

$$D_1\frac{dg_1[P]}{dP}\bigg|_{P=P_{\rm eq}(x)} + D_2\frac{dg_2[P]}{dP}\bigg|_{P=P_{\rm eq}(x)} = \phi(x) + C_1\;, \tag{24}$$

With C_1 being a constant.

Next, we extremize the functional of Equation (18) with respect to the probability, under the constraints of probability normalization and internal-energy definition following Equation (10). For this, we introduce the functional

$$\mathcal{I} = \frac{\Lambda}{k} + \alpha\left(1 - \int_{-\infty}^{\infty} dx\; P(x,t)\right) + \beta\left(U - \int_{-\infty}^{\infty} dx\; \phi(x)P(x,t)\right) , \tag{25}$$

where α and β are Lagrange multipliers. Hence, the extremization, $(\delta\mathcal{I})/(\delta P)|_{P=P_{\rm eq}(x)} = 0$, leads to

$$\frac{D_1}{D}\frac{dg_1[P]}{dP}\bigg|_{P=P_{\rm eq}(x)} + \frac{D_2}{D}\frac{dg_2[P]}{dP}\bigg|_{P=P_{\rm eq}(x)} - \alpha - \beta\phi(x) = 0 . \tag{26}$$

One notices that Equations (24) and (26), resulting from the stationary-state solution of Equation (3) and the extremization of the functional of Equation (18), respectively, which in fact yield stable solutions, coincide if one chooses the Lagrange multipliers $\alpha = C_1/D$ and $\beta = 1/D$. Moreover, one should remind that the relations in Equation (14) were used to get Equation (24).

For reasons that follow next, we impose the Lagrange multiplier $\beta = 1/(D_1 + D_2)$, which implies on $D = D_1 + D_2$; in this case, the functional of Equation (18) becomes

$$\Lambda[\theta_1, \theta_2, P(x,t)] = \frac{\theta_1}{\theta_1 + \theta_2} S_1[P] + \frac{\theta_2}{\theta_1 + \theta_2} S_2[P] , \tag{27}$$

where we have used the temperature definitions of Equation (13). This particular choice for the Lagrange multiplier β yields a functional $\Lambda[\theta_1, \theta_2, P(x,t)]$, given by a linear combination of the two entropic forms $S_1[P]$ and $S_2[P]$ with well-defined coefficients, representing one of the main novelties of this investigation; it presents important properties, listed below.

(i) The corresponding coefficients are both in the interval $[0,1]$ and their sum gives unit, so that they may be interpreted as probabilities related to the contribution of each entropic form to the functional $\Lambda[\theta_1, \theta_2, P(x,t)]$; in this way, the quantity in Equation (27) can be understood as a mean value.

(ii) When one temperature prevails with respect to the other one, e.g., $\theta_1 \gg \theta_2$, the resulting functional becomes essentially the entropic form associated to this temperature, i.e., $\Lambda \approx S_1[P]$, like considered in the physical application explored in [38,43,45,46,65].

(iii) In Equation (17), one has

$$g[\theta_1, \theta_2, P] = \frac{\theta_1}{\theta_1 + \theta_2} g_1[P] + \frac{\theta_2}{\theta_1 + \theta_2} g_2[P] , \tag{28}$$

So that the conditions of Equation (11) yield $g[\theta_1, \theta_2, 0] = g[\theta_1, \theta_2, 1] = 0$, for arbitrary values of θ_1 and θ_2.

(iv) The concavity of the functional $\Lambda[\theta_1, \theta_2, P(x,t)]$ with respect to the probability is well defined; indeed, Equation (16) leads to

$$\frac{d^2 g[\theta_1, \theta_2, P]}{dP^2} = \frac{\theta_1}{\theta_1 + \theta_2}\frac{d^2 g_1[P]}{dP^2} + \frac{\theta_2}{\theta_1 + \theta_2}\frac{d^2 g_2[P]}{dP^2} \leq 0 , \tag{29}$$

where the inequality comes as a direct consequence of the conditions in Equation (11). In addition to this, due to the properties of the coefficients (described in (i) above), the second derivative on the left-hand-side should be in between the two second derivatives on the right-hand side.

4. Physical Application: Type-II Superconducting Vortices

One case of interest in Equation (3) corresponds to the competition of two power-like diffusive terms, given by the NLFPE [46]

$$\frac{\partial P(x,t)}{\partial t} = -\frac{\partial}{\partial x}\{A(x)P(x,t)]\} + D_1 \frac{\partial^2 P^{\mu_1}(x,t)}{\partial x^2} + D_2 \frac{\partial^2 P^{\mu_2}(x,t)}{\partial x^2} , \tag{30}$$

where (D_1, μ_1) and (D_2, μ_2) correspond to coefficients and exponents related to each diffusion contribution; comparing with Equation (3), one has that

$$\begin{aligned}\Psi[P(x,t)] &= P(x,t)\,, \\ \Omega_1[P(x,t)] &= \mu_1[P(x,t)]^{\mu_1-1}\,; \qquad \Omega_2[P(x,t)] = \mu_2[P(x,t)]^{\mu_2-1}\,. \end{aligned} \tag{31}$$

Therefore, the integrations in Equation (14) lead to

$$S_1[P(x,t)] = k\int_{-\infty}^{\infty} dx\,\frac{P(x,t)-[P(x,t)]^{\mu_1}}{\mu_1-1}\,; \qquad S_2[P(x,t)] = k\int_{-\infty}^{\infty} dx\,\frac{P(x,t)-[P(x,t)]^{\mu_2}}{\mu_2-1}\,, \tag{32}$$

Which compose the functional in Equation (27). By comparing this functional with previous ones [e.g., cf. Equation (23) of [46]], the main novelty herein concerns the coefficients of each entropic form, written in the form of probabilities. In this way, the equilibrium equation [cf. Equation (24)] becomes

$$\frac{D_1\mu_1}{\mu_1-1}\,P_{\rm eq}^{\mu_1-1}(x) + \frac{D_2\mu_2}{\mu_2-1}\,P_{\rm eq}^{\mu_2-1}(x) = -\phi(x) + C_1\,. \tag{33}$$

Recently, a special interest was given to a physical application, namely, a system of interacting vortices, used as a suitable model for type-II superconductors [38,43,45,46,65]. For this particular system, in the second diffusive contribution one has $\mu_2 = 2$ and $D_2 = k\theta$, where an effective temperature θ was related to the density of vortices [41]. The first diffusive term comes from a standard uncorrelated thermal noise, leading to the linear contribution ($\mu_1 = 1$) and $D_1 = kT$. The functional of Equation (27) becomes

$$\begin{aligned}\Lambda[\theta,T,P(x,t)] &= \frac{T}{T+\theta}\,S_1[P(x,t)] + \frac{\theta}{T+\theta}\,S_2[P(x,t)] \\ &= -\frac{kT}{T+\theta}\int_{-\infty}^{\infty} dx\,P(x,t)\ln P(x,t) + \frac{k\theta}{T+\theta}\int_{-\infty}^{\infty} dx\,\{P(x,t)-[P(x,t)]^2\}\,,\end{aligned} \tag{34}$$

Which differs from previous ones [e.g., Equation (17) of [38], or Equation (47) of [46]] in the choices for the coefficients. Particularly, with respect to the result of [38], one has now a concrete proposal for the quantity in the denominator of these coefficients, which appeared herein from an appropriate choice of the Lagrange multiplier β.

In the present case, Equation (33) becomes

$$D_1\ln P_{\rm eq}(x) + 2D_2P_{\rm eq}(x) = C'' - \phi(x)\,, \tag{35}$$

Which can be written as

$$\frac{2D_2\,P_{\rm eq}(x)}{D_1}\exp\left(\frac{2D_2}{D_1}\,P_{\rm eq}(x)\right) = \frac{2D_2}{D_1}\exp\left(\frac{C''-\phi(x)}{D_1}\right)\,. \tag{36}$$

where $C'' = C_1 - D_1 + D_2$.

In the equation above one identifies the form $Xe^X = Y$, which defines the implicit W-Lambert function, such that $X = W(Y)$ (see, e.g., [70]). Therefore,

$$P_{\rm eq}(x) = \frac{D_1}{2D_2}\,W\left\{\frac{2D_2}{D_1}\exp\left(\frac{C''-\phi(x)}{D_1}\right)\right\}\,. \tag{37}$$

Choosing a harmonic confining potential, the distribution above interpolates between two well-known limits, namely, the Gaussian distribution ($D_1 \gg D_2$) and the parabola, i.e., q-Gaussian distribution with $q = 2$ ($D_1 \ll D_2$); moreover, both parameters D_1 and D_2 affect directly the width of the distribution, consistently with the temperature definitions of Equation (13), in the sense that larger values of these parameters produce larger widths [46].

Hence, considering $\phi(x) = \alpha x^2/2$ $(\alpha > 0)$, and the temperature definitions of Equation (13) for the present case, i.e., $D_1 = kT$ and $D_2 = k\theta$, g one has

$$P_{\rm eq}(x) = \frac{T}{2\theta} W \left\{ \frac{2\theta}{T} \exp\left(\frac{C''}{kT} - \frac{\alpha x^2}{2kT} \right) \right\} , \tag{38}$$

Which is illustrated in Figure 1 for typical choices of its parameters. The crossover between the parabolic behavior $(T \ll \theta)$ to the Gaussian distribution $(T \gg \theta)$ is shown in Figure 1a, where we present equilibrium distributions for $\alpha = \theta = 1$ and increasing values of T (from top to bottom). The former limit is important for an appropriate description of a type-II superconducting phase, where one finds strongly-interacting vortices [41,43], whereas in the latter limit one approaches the behavior of a system of weakly-interacting particles [45]. In Figure 1b we present equilibrium distributions by considering $T = \theta = 1$ (i.e., in between the two limits mentioned above) and increasing values of α. As expected, the parameter α, which represents the strength of the confining potential, is directly related to the confining of the vortices, affecting the distribution width, in the sense that larger values of α correspond to smaller distribution widths. In the first limiting behavior, a whole consistent thermodynamic framework was developed by neglecting thermal effects, considering the temperature θ, its conjugated entropy S_q (with $q = 2$), the parameter α, and introducing its conjugated parameter σ [41–44,67,68]. At this point, it is important to emphasize that the two temperatures exist in an equilibrium state (there is no flux), presenting different physical meanings. The usual temperature T is related to the thermal noise, and may be changed through heat transfers to (or from) the system, whereas the temperature θ is related to the density of vortices, and can be varied by monitoring an applied magnetic field. Even with these two temperatures, with different physical meanings, there is no nonequilibrium behavior, so that the thermodynamic formalism can be considered.

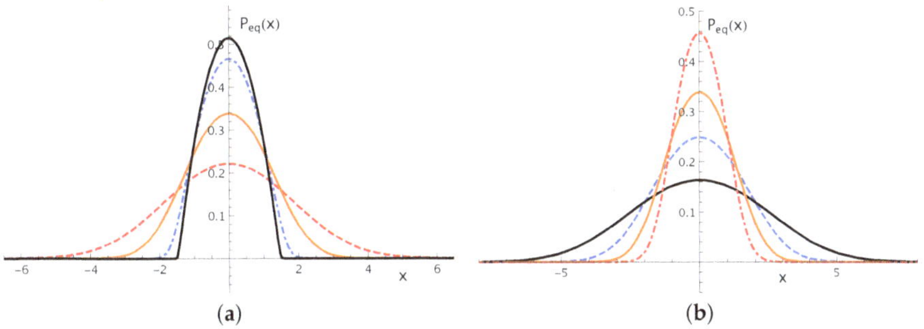

Figure 1. The equilibrium-state distribution $P_{\rm eq}(x)$ of Equation (38) is represented versus x by considering $k = 1$ and special choices of its parameters. (**a**) Fixing $\alpha = \theta = 1$ and increasing values of T [$T = 0.02, 0.2, 1.0$, and 3.0 (from top to bottom)], showing the crossover from the parabolic to the Gaussian behavior; (**b**) Fixing $T = \theta = 1$ and increasing values of α [$\alpha = 0.2, 0.5, 1.0$, and 3.0 (from bottom to top)]. The parameter C'' is found in each case by imposing normalization for $P_{\rm eq}(x)$.

Recently, by analyzing short-range power-law interactions and introducing correlations among particles, a coarse-graining approach has led to a more general NLFPE, extending the previous results to a wider range of values of q [71]. Motivated by this, a consistent thermodynamic framework was proposed for q-Gaussian distributions characterized by a cutoff, under similar conditions [47].

Next, we discuss further the most general physical situation where both temperatures may be of the same order of magnitude. Following previous analyses [42–44,67,68], we consider that the parameter α may be varied continuously, and that these variations are related to a work contribution in a first-law proposal.

5. Discussion and Conclusions

In this section we analyze possible thermodynamical scenarios describing the two-temperature equilibrium state (temperatures θ_1 and θ_2) discussed in Section 3. Considering the free-energy functional of Equation (10), one may define two different types of heat-like contributions, related to variations in each entropic contribution, namely, $\delta Q_1 = \theta_1 dS_1$, and $\delta Q_2 = \theta_2 dS_2$. Moreover, inspired by the physical example of Section 4, one can consider the parameter α as some controllable external field associated with work, which affects directly the volume occupied by the particles; in this way, following previous investigations, we introduce an infinitesimal work contribution, $\delta W = \sigma d\alpha$ [42–44,67,68]. Hence, infinitesimal variations in the internal energy U can be associated to these proposals for infinitesimal work and heat contributions, yielding an equivalent to the first law,

$$dU = \delta Q_1 + \delta Q_2 + \delta W = \theta_1 dS_1 + \theta_2 dS_2 + \sigma d\alpha \,, \tag{39}$$

where the two temperatures may be obtained from

$$\theta_1 = \left(\frac{\partial U}{\partial S_1}\right)_{S_2,\alpha} \;; \qquad \theta_2 = \left(\frac{\partial U}{\partial S_2}\right)_{S_1,\alpha} \,. \tag{40}$$

One should remind that, since the present results follow from a NLFPE, the thermodynamic quantities (U, S_1, S_2, α), presenting infinitesimal changes in Equation (39), refer to one particle of the system. Furthermore, σ corresponds to the parameter thermodynamically conjugated to α, to be determined from an equation of state; considering Equation (39), one has two equivalent ways to calculate σ,

$$\sigma = -\theta_1 \left(\frac{\partial S_1}{\partial \alpha}\right)_{U,S_2} \;; \qquad \sigma = -\theta_2 \left(\frac{\partial S_2}{\partial \alpha}\right)_{U,S_1} \,, \tag{41}$$

By keeping fixed one of the two entropic forms in each case. Summing these two equations, one has

$$\sigma = -\frac{1}{2}\left[\theta_1 \left(\frac{\partial S_1}{\partial \alpha}\right)_{U,S_2} + \theta_2 \left(\frac{\partial S_2}{\partial \alpha}\right)_{U,S_1}\right] \,, \tag{42}$$

Which relates the quantities $(\theta_1, \theta_2, \sigma, \alpha)$, representing the equation of state of the system.

Now, we assume that a given thermodynamical transformation may occur in such a way that variations in the two entropic forms lead to a variation in the functional Λ of Equation (27) following

$$(\theta_1 + \theta_2) d\Lambda = \theta_1 dS_1 + \theta_2 dS_2 = \delta Q_1 + \delta Q_2 \,. \tag{43}$$

In fact, the result above holds in general (not only for a specific transformation) if one considers Λ as a thermodynamic quantity whose natural variables are S_1 and S_2, i.e., $\Lambda \equiv \Lambda(S_1, S_2)$. This is justified by the argument that, in a many-particle system, both entropic forms S_1 and S_2 should be extensive quantities, and so, extensivity of $\Lambda(S_1, S_2)$ follows by imposing Λ to be a homogeneous function of first degree of S_1 and S_2,

$$\Lambda(\lambda S_1, \lambda S_2) = \lambda \Lambda(S_1, S_2) \,, \tag{44}$$

where λ is a positive real number. In this way, one may write the first-law in Equation (39) in the form,

$$dU = (\theta_1 + \theta_2) d\Lambda + \sigma d\alpha \,, \tag{45}$$

From which one identifies $(\theta_1 + \theta_2)$ as the parameter thermodynamically conjugated to Λ. Therefore, another temperature (to be called γ) may be defined, leading to the fundamental equation of thermodynamics,

$$\gamma = \theta_1 + \theta_2 = \left(\frac{\partial U}{\partial \Lambda}\right)_\alpha , \tag{46}$$

Whereas the equation of state may be expressed as

$$\sigma = -\gamma \left(\frac{\partial \Lambda}{\partial \alpha}\right)_U . \tag{47}$$

One should notice that the temperature parameter γ above now appears as a concrete proposal of this quantity in the two-entropic functional of previous works [e.g., Equation (17) of [38]].

Consequently, from the statistical point of view, the dependence of Λ on the entropic forms $S_1[P]$ and $S_2[P]$ imply on $\Lambda \equiv \Lambda[P]$; furthermore, the effective-temperature definition of Equation (46) and the first law in the form of Equation (45) are consistent with a free-energy functional similar to Equation (5),

$$F[P] = U[P] - \gamma\Lambda[P] \; ; \qquad U[P] = \int_{-\infty}^{\infty} dx \; \phi(x) P(x,t) \, , \tag{48}$$

So that results for systems with a single temperature and entropic form apply to the pair of conjugated variables (γ, Λ) defined above. Particularly, the H-theorem for the NLFPE with two diffusive terms in Equation (3) may be proven by considering the functional of Equation (4), leading to the relation in Equation (9), as well as the equivalence of the stationary-state solution of the NLFPE and the extremization of the functional $\Lambda[P]$ (see, e.g., [46]). This means that the stationary-state solution of the NLFPE yields the same solution as the MaxEnt principle, allowing us to assert that this is an equilibrium solution.

To summarize, we have studied a general physical situation of a system characterized by two distinct temperatures, θ_1 and θ_2, which may be of the same order of magnitude, and their thermodynamically conjugated entropic forms, S_1 and S_2, respectively. This study was motivated by recent investigations concerning type-II superconducting vortices, where two entropic forms corresponding to S_q with $q = 2$ and BG entropy S_{BG}, conjugated to temperatures θ and T, appeared. However, in this case, it was shown that an approximated physical description could be developed by neglecting thermal effects, based on the fact that along a typical type-II superconducting phase, the two associated temperatures are significantly different in magnitude, i.e., $\theta \gg T$. In the present work we have assumed that the temperatures should present physically distinct properties, like in the case of type-II superconductors, where they are associated respectively, to fluctuations in velocities and positions of the particles; in this way, an equilibrium state may be attained, with a temperature θ_1 having a different physical meaning than the temperature θ_2. The procedure was based on a nonlinear Fokker-Planck equation, where the temperatures appeared as coefficients of different diffusion contributions. We have proven an H-theorem, relating such a Fokker-Planck equation to a sum of two entropic forms, each of them associated with a given diffusion term. Due to the H-theorem, the corresponding stationary state is considered as an equilibrium state, characterized by two temperature parameters, θ_1 and θ_2, and their associated entropic forms, S_1 and S_2. Particularly, a free-energy functional, together with a first-law proposal, define a four-dimensional space $(\theta_1, \theta_2, S_1, S_2)$ where physical transformations may take place, in such a way to develop a consistent thermodynamical framework. We have also introduced a functional $\Lambda[P] \equiv \Lambda(S_1[P], S_2[P])$, together with a thermodynamically conjugated temperature parameter $\gamma(\theta_1, \theta_2)$, so that an alternative physical description is proposed in terms of these pair of variables. We have shown that the functional $\Lambda[P]$ presents properties characteristic of an entropic form, e.g., it depends only on the probability distribution, and it presents the appropriated concavity sign.

The above-mentioned proposals, and particularly the thermodynamic properties of a system with two distinct temperatures, together with their conjugated entropies, represent open problems of relevant interest that require further investigations from both theoretical and experimental points of view. Among potential candidates in nature, one could mention: (i) Systems of particles interacting repulsively, for which a coarse-graining procedure on the interactions lead to a diffusion contribution in a Fokker-Planck equation of the same order of magnitude as the standard linear diffusion term, associated with an additive uncorrelated thermal noise; (ii) Anomalous-diffusion phenomena in porous media constituted by more than one type of material, or even random porous media.

Acknowledgments: We thank C. Tsallis for fruitful conversations. The partial financial support from CNPq, CAPES, and FAPERJ (Brazilian agencies) is acknowledged.

Conflicts of Interest: The authors declare no conflict of interest.

References

1. Balian, R. *From Microphysics to Macrophysics*; Springer: Berlin, Germany, 1991; Volumes I and II.
2. Reichl, L.E. *A Modern Course in Statistical Physics*, 2nd ed.; John Wiley and Sons: New York, NY, USA, 1998.
3. Balakrishnan, V. *Elements of Nonequilibrium Statistical Mechanics*; CRC Press, Taylor and Francis Group: New York, NY, USA, 2008.
4. Risken, H. *The Fokker-Planck Equation*, 2nd ed.; Springer: Berlin, Germany, 1989.
5. Bouchaud, J.P.; Georges, A. Anomalous diffusion in disordered media: Statistical mechanisms, models and physical applications. *Phys. Rep.* **1990**, *195*, 127–293.
6. Vázquez, J.L. *The Porous Medium Equation*; Oxford University Press: Oxford, UK, 2007.
7. Frank, T.D. *Nonlinear Fokker-Planck Equations: Fundamentals and Applications*; Springer: Berlin, Germay, 2005.
8. Plastino, A.R.; Plastino, A. Non-extensive statistical mechanics and generalized Fokker-Planck equation. *Physica A* **1995**, *222*, 347–354.
9. Tsallis, C.; Bukman, D.J. Anomalous diffusion in the presence of external forces: Exact time-dependent solutions and their thermostatistical basis. *Phys. Rev. E* **1996**, *54*, R2197–R2200.
10. Tsallis, C. *Introduction to Nonextensive Statistical Mechanics*; Springer: New York, NY, USA, 2009.
11. Tsallis, C. An introduction to nonadditive entropies and a thermostatistical approach to inanimate and living matter. *Contemp. Phys.* **2014**, *55*, 179–197.
12. Tsallis, C. Possible generalization of Boltzmann-Gibbs statistics. *J. Stat. Phys.* **1988**, *52*, 479–487.
13. Abe, S. A note on the q-deformation-theoretic aspect of the generalized entropies in nonextensive physics. *Phys. Lett. A* **1997**, *224*, 326–330.
14. Borges, E.P.; Roditi, I. A family of nonextensive entropies. *Phys. Lett. A* **1998**, *246*, 399–402.
15. Anteneodo, C.; Plastino, A.R. Maximum entropy approach to stretched exponential probability distributions. *J. Phys. A* **1999**, *32*, 1089–1097.
16. Curado, E.M.F. General Aspects of the Thermodynamical Formalism. *Braz. J. Phys.* **1999**, *29*, 36–45.
17. Curado, E.M.F.; Nobre, F.D. On the stability of analytic entropic forms. *Physica A* **2004**, *335*, 94–106.
18. Kaniadakis, G. Non-linear kinetics underlying generalized statistics. *Physica A* **2001**, *296*, 405–425.
19. Kaniadakis, G. Statistical mechanics in the context of special relativity. *Phys. Rev. E* **2002**, *66*, 056125.
20. Kaniadakis, G. Statistical mechanics in the context of special relativity II. *Phys. Rev. E* **2005**, *72*, 036108.
21. Hanel, R.; Thurner, S. A comprehensive classification of complex statistical systems and an axiomatic derivation of their entropy and distribution functions. *Europhys. Lett.* **2011**, *93*, 20006.
22. Hanel, R.; Thurner, S. When do generalized entropies apply? How phase space volume determines entropy. *Europhys. Lett.* **2011**, *96*, 50003.
23. Hanel, R.; Thurner, S. Generalized (c,d)-entropy and aging random walks. *Entropy* **2013**, *15*, 5324–5337.
24. Yamano, T. On a simple derivation of a family of nonextensive entropies from information content. *Entropy* **2004**, *6*, 364–374.
25. Kaniadakis, G. H-theorem and generalized entropies within the framework of nonlinear kinetics. *Phys. Lett. A* **2001**, *288*, 283–291.
26. Shiino, M. Free energies based on generalized entropies and H-theorems for nonlinear Fokker-Planck equations. *J. Math. Phys.* **2001**, *42*, 2540–2553.
27. Frank, T.D.; Daffertshofer, A. H-theorem for nonlinear Fokker-Planck equations related to generalized thermostatistics. *Physica A* **2001**, *295*, 455–474.
28. Frank, T.D. Generalized Fokker-Planck equations derived from generalized linear nonequilibrium thermodynamics. *Physica A* **2002**, *310*, 397–412.

29. Shiino, M. Stability analysis of mean-field-type nonlinear Fokker-Planck equations associated with a generalized entropy and its application to the self-gravitating system. *Phys. Rev. E* **2003**, *67*, 056118.
30. Chavanis, P.-H. Generalized thermodynamics and Fokker-Planck equations: Applications to stellar dynamics and two-dimensional turbulence. *Phys. Rev. E* **2003**, *68*, 036108.
31. Chavanis, P.-H. Generalized Fokker-Planck equations and effective thermodynamics. *Physica A* **2004**, *340*, 57–65.
32. Schwämmle, V.; Nobre, F.D.; Curado, E.M.F. Consequences of the *H* theorem from nonlinear Fokker-Planck equations. *Phys. Rev. E* **2007**, *76*, 041123.
33. Schwämmle, V.; Curado, E.M.F.; Nobre, F.D. A general nonlinear Fokker-Planck equation and its associated entropy. *Eur. Phys. J. B* **2007**, *58*, 159–165.
34. Chavanis, P.-H. Nonlinear mean field Fokker-Planck equations. Application to the chemotaxis of biological population. *Eur. Phys. J. B* **2008**, *62*, 179–208.
35. Schwämmle, V.; Curado, E.M.F.; Nobre, F.D. Dynamics of normal and anomalous diffusion in nonlinear Fokker-Planck equations. *Eur. Phys. J. B* **2009**, *70*, 107–116.
36. Shiino, M. Nonlinear Fokker-Planck equations associated with generalized entropies: Dynamical characterization and stability analyses. *J. Phys. Conf. Ser.* **2010**, *201*, 012004.
37. Ribeiro, M.S.; Nobre, F.D.; Curado, E.M.F. Classes of *N*-Dimensional Nonlinear Fokker-Planck Equations Associated to Tsallis Entropy. *Entropy* **2011**, *13*, 1928–1944.
38. Andrade, J.S., Jr.; da Silva, G.F.T.; Moreira, A.A.; Nobre, F.D.; Curado, E.M.F. Thermostatistics of overdamped motion of interacting particles. *Phys. Rev. Lett.* **2010**, *105*, 260601.
39. Ribeiro, M.S.; Nobre, F.D.; Curado, E.M.F. Time evolution of interacting vortices under overdamped motion. *Phys. Rev. E* **2012**, *85*, 021146.
40. Ribeiro, M.S.; Nobre, F.D.; Curado, E.M.F. Overdamped motion of interacting particles in general confining potentials: time-dependent and stationary-state analyses. *Eur. Phys. J. B* **2012**, *85*, 399.
41. Nobre, F.D.; Souza, A.M.C.; Curado, E.M.F. Effective-temperature concept: A physical application for nonextensive statistical mechanics. *Phys. Rev. E* **2012**, *86*, 061113.
42. Curado, E.M.F.; Souza, A.M.C.; Nobre, F.D.; Andrade, R.F.S. Carnot cycle for interacting particles in the absence of thermal noise. *Phys. Rev. E* **2014**, *89*, 022117.
43. Nobre, F.D.; Curado, E.M.F.; Souza, A.M.C.; Andrade, R.F.S. Consistent thermodynamic framework for interacting particles by neglecting thermal noise. *Phys. Rev. E* **2015**, *91*, 022135.
44. Ribeiro, M.S.; Casas, G.A.; Nobre, F.D. Second law and entropy production in a nonextensive system. *Phys. Rev. E* **2015**, *91*, 012140.
45. Ribeiro, M.S.; Nobre, F.D.; Curado, E.M.F. Comment on "Vortex distribution in a confining potential". *Phys. Rev. E* **2014**, *90*, 026101.
46. Ribeiro, M.S.; Casas, G.A.; Nobre, F.D. Multi-diffusive nonlinear Fokker-Planck equation. *J. Phys. A* **2017**, *50*, 065001.
47. Souza, A.M.C.; Andrade, R.F.S.; Nobre, F.D.; Curado, E.M.F. Thermodynamic framework for compact q-Gaussian distributions. *Physica A* **2018**, *491*, 153–166.
48. Malacarne, L.C.; Mendes, R.S.; Pedron, I.T.; Lenzi, E.K. Nonlinear equation for anomalous diffusion: Unified power-law and stretched exponential exact solution. *Phys. Rev. E* **2001**, *63*, 030101.
49. Malacarne, L.C.; Mendes, R.S.; Pedron, I.T.; Lenzi, E.K. N-dimensional nonlinear Fokker-Planck equation with time-dependent coefficients. *Phys. Rev. E* **2002**, *65*, 052101.
50. Da Silva, L.R.; Lucena, L.S.; da Silva, P.C.; Lenzi, E.K.; Mendes, R.S. Multidimensional nonlinear diffusion equation: Spatial time dependent diffusion coefficient and external forces. *Physica A* **2005**, *357*, 103–108.
51. Nobre, F.D.; Curado, E.M.F.; Rowlands, G. A procedure for obtaining general nonlinear Fokker-Planck equations. *Physica A* **2004**, *334*, 109–118.
52. Curado, E.M.F.; Nobre, F.D. Derivation of nonlinear Fokker-Planck equations by means of approximations to the master equation. *Phys. Rev. E* **2003**, *67*, 021107.
53. Boon, J.P.; Lutsko, J.F. Nonlinear diffusion from Einstein's master equation. *Europhys. Lett.* **2007**, *80*, 60006.
54. Lutsko, J.F.; Boon, J.P. Generalized diffusion: A microscopic approach. *Phys. Rev. E* **2008**, *77*, 051103.
55. Zand, J.; Tirnakli, U.; Jensen, H.J. On the relevance of *q*-distribution functions: The return time distribution of restricted random walker. *J. Phys. A* **2015**, *48*, 425004.
56. Borland, L. Microscopic dynamics of the nonlinear Fokker-Planck equation: A phenomenological model. *Phys. Rev. E* **1998**, *57*, 6634–6642.

57. Borland, L. Ito-Langevin equations within generalized thermostatistics. *Phys. Lett. A* **1998**, *245*, 67–72.
58. Beck, C. Dynamical Foundations of Nonextensive Statistical Mechanics. *Phys. Rev. Lett.* **2001**, *87*, 180601.
59. Anteneodo, C.; Tsallis, C. Multiplicative noise: A mechanism leading to nonextensive statistical mechanics. *J. Math. Phys.* **2003**, *44*, 5194–5203.
60. Fuentes, M.A.; Cáceres, M.O. Computing the non-linear anomalous diffusion equation from first principles. *Phys. Lett. A* **2008**, *372*, 1236–1239.
61. dos Santos, B.C.; Tsallis, C. Time evolution towards q-Gaussian stationary states through unified Itô-Stratonovich stochastic equation. *Phys. Rev. E* **2010**, *82*, 061119.
62. Casas, G.A.; Nobre, F.D.; Curado, E.M.F. Entropy production and nonlinear Fokker-Planck equations. *Phys. Rev. E* **2012**, *86*, 061136.
63. Arenas, Z.G.; Barci, D.G.; Tsallis, C. Nonlinear inhomogeneous Fokker-Planck equation within a generalized Stratonovich prescription. *Phys. Rev. E* **2014**, *90*, 032118.
64. Bianucci, M. Large Scale Emerging Properties from Non Hamiltonian Complex Systems. *Entropy* **2017**, *19*, 302.
65. Zapperi, S.; Moreira, A.A.; Andrade, J.S., Jr. Thermostatistics of overdamped motion of interacting particles. *Phys. Rev. Lett.* **2001**, *87*, 180601.
66. Lenzi, E.K.; Mendes, R.S.; Tsallis, C. Crossover in diffusion equation: Anomalous and normal behaviors. *Phys. Rev. E* **2003**, *67*, 031104.
67. Andrade, R.F.S.; Souza, A.M.C.; Curado, E.M.F.; Nobre, F.D. A thermodynamical formalism describing mechanical interactions. *Europhys. Lett.* **2014**, *108*, 20001.
68. Ribeiro, M.S.; Nobre, F.D. Repulsive particles under a general external potential: Thermodynamics by neglecting thermal noise. *Phys. Rev. E* **2016**, *94*, 022120.
69. Khinchin, A.I. *Mathematical Foundations of Information Theory*; Dover Publications: New York, NY, USA, 1957.
70. Valluri, S.R.; Gil, M.; Jeffrey, D.J.; Basu, S. The Lambert *W* function and quantum statistics. *J. Math. Phys.* **2009**, *50*, 102103.
71. Vieira, C.M.; Carmona, H.A.; Andrade, J.S. Jr.; Moreira, A.A. General continuum approach for dissipative systems of repulsive particles. *Phys. Rev. E* **2016**, *93*, 060103.

entropy

MDPI

Article

Information-Length Scaling in a Generalized One-Dimensional Lloyd's Model

J. A. Méndez-Bermúdez [1,*] and R. Aguilar-Sánchez [2]

[1] Instituto de Física, Benemérita Universidad Autónoma de Puebla, Puebla 72570, Mexico
[2] Facultad de Ciencias Químicas, Benemérita Universidad Autónoma de Puebla, Puebla 72570, Mexico; ras747698@gmail.com
* Correspondence: jmendezb@ifuap.buap.mx; Tel.: +52-222-229-5610

Received: 27 December 2017; Accepted: 8 April 2018; Published: 20 April 2018

Abstract: We perform a detailed numerical study of the localization properties of the eigenfunctions of one-dimensional (1D) tight-binding wires with on-site disorder characterized by long-tailed distributions: For large ϵ, $P(\epsilon) \sim 1/\epsilon^{1+\alpha}$ with $\alpha \in (0,2]$; where ϵ are the on-site random energies. Our model serves as a generalization of 1D Lloyd's model, which corresponds to $\alpha = 1$. In particular, we demonstrate that the information length β of the eigenfunctions follows the scaling law $\beta = \gamma x/(1+\gamma x)$, with $x = \xi/L$ and $\gamma \equiv \gamma(\alpha)$. Here, ξ is the eigenfunction localization length (that we extract from the scaling of Landauer's conductance) and L is the wire length. We also report that for $\alpha = 2$ the properties of the 1D Anderson model are effectively reproduced.

Keywords: Lloyd model; scaling laws; information length; one-dimensional disordered systems

1. Introduction

There is a class of disordered systems characterized by random variables $\{\epsilon\}$ whose density distribution function exhibits a slow decaying tail:

$$P(\epsilon) \sim \frac{1}{|\epsilon|^{1+\alpha}}, \tag{1}$$

for large $|\epsilon|$, with $0 < \alpha < 2$. The study of this class of disordered systems dates back to Lloyd [1], who studied spectral properties of a three-dimensional (3D) lattice described by a 3D tight-binding Hamiltonian with Cauchy-distributed on-site potentials (which corresponds to the particular value $\alpha = 1$ in Equation (1)). Since then, a considerable number of works have been devoted to the study of spectral, eigenfunction, and transport properties of Lloyd's model in its original 3D setup [2–12] and in lower dimensional versions [11–30]. Consequently, disorder characterized by Equation (1) is commonly known as Lévy-type disorder. In addition, the recent experimental realizations of the so-called Lévy glasses [31] as well as Lévy waveguides [32,33] has refreshed the interest in the study of systems characterized by Lévy-type disorder; see some examples in Refs. [34–50].

It is important to point out that one-dimensional (1D) tight-binding wires with power-law distributed random on-site potentials, characterized by power-laws different from $\alpha = 1$ (which corresponds to the 1D Lloyd's model), have been scarcely studied; for prominent exceptions see Refs. [26,27]. Thus, in this paper, we perform a detailed numerical study of the localization properties of the eigenfunctions of disordered wires defined as a generalization of the 1D Lloyd's model as follows. We shall study 1D wires described by the Hamiltonian

$$H = \sum_{n=1}^{L} \left[\epsilon_n \mid n \rangle \langle n \mid - v_{n,n+1} \mid n \rangle \langle n+1 \mid - v_{n,n-1} \mid n \rangle \langle n-1 \mid \right]; \tag{2}$$

where L is the length of the wire given as the total number of sites n, ϵ_n are random on-site potentials, and $v_{n,m}$ are the hopping integrals between nearest neighbours. In particular, we set $v_{n,m} = v = 1$ and consider the on-site potentials ϵ_n following the distribution of Equation (1) with $0 < \alpha \leq 2$. We note that when $\alpha = 1$ we recover the original 1D Lloyd's model.

Of particular interest is the comparison between the 1D Anderson model [51] and our generalization of 1D Lloyd's model, since the former represents the most prominent model of disordered wires [52]. Indeed, the 1D Anderson model is also described by the tight-binding Hamiltonian of Equation (2). However, while for the standard 1D Anderson model (with white-noise on-site disorder $\langle \epsilon_n \epsilon_m \rangle = \sigma^2 \delta_{nm}$ and $\langle \epsilon_n \rangle = 0$) the on-site potentials are characterized by a finite variance $\sigma^2 = \langle \epsilon_n^2 \rangle$ (in most cases the corresponding probability distribution function $P(\epsilon)$ is chosen as a box or a Gaussian distribution), in the generalized 1D Lloyd's model the second moment of the random on-site energies ϵ_n diverges for $0 < \alpha < 2$, and if $0 < \alpha < 1$ also the first moment diverges. Moreover, when $\alpha = 2$ the properties of the generalized 1D Lloyd's model are expected to be similar to those of the 1D Anderson model, since the variance of Equation (1) becomes finite.

It is relevant to recall that the eigenstates Ψ of the *infinite* 1D Anderson model are exponentially localized around a site position n_0 [52]:

$$|\Psi_n| \sim \exp\left(-\frac{|n - n_0|}{\xi} \right) ; \tag{3}$$

where ξ is the eigenfunction localization length. Moreover, for weak disorder ($\sigma^2 \ll 1$), the only relevant parameter for describing the statistical properties of the transmission of the *finite* 1D Anderson model is the ratio L/ξ [53], a fact known as single parameter scaling. The above exponential localization of eigenfunctions strongly affects the scattering and transport properties of the corresponding open wire. In particular, the transmission or dimensionless conductance G, that within a scattering approach to the electronic transport is given as [54–57]

$$G = |t|^2 \tag{4}$$

(where t is the transmission amplitude of the 1D wire), becomes exponentially small [58]:

$$\langle -\ln G \rangle = \frac{2}{x} \tag{5}$$

with

$$x = \frac{\xi}{L} . \tag{6}$$

Thus, relation (5) can be used to obtain the localization length ξ from the transmission of the disordered wire. Remarkably, it has been shown that Equation (5) is also valid for the generalized 1D Lloyd's model [26,27] implying the single parameter scaling, see also [22,23].

Moreover, outstandingly, it has been found [30] that the eigenfunction properties of both the 1D Anderson model and the 1D Lloyd's model (i.e., our generalized 1D Lloyd's model with $\alpha = 1$), characterized by the *information length* β (defined in Equation (12) below), are *universal* for the fixed ratio of Equation (6). More specifically, it was numerically shown that the scaling function

$$\beta = \frac{\gamma x}{1 + \gamma x} , \tag{7}$$

with $\gamma \sim 1$, holds; see also [59].

Thus, below we explore the validity of scaling (7) for the eigenfunctions of the generalized 1D Lloyd's model. Of particular relevance is the regime $0 < \alpha < 1$ where the coexistence of insulating and ballistic regimes has been reported [27] (evidenced by well defined peaks in $P(G)$ at $G = 0$ and

$G = 1$); a signature of the coexistence of localized and extended eigenfunctions. In our study, we use the 1D Anderson model as reference.

Alongside this work we generate the random variables $\{\epsilon\}$ needed to construct the generalized 1D Lloyd's model by the use of the algorithm introduced in Ref. [60]. That algorithm was designed to produce random variables having the probability density function

$$\mathcal{P}(\epsilon) = \frac{\mathcal{C}}{\epsilon^{1+\alpha}}, \tag{8}$$

which has a lower cut-off at $\epsilon_0 = (\mathcal{C}/\alpha)^{1/\alpha}$ such that $\int_{\epsilon_0}^{\infty} \mathcal{P}(\epsilon) d\epsilon = 1$. In addition, we randomize the signs of the obtained variables $\{\epsilon\}$ in order to set the band center of the generalized 1D Lloyd's model to $E = 0$. Moreover, we have verified that our conclusions do not depend on our choice of $\mathcal{P}(\epsilon)$. Indeed, we obtained similar results (not shown here) with $\mathcal{P}(\epsilon) = \alpha/(1+\epsilon)^{1+\alpha}$ and $\mathcal{P}(\epsilon) = [1/\Gamma(\alpha)]2^{-\alpha}\exp(-1/2\epsilon)/\epsilon^{1+\alpha}$ (where Γ is the Euler gamma function), that we also used in [27].

2. Results

2.1. Extraction of the Localization Length ξ

We obtain the localization length ξ for the generalized 1D Lloyd's model as follows: We open the isolated wires by attaching two semi-infinite single channel leads to the border sites at opposite sides. Each lead is described by a 1D semi-infinite tight-binding Hamiltonian. Using standard methods, see e.g., [61–64], we can write the transmission amplitude through the disordered wires as

$$t = -2i\sin(k)\, \mathcal{W}^T \frac{1}{E - \mathcal{H}_{\text{eff}}} \mathcal{W}, \tag{9}$$

where $k = \arccos(E/2)$ is the wave vector supported in the leads and $\mathcal{H}_{\text{eff}}$ is an effective non-hermitian Hamiltonian given by $\mathcal{H}_{\text{eff}} = H - e^{ik}\mathcal{W}\mathcal{W}^T$. Here, $\mathcal{W}$ is a $L \times 1$ vector that specifies the positions of the attached leads to the wire. In our setup, all elements of $\mathcal{W}$ are equal to zero except $\mathcal{W}_{11}$ and $\mathcal{W}_{L1}$ which we set to unity (i.e., the leads are attached to the wire with a strength equal to the inter-site hopping amplitudes: $v = 1$). Also, we have fixed the energy at $E = 0$. Therefore, we use Equation (4) to compute G.

Then, in Figure 1a we present the ensemble average $\langle -\ln G\rangle$ as a function of L for the generalized 1D Lloyd's model for several values of α. It is clear from this figure that $\langle -\ln G\rangle \propto L$ for all the values of α we consider here. Therefore, we can extract the localization length ξ by fitting the curves $\langle -\ln G\rangle$ vs. L with Equation (5); see dashed lines in Figure 1a. We have observed that it is possible to tune ξ for the same value of α by moving $\mathcal{C}$ in Equation (8). Thus, in Figure 1b we report the values of ξ, extracted from the curves $\langle -\ln G\rangle$ vs. L, as a function of $\mathcal{C}$. Moreover, note that

$$\xi \propto \mathcal{C}^{-\alpha}, \tag{10}$$

as shown with the dashed lines in Figure 1b.

In the following Subsection, we use the values of ξ obtained here to construct the ratio of Equation (6) and verify scaling (7) accordingly.

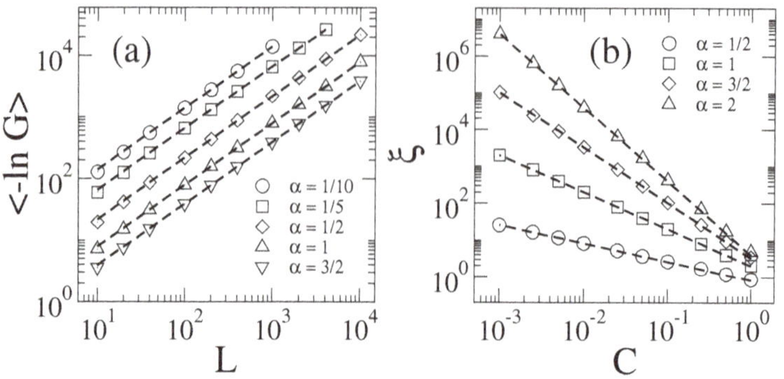

Figure 1. (**a**) Average logarithm of the conductance $\langle -\ln G \rangle$ as a function of L for the generalized 1D Lloyd's model characterized by the values of α indicated on the figure (symbols). $C = 1$ and $E = 0$ were used. Each point was calculated using 10^4 disorder realizations. Dashed lines are the fittings of the data with Equation (5). The obtained values of ξ from the fittings are: $\xi \approx 0.143$ (for $\alpha = 1/10$), $\xi \approx 0.304$ (for $\alpha = 1/5$), $\xi \approx 0.919$ (for $\alpha = 1/2$), $\xi \approx 2.535$ (for $\alpha = 1$), and $\xi \approx 5.144$ (for $\alpha = 3/2$). (**b**) ξ as a function of the constant C. Dashed lines are the fittings of the data with Equation (10). The obtained values for the proportionality constant c from the fittings are: $c \approx 0.82$ (for $\alpha = 1/2$), $c \approx 2.01$ (for $\alpha = 1$), $c \approx 3.25$ (for $\alpha = 3/2$), and $c \approx 4.01$ (for $\alpha = 2$). Error bars on both panels are not shown since they are much smaller than symbol size.

2.2. Calculation of the Information Length β

We compute the information length β for the generalized 1D Lloyd's model by the use of the Shannon entropy S of the corresponding eigenfunctions: With the Shannon entropy for the eigenfunction Ψ^m, which is given as

$$S = -\sum_{n=1}^{L} (\Psi_n^m)^2 \ln(\Psi_n^m)^2 , \tag{11}$$

we write β (see e.g., [65]) as

$$\beta = \exp\left[-\left(S_{\text{GOE}} - \langle S \rangle\right)\right] . \tag{12}$$

Here $S_{\text{GOE}} \approx \ln(L/2.07)$ is the entropy of a random eigenfunction with Gaussian distributed amplitudes (i.e., an eigenfunction of the Gaussian Orthogonal Ensemble [66]). With this definition for S (recall that S provides the number of principal components of an eigenfunction in a given basis) and β, when the eigenfunctions are localized $\langle S \rangle \approx 0$ and $\beta \approx 2.07/L$, which tends to zero for large L. On the other hand, when *fully chaotic* eigenfunctions extend over the L available basis states $\langle S \rangle = S_{\text{GOE}}$ and $\beta = 1$. Therefore, β can take values in the range $(0, 1]$.

Below we use exact numerical diagonalization to obtain the eigenfunctions Ψ^m ($m = 1 \ldots L$) of large ensembles of the generalized 1D Lloyd's model characterized by the parameters L, C, and α. We perform the average $\langle S \rangle$ taking half of the eigenfunctions, around the band center, for each disordered wire length such that $\langle S \rangle$ is computed with 10^5 data values. For example, since for a wire of length $L = 10^3$ we extract $10^3/2$ Shannon entropies only, we construct an ensemble of 2×10^2 disordered wires of that length to compute $\langle S \rangle$ from a double average; that is, we average over a subset of the eigenfunctions and over wire realizations. We use half of the eigenfunctions around the band center to be consistent with the fact that we have fixed $E = 0$ in Equation (9), however we have verified that our conclusions do not depend on this choice: i.e., we could reduce the energy window around $E = 0$ or even consider all eigenfunctions to compute $\langle S \rangle$ obtaining equivalent results.

In Figure 2a we present β as a function of ξ/L, see Equations (6) and (7), for the generalized 1D Lloyd's model with $\alpha = 1$, i.e., for the actual 1D Lloyd's model. Recall that to construct the ratios

ξ/L we are using the values of ξ obtained in the previous Subsection from the scaling of Landauer's conductance; see Figure 1a. In addition, in Figure 2b the logarithm of $\beta/(1-\beta)$ as a function of $\ln(\xi/L)$ is presented. The quantity $\beta/(1-\beta)$ is useful in the study of the scaling properties of β because

$$\frac{\beta}{1-\beta} = \gamma x \,, \tag{13}$$

which is obtained directly by equating γx from (7), implies that a plot of $\ln[\beta/(1-\beta)]$ vs. $\ln(x)$ (with $x = \xi/L$) is a straight line with unit slope. This fact applies for 1D Lloyd's model as can be easily verified by comparing the black symbols with the black dashed line in Figure 2b. Note that to construct Figure 2 we have used wires with lengths in the range $10^2 \leq L \leq 10^3$ only, however by changing C (different symbols indicate different values of C) it is possible to span a large range of ξ/L values.

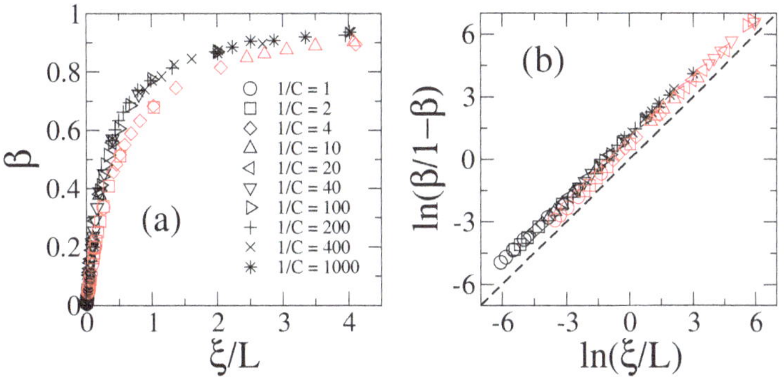

Figure 2. (**a**) Information length β as a function of ξ/L [see Equations (6) and (7)] for the generalized 1D Lloyd's model with $\alpha = 1$ (black symbols) and the 1D Anderson model (red symbols). (**b**) Logarithm of $\beta/(1-\beta)$ as a function of $\ln(\xi/L)$ [see Equation (13)]. Different symbols correspond to disorder distributions with different constants C, as indicated in (**a**). Each symbol is computed by averaging over 10^5 eigenstates. Wires with $10^2 \leq L \leq 10^3$ were used. The identity, black dashed line in (**b**), is shown to guide the eye.

In Figure 2 we include, in red symbols, the corresponding data for the 1D Anderson model. Note that the functional forms in both panels, for both data sets, are very similar. With this, we confirm that scaling (7) is valid for the 1D Lloyd's model and the 1D Anderson model; as already shown in [30].

Now, in Figure 3 we plot curves of β vs. ξ/L (left panels) and $\ln[\beta/(1-\beta)]$ vs. $\ln(\xi/L)$ (right panels) for several values of α. Again, as reference, we include curves for the 1D Anderson model (red dashed lines). In particular, we present representative values of α in the interval $0 < \alpha < 1$ ($\alpha = 1/2$ and $3/4$), one value of α in the interval $1 \leq \alpha < 2$ ($\alpha = 3/2$), and $\alpha = 2$. We stress the choice of the values of α in Figure 3 because it is known that the properties of systems with Levy-type disorder can be significantly different when the disorder is characterized by values of α in the intervals $0 < \alpha < 1$ or $1 \leq \alpha < 2$. However, as shown in Figure 3 there is no major difference between different values of α except for the evolution of the curves towards that for the 1D Anderson model when increasing α. In fact, once $\alpha = 2$ the 1D Anderson model is effectively reproduced, see the lower panels of Figure 3.

Thus, we have verified that scaling (7) describes remarkably well all data sets. Indeed, in left panels of Figure 3 we show fittings of the data with Equation (7); see full black lines. Finally, in Figure 4 we report the values of the coefficients γ, obtained from the fittings of the data using Equation (7), for the values of α considered here. Clearly, when $\alpha = 2$, the coefficient γ is already very close to that for the 1D Anderson model ($\gamma \approx 2$); while, in the opposite limit (i.e., when $\alpha \to 0$) the value of γ becomes relatively large.

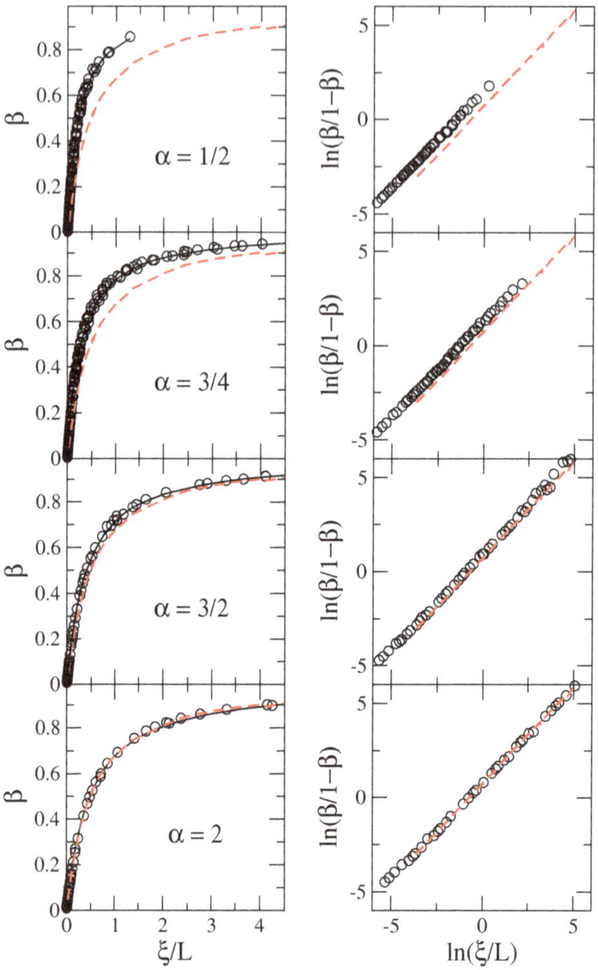

Figure 3. (**Left panels**) Information length β as a function of ξ/L [see Equations (6) and (7)] for ensembles of 1D disordered wires. (**Right panels**) Logarithm of $\beta/(1-\beta)$ as a function of $\ln(\xi/L)$ [see Equation (13)]. Black circles correspond to the generalized 1D Lloyd's model with $\alpha = 1/2$, $3/4$, $3/2$, and 2 (from top to bottom); while the Anderson model is represented by red dashed lines. Each symbol is computed by averaging over 10^5 (10^6) eigenstates when $L \geq 10^2$ ($L < 10^2$). Here, as in Figure 2, we change $\mathcal{C}$ to span a large range of ξ/L values however we do not use different symbols to indicate different values of $\mathcal{C}$ to avoid figure saturation. Full black lines in left panels correspond to fittings of the data with Equation (7) with $\gamma \approx 4.3798$ (for $\alpha = 1/2$), $\gamma \approx 3.5453$ (for $\alpha = 3/4$), $\gamma \approx 2.4891$ (for $\alpha = 3/2$), and $\gamma \approx 2.069$ (for $\alpha = 2$).

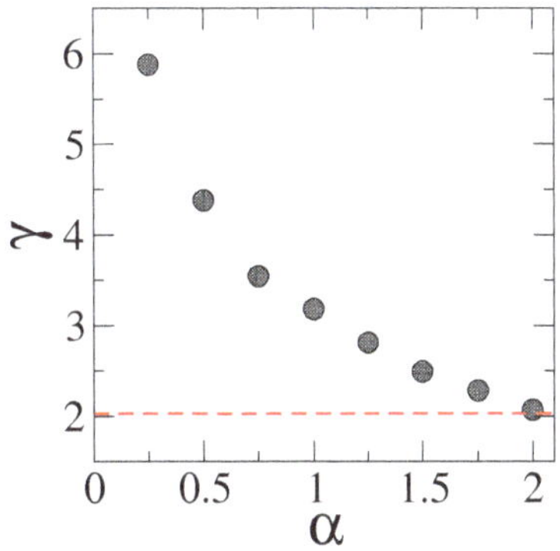

Figure 4. Coefficient γ as a function of α, obtained from the fittings of the curves β vs. ξ/L with Equation (7); see left panels in Figure 3. Black circles correspond to the generalized 1D Lloyd's model, while the 1D Anderson model is represented by the red dashed line at $\gamma \approx 2$. Error bars are not shown since they are much smaller than symbol size.

3. Discussion

In this paper, by the use of extensive numerical simulations, we demonstrate that the information length β of the eigenfunctions of our generalization of 1D Lloyd's model scales with the ratio $x = \xi(\alpha)/L$ as $\gamma x/(1+\gamma x)$, where $\xi(\alpha)$ is the eigenfunction localization length, $\gamma \equiv \gamma(\alpha)$, and L is the wire length. Here α is the power-law decay of the long-tailed distributions, $P(\epsilon) \sim 1/\epsilon^{1+\alpha}$, characterizing the on-site random energies of 1D tight-binding wires.

It is particularly relevant that scaling (7) describes the eigenfunction properties of our generalization of 1D Lloyd's model for all values of α, since it has been shown that transport properties of this model are significatively different in the intervals $0 < \alpha < 1$ and $1 \leq \alpha < 2$ [27]. Moreover, we have shown that for $\alpha = 2$ the generalized 1D Lloyd's model already reproduces the properties of the 1D Anderson model.

It is pertinent to add that scaling (7) is also valid for the eigenfunctions of other disordered models (when the scaling parameter x is properly defined): the banded random matrix (BRM) model [67–74], the kicked-rotator model [65,71,75] (a quantum-chaotic system characterized by a random-like banded Hamiltonian matrix), the diluted BRM model [76], and multiplex and multilayer random networks [77]. Thus, we include our generalization of 1D Lloyd's model to the family of complex systems described by scaling (7).

Finally, we want to recall that scaling (7), can be rewritten in a "model independent" form as a relation between properly-defined inverse lengths [30,67]:

$$\frac{1}{d(L,\xi)} = \frac{1}{d(\infty,\xi)} + \frac{1}{d(L,0)}, \tag{14}$$

were $d(L,\xi) \equiv \exp[\langle S(L,\xi)\rangle]$, which is also applicable to our generalization of 1D Lloyd's model.

Acknowledgments: This work was supported by VIEP-BUAP (Grant Nos. MEBJ-EXC17-I and AGSR-NAT17-I), Fondo Institucional PIFCA (Grant Nos. BUAP-CA-169 and BUAP-CA-40), and CONACyT (Grant Nos. CB-2013/220624 and CB-2014/243030).

Author Contributions: The authors contributed equally to this work. J. A. Mendez-Bermudez and R. Aguilar-Sanchez conceived, designed and performed the experiments; J. A. Mendez-Bermudez and R. Aguilar-Sanchez analyzed the data and wrote the paper.

Conflicts of Interest: The authors declare no conflict of interest.

References

1. Lloyd, P. Exactly solvable model of electronic states in a three-dimensional disordered Hamiltonian: non-existence of localized states. *J. Phys. C* **1969**, *2*, 1717–1725. [CrossRef]
2. Saitoh, M. Existence of localization in Lloyd's disordered lattice model. *Phys. Lett. A* **1970**, *33*, 44–45. [CrossRef]
3. Saitoh, M. Electric Conductivity for Lloyd's Disordered Lattice Model. *Progr. Theor. Phys.* **1971**, *45*, 746–755. [CrossRef]
4. Kumar, A.P.; Baskaran, G. Coherent potential approximation, averaged T-matrix approximation and Lloyd's model. *J. Phys. C* **1973**, *6*, L399–L401. [CrossRef]
5. Hoshino, K. Electrical conductivity and electron localization for the Lloyd model. *Phys. Lett. A* **1976**, *56*, 133–134. [CrossRef]
6. Bandy, W.R.; Glick, A.J. Tight-binding Green's-function calculations of electron tunneling. II. Diagonal disorder in the one-dimensional two-band model. *Phys. Rev. B* **1977**, *16*, 2346–2349. [CrossRef]
7. Kivelson, S.; Gelatt, C.D. Impurity states in a disordered insulator: The Lloyd model. *Phys. Rev. B* **1979**, *20*, 4167–4170. [CrossRef]
8. Simon, B. Equality of the density of states in a wide class of tight-binding Lorentzian random models. *Phys. Rev. B* **1983**, *27*, 3859–3860. [CrossRef]
9. Rodrigues, D.E.; Weisz, J.F. Generalization of the Lloyd model for calculation of electronic structure at disordered interfaces. *Phys. Rev. B* **1986**, *34*, 2306–2310. [CrossRef]
10. Kolley, E.; Kolley, W. Conductivity in Anderson-type models: a comparative study of critical disorder. *J. Phys. C* **1988**, *21*, 6099–6109. [CrossRef]
11. Johnston, R.; Kunz, H. A method for calculating the localisation length, with an analysis of the Lloyd model. *J. Phys. C* **1983**, *16*, 4565–4580. [CrossRef]
12. Rodrigues, D.E.; Pastawski, H.M.; Weisz, J.F. Localization and phase coherence length in the Lloyd model. *Phys. Rev. B* **1986**, *34*, 8545–8549. [CrossRef]
13. Thouless, D.J. A relation between the density of states and range of localization for one dimensional random systems. *J. Phys. C* **1972**, *5*, 77–81. [CrossRef]
14. Ishii, K. Localization of eigenstates and transport phenomena in the one-dimensional disordered system. *Suppl. Progr. Theor. Phys.* **1973**, *53*, 77–138. [CrossRef]
15. Abou-Chacra, R.; Thouless, D.J. Self-consistent theory of localization. II. Localization near the band edges. *J. Phys. C* **1974**, *7*, 65–75. [CrossRef]
16. Thouless, D.J. Localisation in the Lloyd model. *J. Phys. C* **1983**, *16*, L929–L931. [CrossRef]
17. MacKinnon, A. Localisation in the Lloyd model of a disordered solid. *J. Phys. C* **1984**, *17*, L289–L291. [CrossRef]
18. Robbins, M.O.; Koiller, B. Localization properties of random and partially ordered one-dimensional systems. *Phys. Rev. B* **1985**, *32*, 4576–4583. [CrossRef]
19. Shepelyansky, D.L. Localization of quasienergy eigenfunctions in action space. *Phys. Rev. Lett.* **1986**, *56*, 677–680. [CrossRef]
20. Fishman, S.; Prange, R.E.; Griniasty, M. Scaling theory for the localization length of the kicked rotor. *Phys. Rev. A* **1989**, *39*, 1628–1633. [CrossRef]
21. Murdy, C.; Brouwer, P.W.; Halperin, B.I.; Gurarie, V.; Zee, A. Density of states in the non-Hermitian Lloyd model. *Phys. Rev. B* **1998**, *58*, 13539–13543. [CrossRef]
22. Deych, L.I.; Lisyansky, A.A.; Altshuler, B.L. Single parameter scaling in one-dimensional localization revisited. *Phys. Rev. Lett.* **2000**, *84*, 2678–2681. [CrossRef]
23. Deych, L.I.; Lisyansky, A.A.; Altshuler, B.L. Single-parameter scaling in one-dimensional Anderson localization: Exact analytical solution. *Phys. Rev. B* **2001**, *64*, 224202. [CrossRef]
24. Gangardt, D.M.; Fishman, S. Localization of eigenstates in a modified Tomonaga-Luttinger model. *Phys. Rev. B* **2001**, *63*, 045106. [CrossRef]
25. Fuchs, C.; Baltz, R.V. Optical properties of quantum wires: Disorder scattering in the Lloyd model. *Phys. Rev. B* **2001**, *63*, 085318. [CrossRef]
26. Titov, M.; Schomerus, H. Anomalous wave function statistics on a one-dimensional lattice with power-law disorder. *Phys. Rev. Lett.* **2003**, *91*, 176601. [CrossRef]

27. Mendez-Bermudez, J.A.; Martinez-Mendoza, A.J.; Gopar, V.A.; Varga, I. Lloyd-model generalization: Conductance fluctuations in one-dimensional disordered systems. *Phys. Rev. E* **2016**, *93*, 012135. [CrossRef]
28. Roy, D.; Kumar, N. Random-phase reservoir and a quantum resistor: The Lloyd model. *Phys. Rev. B* **2007**, *76*, 092202. [CrossRef]
29. Kozlov, G.G. Spectral dependence of the localization degree in the one-dimensional disordered Lloyd model. *Theor. Math. Phys.* **2012**, *171*, 531–540. [CrossRef]
30. Casati, G.; Guarneri, I.; Izrailev, F.M.; Fishman, S.; Molinari, L. Scaling of the information length in 1D tight-binding models. *J. Phys. Condens. Matter* **1992**, *4*, 149–156. [CrossRef]
31. Barthelemy, P.; Bertolotti, J.; Wiersma, D.S. A Levy flight for light. *Nature* **2008**, *453*, 495–498. [CrossRef]
32. Fernandez-Marin, A.A.; Mendez-Bermudez, J.A.; Carbonell, J.; Cervera, F.; Sanchez-Dehesa, J.; Gopar, V.A. Beyond anderson localization in 1D: Anomalous localization of microwaves in random waveguides. *Phys. Rev. Lett.* **2014**, *113*, 233901. [CrossRef]
33. Fernandez-Marin, A.A.; Mendez-Bermudez, J.A.; Carbonell, J.; Cervera, F.; Sanchez-Dehesa, J.; Gopar, V.A. Beyond Anderson localization: Anomalous transmission of waves through media with Levy disorder. In Proceedings of the 9th International Congress on Advanced Electromagnetic Materials in Microwaves and Optics (Metamaterials 2015), Oxford, UK, 7–12 September 2015; pp. 409–411.
34. Beenakker, C.W.J.; Groth, C.W.; Akhmerov, A.R. Nonalgebraic length dependence of transmission through a chain of barriers with a Levy spacing distribution. *Phys. Rev. B* **2009**, *79*, 024204. [CrossRef]
35. Burioni, R.; Caniparoli, L.; Vezzani, A. Levy walks and scaling in quenched disordered media. *Phys. Rev. E* **2010**, *81*, 060101(R). [CrossRef]
36. Eisfeld, A.; Vlaming, S.M.; Malyshev, V.A.; Knoester, J. Excitons in molecular aggregates with Levy-type disorder: Anomalous localization and exchange broadening of optical spectra. *Phys. Rev. Lett.* **2010**, *105*, 137402. [CrossRef]
37. Bertolotti, J.; Vynck, K.; Pattelli, L.; Barthelemy, P.; Lepri, S.; Wiersma, D.S. Engineering disorder in superdiffusive Levy glasses. *Adv. Funct. Mater.* **2010**, *20*, 965–968. [CrossRef]
38. Barthelemy, P.; Bertolotti, J.; Vynck, K.; Lepri, S.; Wiersma, D.S. Role of quenching on superdiffusive transport in two-dimensional random media. *Phys. Rev. E* **2010**, *82*, 011101. [CrossRef]
39. Burresi, M.; Radhalakshmi, V.; Savo, R.; Bertolotti, J.; Vynck, K.; Wiersma, D.S. Weak localization of light in superdiffusive random systems. *Phys. Rev. Lett.* **2012**, *108*, 110604. [CrossRef]
40. Groth, C.W.; Akhmerov, A.R.; Beenakker, C.W.J. Transmission probability through a Levy glass and comparison with a Levy walk. *Phys. Rev. E* **2012**, *85*, 021138. [CrossRef]
41. Burioni, R.; diSanto, S.; Lepri, S.; Vezzani, A. Scattering lengths and universality in superdiffusive Levy materials. *Phys. Rev. E* **2012**, *86*, 031125. [CrossRef]
42. Vlaming, S.M.; Malyshev, V.A.; Eisfeld, A.; Knoester, J. Subdiffusive exciton motion in systems with heavy-tailed disorder. *J. Chem. Phys.* **2013**, *138*, 214316. [CrossRef]
43. Burioni, R.; Ubaldi, E.; Vezzani, A. Superdiffusion and transport in two-dimensional systems with Levy-like quenched disorder. *Phys. Rev. E* **2014**, *89*, 022135. [CrossRef]
44. Bernabo, P.; Burioni, R.; Lepri, S.; Vezzani, A. Anomalous transmission and drifts in one-dimensional Levy structures. *Chaos Solitons Fractals* **2014**, *67*, 11–19. [CrossRef]
45. Zakeri, S.S.; Lepri, S.; Wiersma, D.S. Localization in one-dimensional chains with Levy-type disorder. *Phys. Rev. E* **2015**, *91*, 032112. [CrossRef]
46. Ardakani, A.G.; Nezhadhaghighi, M.G. Controlling Anderson localization in disordered heterostrctures with Levy-type distribution. *J. Opt.* **2015**, *17*, 105601. [CrossRef]
47. Falceto, F.; Gopar, V.A. Conductance through quantum wires with Levy-type disorder: Universal statistics in anomalous quantum transport. *Europhys. Lett.* **2010**, *92*, 57014. [CrossRef]
48. Fernandez-Marin, A.A.; Mendez-Bermudez, J.A.; Gopar, V.A. Photonic heterostructures with Levy-type disorder: Statistics of coherent transmission. *Phys. Rev. A* **2012**, *85*, 035803. [CrossRef]
49. Amanatidis, I.; Kleftogiannis, I.; Falceto, F.; Gopar, V.A. Conductance of one-dimensional quantum wires with anomalous electron wave-function localization. *Phys. Rev. B* **2012**, *85*, 235450. [CrossRef]
50. Amanatidis, I.; Kleftogiannis, I.; Falceto, F.; Gopar, V.A. Coherent wave transmission in quasi-one-dimensional systems with Lévy disorder. *Phys. Rev. E* **2017**, *96*, 062141. [CrossRef]
51. Anderson, P.W. Absence of diffusion in certain random lattices. *Phys. Rev.* **1958**, *109*, 1492–1505. [CrossRef]
52. Abrahams, E. (Ed.) *50 Years of Anderson Localization*; World Scientific, Singapore, Singapore, 2010.

53. Anderson, P.W.; Thouless, D.J.; Abrahams, E.; Fisher, D.S. New method for a scaling theory of localization. *Phys. Rev. B* **1980**, *22*, 3519–3526. [CrossRef]
54. Landauer, R. Spatial variation of currents and fields due to localized scatterers in metallic conduction. *IBM J. Res. Dev.* **1957**, *1*, 223–231. [CrossRef]
55. Landauer, R. Spatial variation of currents and fields due to localized scatterers in metallic conduction. *IBM J. Res. Dev.* **1988**, *32*, 306–316. [CrossRef]
56. Buttiker, M. Four-terminal phase-coherent conductance. *Phys. Rev. Lett.* **1986**, *57*, 1761–1764. [CrossRef]
57. Buttiker, M. Symmetry of electrical conduction. *IBM J. Res. Dev.* **1988**, *32*, 317–334. [CrossRef]
58. Lifshits, I.M.; Gredeskul, S.A.; Pastur, L.A. *Introduction to the Theory of Disordered Systems*; Willey: New York, NY, USA, 1988.
59. Varga, I.; Pipek, J. Information length and localization in one dimension. *J. Phys. Condens. Matter* **1994**, *6*, L115–L122. [CrossRef]
60. Mendez-Bermudez, J.A.; deOliveira, J.A.; Leonel, E.D. Two-dimensional nonlinear map characterized by tunable Levy fights. *Phys. Rev. E* **2014**, *90*, 042138.
61. Mahaux, C.; Weidenmüller, H.A. *Shell Model Approach in Nuclear Reactions*; North-Holland Pub. Co.: Amsterdam, The Netherlands, 1969.
62. Verbaarschot, J.J.M.; Weidenmüller, H.A.; Zirnbauer, M.R. Grassmann integration in stochastic quantum physics: The case of compound-nucleus scattering. *Phys. Rep.* **1985**, *129*, 367–438. [CrossRef]
63. Rotter, I. A continuum shell model for the open quantum mechanical nuclear system. *Rep. Prog. Phys.* **1991**, *54*, 635–682. [CrossRef]
64. Auerbach, N.; Zelevinsky, V. Super-radiant dynamics, doorways and resonances in nuclei and other open mesoscopic systems. *Rep. Prog. Phys.* **2011**, *74*, 106301. [CrossRef]
65. Izrailev, F.M. Simple models of quantum chaos: Spectrum and eigenfunctions. *Phys. Rep.* **1990**, *196*, 299–392. [CrossRef]
66. Metha, M.L. *Random Matrices*; Elsevier: Amsterdam, The Netherlands, 2004.
67. Fyodorov, Y.F.; Mirlin, A.D. Analytical derivation of the scaling law for the inverse participation ratio in quasi-one-dimensional disordered systems. *Phys. Rev. Lett.* **1992**, *69*, 1093–1096. [CrossRef]
68. Casati, G.; Molinari, L.; Izrailev, F.M. Scaling properties of band random matrices. *Phys. Rev. Lett.* **1990**, *64*, 1851–1854. [CrossRef]
69. Evangelou, S.N.; Economou, E.N. Eigenvector statistics and multifractal scaling of band random matrices. *Phys. Lett. A* **1990**, *151*, 345–348. [CrossRef]
70. Fyodorov, Y.F.; Mirlin, A.D. Scaling properties of localization in random band matrices: A σ-model approach. *Phys. Rev. Lett.* **1991**, *67*, 2405–2409. [CrossRef]
71. Izrailev, F.M. Scaling properties of spectra and eigenfunctions for quantum dynamical and disordered systems. *Chaos Solitons Fractals* **1995**, *5*, 1219–1234. [CrossRef]
72. Mirlin, A.D.; Fyodorov, Y.F. The statistics of eigenvector components of random band matrices: Analytical results. *J. Phys. A Math. Gen.* **1993**, *26*, L551–L558. [CrossRef]
73. Fyodorov, Y.F.; Mirlin, A.D. Level-to-level fluctuations of the inverse participation ratio in finite quasi 1D disordered systems. *Phys. Rev. Lett.* **1993**, *71*, 412–415. [CrossRef]
74. Fyodorov, Y.F.; Mirlin, A.D. Statistical properties of eigenfunctions of random quasi 1D one-particle Hamiltonians. *Int. J. Mod. Phys. B* **1994**, *8*, 3795–3842. [CrossRef]
75. Casati, G.; Guarneri, I.; Izrailev, F.M.; Scharf, R. Scaling behavior of localization in quantum chaos. *Phys. Rev. Lett.* **1990**, *64*, 5–8. [CrossRef]
76. Mendez-Bermudez, J.A.; Ferraz de Arruda, G.; Rodrigues, F.A.; Moreno, Y. Diluted banded random matrices: scaling behavior of eigenfunction and spectral properties. *J. Phys. A Math. Theor.* **2017**, *50*, 495205. [CrossRef]
77. Mendez-Bermudez, J.A.; Ferraz de Arruda, G.; Rodrigues, F.A.; Moreno, Y. Scaling properties of multilayer random networks. *Phys. Rev. E* **2017**, *96*, 012307. [CrossRef]

entropy

MDPI

Article

Oscillations in Multiparticle Production Processes †

Grzegorz Wilk [1,*] and Zbigniew Włodarczyk [2]

1 Department of Fundamental Research, National Centre for Nuclear Research, 00-681 Warsaw, Poland
2 Institute of Physics, Jan Kochanowski University, 25-406 Kielce, Poland; zbigniew.wlodarczyk@ujk.edu.pl
* Correspondence: grzegorz.wilk@ncbj.gov.pl; Tel.: +48-22-621-6085

† This paper is an extended version of conference paper given at the SigmaPhi 2017, Corfu, Greece, 10–14 July 2017.

Received: 22 November 2017; Accepted: 4 December 2017; Published: 6 December 2017

Abstract: We discuss two examples of oscillations apparently hidden in some experimental results for high-energy multiparticle production processes: (i) the log-periodic oscillatory pattern decorating the power-like Tsallis distributions of transverse momenta; (ii) the oscillations of the modified combinants obtained from the measured multiplicity distributions. Our calculations are confronted with pp data from the Large Hadron Collider (LHC). We show that in both cases, these phenomena can provide new insight into the dynamics of these processes.

Keywords: scale invariance; log-periodic oscillation; complex nonextensivity parameter; self-similarity; combinants; compound distributions

1. Introduction

In this work, we argue that closer scrutiny of the available experimental results from Large Hadron Collider (LHC) experiments can result in some new, so-far unnoticed (or underrated) features, which can provide new insight into the dynamics of the processes under consideration. More specifically, we shall concentrate on multiparticle production processes at high energies and on two examples of hidden oscillations apparently visible there: (i) the log-periodic oscillations in data on the large transverse momenta spectra, $f(p_T)$ (presented in Section 2); (ii) the oscillations of some coefficients in the recurrence relation defining the multiplicity distributions, $P(N)$ (presented in Section 3).

As will be seen, the first phenomenon is connected with the fact that large p_T distributions follow a quasi power-like pattern, which is best described by the Tsallis distribution [1–3]:

$$f(p_T) = C\left(1 + \frac{p_T}{nT}\right)^{-n}, \qquad n = \frac{1}{q-1} \tag{1}$$

This is a quasi power-law distribution with two parameters: power index n (connected with the nonextensivity parameter q) and scale parameter T (in many applications identified with temperature) [3]. In Figure 1, we present examples of applications of the nonextensive approach to multiparticle distributions represented by Equation (1). Figure 1a,b show the high-quality Tsallis fit as well as a kind of self-similarity of p_T distributions of jets and hadrons [4]. Figure 1c demonstrates that the values of the nonextensivity parameters q for particles in jets correspond rather closely to values of q obtained from the inclusive distributions measured in pp collisions [4,5]. (This observation should be connected with the fact that multiplicity distributions, $P(N)$, are closely connected with the nonextensive approach, and that $q-1 = Var(N)/<N>^2 - 1/<N>$ for negative binomial distributions (NBDs), Poisson distributions (PDs) and binomial distributions (BDs) [5]). In general, one observes the self-similar characteristics of the production process, in both cases originating from their cascading characteristic, which always results in a Tsallis distribution. (In fact, this is a very old idea, introduced by Hagedorn in [6,7], that the production of hadrons proceeds through the formation of *fireballs*, which produces a statistical equilibrium of an undetermined number of all kinds of fireballs, each of which in turn is considered to be a fireball. This idea returned recently in the form of thermofractals introduced in [8]. We note that the NBD discussed below also has a self-similar character [9]. We note also that the self-similarity and fractality features of the multiparticle production process were exhaustively discussed in [10,11].). In the pure dynamical QCD approach to hadronization, one could think of partons fragmenting into final-state hadrons through multiple sub-jet production [11].

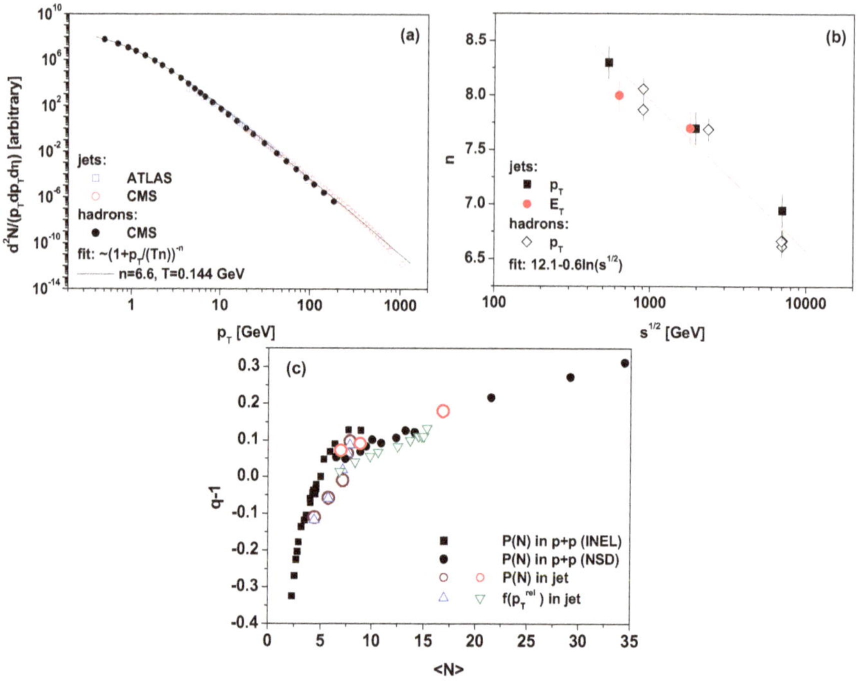

Figure 1. (**a**) Transverse momentum spectra p_T for jets (open symbols) compared with p_T spectra for hadrons (full symbols) from proton–proton collisions at $\sqrt{s} = 7$ TeV (with arbitrary normalization at $p_T = 20$ GeV; data are from [12–15]). The dashed line shows the fit using Equation (1). (**b**) The power index n extracted from fits to the jet spectra shown in (**a**) (full symbols) compared with n from p_T distributions of charged particles (open symbols, data are from [14–16]. (**c**) Compilation of values of the parameters q obtained from the p_T spectra (triangles) and from the multiplicity distributions (circles). Triangles at small $\langle N \rangle$ are obtained from [12,13]; those for larger $\langle N \rangle$ are from [17]. Full squares and circles are from data on multiparticle production in $p + p$ collisions: squares (inelastic data) are from the compilation for beam energy 3.7–303 GeV presented in [18], full circles (non-single diffractive data) are from the compilation in [19], and open-red circ les are from [20].

2. Log-Periodic Oscillations in Data on Large p_T Momenta Distributions

To start with the first example, we note that, despite the exceptional quality of the Tsallis fit presented in Figure 1a (in fact, only such a two-parameter quasi-power-like formula can fit the data over the whole range of p_T), the ratio $R = data/fit$ (which is expected to be a flat function of p_T, $R(p_T) \sim 1$) presented in Figure 2a shows clear log-periodic oscillatory behaviour as a function of the transverse momentum p_T. In fact, it turns out that such behaviour occurs (at the same strength) in data from all LHC experiments, at all energies (provided that the range of the measured p_T is large enough), and that its amplitude increases substantially for the nuclear collisions [21]. These observations strongly suggest that closer scrutiny should be undertaken to understand its possible physical origin.

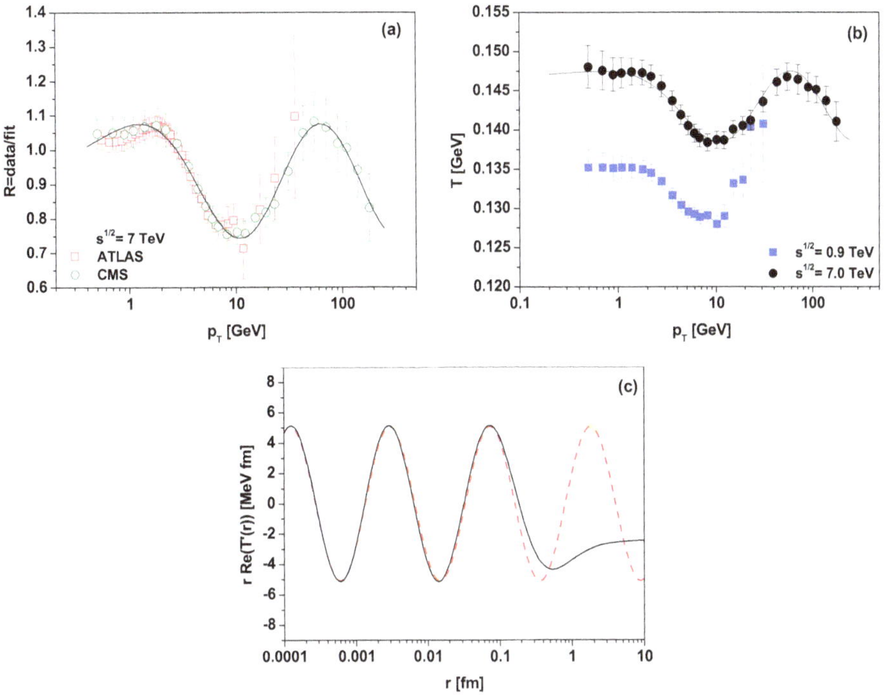

Figure 2. (**a**) The $R = data/fit$ ratio for pp collisions at TeV from the CMS [14,15,22] and ATLAS [12] experiments fitted by Equation (2) with parameters $a = 0.909$, $b = 0.166$, $c = 1.86$, $d = 0.948$ and $f = -1.462$. (**b**) The log-periodic oscillations of $T = T(p_T)$ fitting $R(p_T)$ from panel (**a**) for pp collisions at 0.9 and 7 TeV from the CMS experiment [14,15,22]. These can be fitted by Equation (3) with parameters $\tilde{a} = 0.132$, $\tilde{b} = 0.0035$, $\tilde{c} = 2.2$, $\tilde{d} = 2.0$, and $\tilde{f} = -0.5$ for 0.9 TeV and $\tilde{a} = 0.143$, $\tilde{b} = 0.0045$, $\tilde{c} = 2.0$, $\tilde{d} = 2.0$, and $\tilde{f} = -0.4$ for 7 TeV. (**c**) The results of the Fourier transform of $T(p_T)$ from Figure 2b. The continuous line represents $rT'(r)$ versus r, and the dashed line denotes the function $rT'(r) = 5.1\sin[(2\pi/3.2)\ln(1.24r)]$ fitted for small values of r.

We note that we have two parameters in the Tsallis formula (Equation (1)), n and T, and each of them (or both, but we do not consider such a situation) could be a priori responsible for the observed effect. We start with the power index n. In this case, the observed oscillations may be related to some scale invariance present in the system and are an immanent feature of any power-like distribution [23]. In [24], we showed that they also appear in quasi-power-like distributions of the Tsallis type. In general, they are attributed to a discrete scale invariance (connected with a possible fractal structure of the process under consideration) and are described by introducing a complex power index n (or q). This, in turn, has a number of interesting consequences [25–27], such as a complex heat capacity of the system or complex probability and complex multiplicative noise, all of these known already from other branches of physics. In short, one relies on the fact that power-like distributions, say $O(x) = Cx^{-m}$, exhibit scale invariant behaviour, $Q(\lambda x) = \mu O(x)$, where parameters λ and μ are, in general, related by the condition that $\mu\lambda^m = 1 = \exp(2\pi i k)$, $k = 0, 1, 2, \ldots$, which means that the power index m can take complex values: $m = -\frac{\ln\mu}{\ln\lambda} + i\frac{2\pi k}{\ln\lambda}$. In the case of a Tsallis distribution, this means that Equation (1) is decorated by some oscillating factor $R(p_T)$, the form of which (when keeping only the $k = 0$ and $k = 1$ terms) is

$$R(p_T) = a + b\left[\cos\left(p_T + d\right) + f\right] \tag{2}$$

As one can see in Figure 2, this perfectly fits the observed log-oscillatory pattern.

The second possibility is to keep n constant but allow the scale parameter T to vary with p_T in such a way as to allow a fit to the data on $R\left(p_T\right)$ [27]. The result is shown in Figure 2b. The resulting $T\left(p_T\right)$ has the form of log-periodic oscillations in p_T, which can be parameterized by

$$T\left(p_T\right) = \tilde{a} + \tilde{b}\left[\sin\left(p_T + \tilde{d}\right) + \tilde{f}\right] \tag{3}$$

Such behaviour of $T\left(p_T\right)$ can originate from the well-known stochastic equation for the temperature evolution, which in the Langevin formulation has the following form:

$$\frac{dT}{dt} + \frac{1}{\tau}T + \xi(t) = \Phi \tag{4}$$

where τ is the relaxation time and $\xi(t)$ is time-dependent white noise. Assuming additionally that we have time-dependent transverse momentum $p_T = p_T(t)$ increasing in a way following the scenario of the preferential growth of networks [28],

$$\frac{dp_T}{dt} = \frac{1}{\tau_0}\left(\frac{p_T}{n} \pm T\right) \tag{5}$$

where n is the power index and τ_0 is some characteristic time-step, one may write

$$\frac{1}{\tau_0}\left(\frac{p_T}{n} \pm T\right)\frac{dT}{dp_T} + \frac{1}{\tau}T + \xi(t) = \Phi \tag{6}$$

Equation (3) is obtained in two cases: (i) The noise term increases logarithmically with p_T while the relaxation time τ remains constant:

$$\xi\left(t, p_T\right) = \xi_0(t) + \frac{\omega^2}{n}\ln\left(p_T\right) \tag{7}$$

(ii) The white noise is constant, $\xi\left(t, p_T\right) = \xi_0(t)$, but the relaxation time becomes p_T-dependent, for example,

$$\tau = \tau\left(p_T\right) = \frac{n\tau_0}{n + \omega^2\ln\left(p_T\right)} \tag{8}$$

In both cases ω is some new parameter [3,27]. (To fit data, one needs only a rather small admixture of the stochastic processes with noise depending on p_T. The main contribution comes from the usual energy-independent Gaussian white noise. We note that whereas each of the proposed approaches is based on a different dynamical picture, they are numerically equivalent.).

However, we can use Equation (3) in a way that allows us to look deeper into our dynamical process. To this end, we calculate the Fourier transform of $T\left(p_T\right)$ presented there:

$$T(r) = \sqrt{\frac{2}{\pi}}\int_0^\infty T\left(p_T\right)e^{ip_T r}dp_T = T_0 + T'(r) \tag{9}$$

One can see now how the temperature T (more exactly, its part that varies with r, $T'(r)$) changes with distance r from the collision axis (defined in the plane perpendicular to the collision axis and located at the collision point). The result of this operation is seen in Figure 2c as specific log-periodic oscillations in p_T. Such behaviour can be studied by considering the flow of a compressible fluid in a cylindrical source. Assuming oscillations with small amplitude and velocity v, and introducing the velocity potential f such that $v = \boldsymbol{grad} f$, one finds that f must satisfy the following cylindrical wave equation:

$$\frac{1}{r}\frac{\partial}{\partial r}\left(r\frac{\partial f}{\partial r}\right) - \frac{1}{c^2}\frac{\partial^2 f}{\partial t^2} = 0 \tag{10}$$

It can be shown that this represents a travelling sound wave with velocity v in the direction of propagation. Because the oscillating part of the temperature, T', is related to the velocity v, then

$$T' = \frac{c\kappa T}{c_P} v \tag{11}$$

where $\kappa = \frac{1}{V}\left(\frac{\partial V}{\partial T}\right)_P$ is the coefficient of thermal expansion and c_P denotes the specific heat at constant pressure [29]; in the case of a monochromatic wave where $f(r,t) = f(r)\exp(-i\omega t)$, we have that

$$\frac{\partial^2 f(r)}{\partial r^2} + \frac{1}{r}\frac{\partial f(r)}{\partial r} + K^2 f(r) = 0, \qquad K = K(r) = \frac{\omega}{c(r)} \tag{12}$$

with K being the wave number depending in general on r. For

$$K(r) = \frac{\alpha}{r} \tag{13}$$

the solution of Equation (12) takes the form of some log-periodic oscillation:

$$f(r) \propto \sin[\alpha \ln(r)] \tag{14}$$

Because in our case $f(r) \propto vr$, using Equation (11), we can write that

$$rT'(r) \propto \frac{c\kappa T_0}{c_P} f(r) = \frac{c\kappa T_0}{c_P}\sin[\alpha \ln(r)] \tag{15}$$

which is what we have used in describing the $T'(r)$ presented in panel (c) of Figure 2. The space picture of the collision (in the plane perpendicular to the collision axis and located at the collision point) that emerges is some regular logarithmic structure for small distances, which disappears when r reaches the dimension of the nucleon, that is, for $r \sim 1$ fm. Whether it is connected with the parton structure of the nucleon remains for the moment an open question.

To end this section, we note that Equation (12) with $K(r)$ given by Equation (13) is *scale invariant* and that $f(\lambda r) = f(r)$. We note also that in the variable $\xi = \ln r$, Equation (12) is also *self-similar* because in this variable, it takes the form of the *traveling wave equation*:

$$\frac{\partial^2 F(\xi)}{\partial \xi^2} + \alpha^2 F(\xi) = 0 \tag{16}$$

which has the self-similar solution

$$F(\xi) \propto \cos[\alpha\xi] \tag{17}$$

for which $F(\xi + \ln\lambda) = F(\xi)$ (with $\alpha = \frac{2\pi k}{\ln \lambda}$ or $\lambda = \exp\left(\frac{2\pi k}{\alpha}\right)$, where $k = 1,2,3,\ldots$). This is the *self-similar solution of the second kind*, usually encountered in the description of the intermediate asymptotic. Such asymptotics are observed in phenomena that do not depend on the initial conditions because sufficient time has already passed; nevertheless the system considered is still out of equilibrium [30].

3. Oscillations Hidden in the Multiplicity Distributions Data

Whereas the previous section is concerned with oscillations hidden in the distributions of produced particles in transverse momenta $f(p_T)$ (which by using the Fourier transformation allows us to gain some insight into the space picture of the interaction process), in this section, we concentrate on another important characteristic of the multiparticle production process, namely, on the question of how many particles are produced and with what probability, that is, on the multiplicity distribution function $P(N)$, where N is the observed number of particles. This is usually one of the first observables measured in any multiparticle production experiment [11].

At first we note that any $P(N)$ can be defined in terms of some recurrence relation, the most popular takes the following form:

$$(N+1)P(N+1) = g(N)P(N) \qquad \text{where} \qquad g(N) = \alpha + \beta N \tag{18}$$

Such a linear form of $g(N)$ leads to a NBD, BD or PD:

$$NBD: \quad P(N) = \frac{\Gamma(N+k)}{\Gamma(N+1)\Gamma(k)} p^N (1-p)^k \qquad \text{with} \quad \alpha = kp, \qquad \beta = \frac{\alpha}{k} \tag{19}$$

$$BD: \quad P(N) = \frac{K!}{N!(K-N)!} p^N (1-p)^{K-N} \qquad \text{with} \quad \alpha = \frac{Kp}{1-p}, \quad \beta = -\frac{\alpha}{K} \tag{20}$$

$$PD: \quad P(N) = \frac{\lambda^N}{N!} \exp(-\lambda) \qquad \text{with} \quad \alpha = \lambda, \qquad \beta = 0 \tag{21}$$

Suitable modifications of $g(N)$ result in more involved distributions $P(N)$ (cf. [31] for references).

The most popular form of $P(N)$ is the NBD type of distribution, Equation (19). However, with a growing energy and number of produced secondaries, the NBD starts to deviate from data for large N, and one has to use combinations of NBDs (two [32], three [33] and multicomponent NBDs [34] were proposed) or try to use some other form of $P(N)$ [11,35,36]. For example, in Figure 3a, a single NBD is compared with a two-NBD. However, as shown there, the improvement, although substantial, is not completely adequate. It is best seen when looking at the ratio $R = P_{CMS}(N)/P_{fit}(N)$, which still shows some wiggly structure (albeit substantially weaker than in the case of using only a single NBD to fit the data). Taken seriously, this observation suggests that there is some additional information hidden in the $P(N)$. The question of how to retrieve this information was addressed in [31] by resorting to a more general form of Equation (18) usually used in counting statistics when dealing with cascade stochastic processes [37] in which all multiplicities are connected. In this case, one has coefficients C_j defining the corresponding $P(N)$ in the following way:

$$(N+1)P(N+1) = \langle N \rangle \sum_{j=0}^{N} C_j P(N-j) \tag{22}$$

These coefficients contain the memory of particle $N+1$ about all $N-j$ previously produced particles. Assuming now that all $P(N)$ are given by experiment, one can reverse Equation (22) and obtain a recurrence formula for the coefficients C_j:

$$\langle N \rangle C_j = (j+1)\left[\frac{P(j+1)}{P(0)}\right] - \langle N \rangle \sum_{i=0}^{j-1}\left[\frac{P(j-i)}{P(0)}\right] \tag{23}$$

As can be seen in Figure 3c, the coefficients C_j obtained from the data presented in Figure 3a show oscillatory behaviour (with period roughly equal to 16) gradually disappearing with N. They can be fitted by the following formula:

$$\langle N \rangle C_j = \left(a^2 + b^2\right)^{j/2} \sin\left[c + j \arctan(b/a)\right] + d^j \tag{24}$$

with parameters $a = 0.89$, $b = 0.37$, $c = 5.36$, and $d = 0.95$. Such oscillations do not appear in the single NBD fit presented in Figure 3a, and there is only a small trace of oscillations for the two-NBD fit presented in Figure 3a. This is because for a single NBD, one has a smooth exponential dependence of the corresponding C_j on the rank j:

$$C_j = \frac{k}{\langle N \rangle} p^{j+1} = \frac{k}{k+m} \exp(j \ln p) \tag{25}$$

and one can expect any structure only for the multi-NBD cases [31].

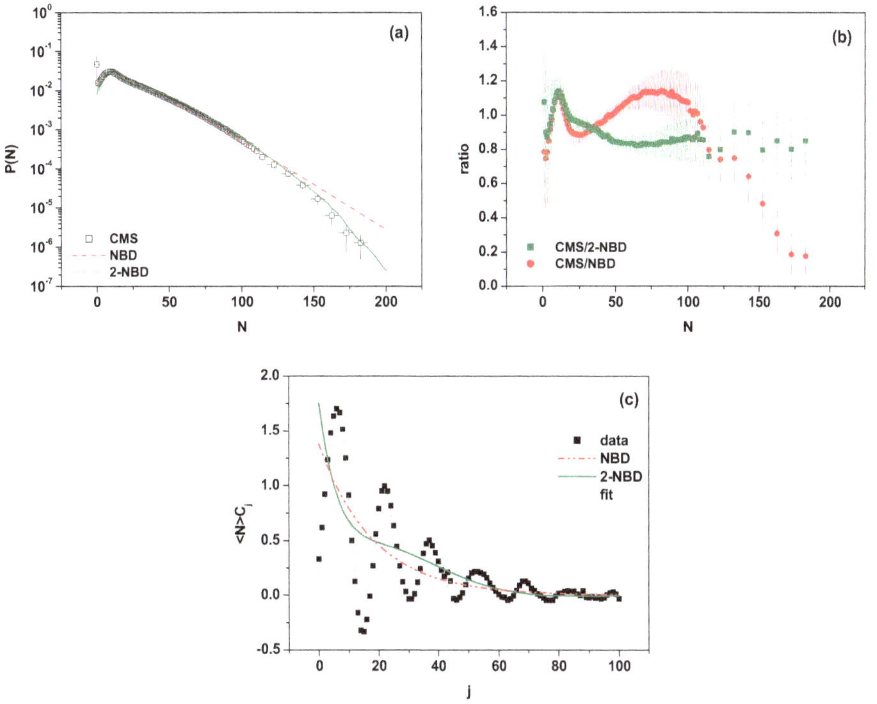

Figure 3. (**a**) Charged hadron multiplicity distributions for the pseudorapidity range $|\eta| < 2$ at $\sqrt{s} =$ 7 TeV, as given by the CMS experiment [38] (points), compared with the negative binomial distribution (NBD) for parameters $\langle N \rangle = 25.5$ and $k = 1.45$ (dashed line) and with the two-component NBD (solid line) with parameters from [39]. (**b**) Multiplicity dependence of the ratio $R = P_{CMS}(N)/P_{fit}(N)$ for the NBD (red circles) and for the two-component NBD for the same data as in panel (**a**) (green squares). (**c**) Coefficients C_j emerging from the data and NBD fits presented in panel (**a**). The data points are fitted by Equation (24); see text for details.

Before proceeding further we note that the coefficients C_j are closely related to the *combinants* $C_j^{\star}$, which were introduced in [40] (see also [11,41,42]) and are defined in terms of the generating function $G(z)$ as

$$C_j^{\star} = \frac{1}{j!} \frac{d^j \ln G(z)}{dz^j}\bigg|_{z=0}, \qquad \text{where} \qquad G(z) = \sum_{N=0}^{\infty} P(N) z^N \tag{26}$$

or by the following relation [31]:

$$\ln G(z) = \ln P(0) + \sum_{j=1}^{\infty} C_j^{\star} z^j \tag{27}$$

From the above, one can deduce the following [31]:

$$C_j = \frac{j+1}{\langle N \rangle} C^{\star}_{j+1} \tag{28}$$

This means that one can rewrite the recurrence relation, Equation (22), in terms of the combinants $C^{\star}_j$:

$$(N+1)P(N+1) = \sum_{j=0}^{N} (j+1)C^{\star}_j P(N-j) \tag{29}$$

When compared with Equation (22), this allows us to express our coefficients C_j, which henceforth we shall call *modified combinants*, by the generating function $G(z)$ of $P(N)$:

$$\langle N \rangle C_j = \frac{1}{j!} \frac{d^{j+1} \ln G(z)}{dz^{j+1}} \bigg|_{z=0} \tag{30}$$

This is the relation we use in the following when calculating the C_j from distributions defined by some $G(z)$.

To continue our reasoning, we note first that whereas a single NBD does not lead to oscillatory behaviour of the modified combinants C_j, there is a distribution for which the corresponding C_j oscillate in a maximum way. This is the BD for which the modified combinants are given by the following formula:

$$C_j = (-1)^j \frac{K}{\langle N \rangle} \left(\frac{\langle N \rangle}{K - \langle N \rangle} \right)^{(j+1)} = \frac{(-1)^j}{1-p} \left(\frac{p}{1-p} \right)^j \tag{31}$$

which oscillates rapidly with a period equal to 2. In Figure 4a, one can see that the amplitude of these oscillations depends on p; generally the C_j increase with rank j for $p > 0.5$ and decrease for $p < 0.5$. However, their general shape lacks the fading-down feature of the C_j observed experimentally. This suggests that the BD alone is not enough to explain the data but must be somehow combined with some other distribution. (In fact, in [43], we have already used a combination of *elementary emitting cells* (EECs) producing particles following a geometrical distribution (our aim at that time was to explain the phenomenon of Bose–Einstein correlations). For a constant number k of EECs, one obtains the NBD as the resultant $P(N)$, whereas for k distributed according to the BD, the resulting $P(N)$ is a modified NBD. However, we could not find a set of parameters providing both the observed $P(N)$ and oscillating C_j. We note that originally, the NBD was seen as a compound PD, with the number of clusters given by a PD and the particle inside the clusters distributed according to a logarithmic distribution [44].).

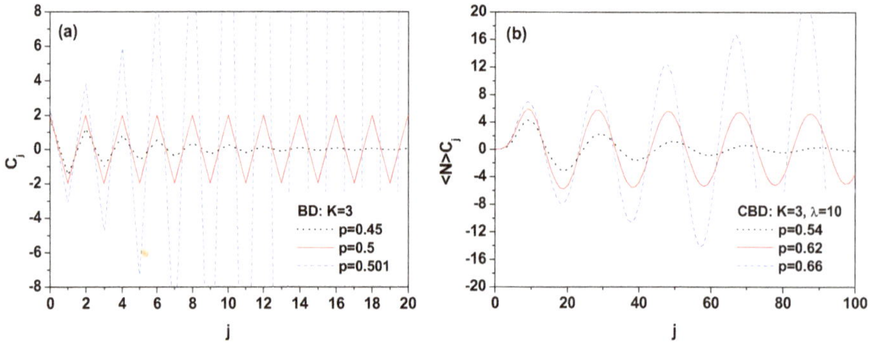

Figure 4. Examples of C_j for binomial distributions (**a**), and compound binomial distributions (**b**) from Equation (37).

We resort therefore to the idea of *compound distributions* (CDs) [45], which, from the point of view of the physics involved in our case, could for example describe a production process in which a number M of some objects (clusters, fireballs, etc.) are produced following, in general, some distribution $f(M)$ (with generating function $F(z)$), and subsequently decay independently into a number of secondaries, $n_{i=1,\ldots,M}$, always following some other (the same for all) distribution $g(n)$ (with a generating function $G(z)$). The distribution $h(N)$, where

$$N = \sum_{i=0}^{M} n_i \tag{32}$$

is a CD of f and g: $h = f \otimes g$. For CDs, we have that

$$\langle N\rangle = \langle M\rangle\langle n\rangle \qquad \text{and} \qquad Var(N) = \langle M\rangle Var(n) + Var(M)\langle n\rangle^2 \tag{33}$$

and its generating function $H(z)$ is equal to

$$H(z) = F[G(z)] \tag{34}$$

It should be mentioned that for the class of distributions of M that satisfy our recursion relation, Equation (18), the CD $h = f \otimes g$ is given by Panjer's recursion relation [46]:

$$Nh(N) = \sum_{j=1}^{N} [\beta N + (\alpha - \beta)j] g(j) h(N-j) = \sum_{j=1}^{N} C_j^{(P)}(N) h(N-j) \tag{35}$$

with the initial value $h(0) = f(0)$. However, the coefficients $C_j^{(P)}$ occurring here depend on N, contrary to our recursion given by Equation (22), for which the modified combinants, C_j, are independent of N. Moreover, Equation (22) is not limited to the class of distributions satisfying Equation (18) but is valid for any distribution $P(N)$. For this reason, the recursion relation, Equation (35), is not suitable for us.

To visualize the CD in action, we take for f a BD with generating function $F(z) = (pz + 1 - p)^K$, and for g we take a PD with generating function $G(z) = \exp[\lambda(z-1)]$. The generating function of the resultant distribution is now equal to

$$H(z) = \{p \exp[\lambda(z-1)] + 1 - p\}^K \tag{36}$$

and the corresponding modified combinants are

$$\begin{aligned}\langle N\rangle C_j &= \frac{K\lambda^{j+2}\exp(-\lambda)}{j!} \sum_{i=1}^{j+2} \left[\frac{p}{1-p+p\exp(-\lambda)}\right]^i \frac{1}{i} \sum_{k=0}^{i} (-1)^{k+1} \binom{i}{k} k^{j+1} = \\ &= \frac{K\lambda^{j+2}\exp(-\lambda)}{j!} \sum_{i=1}^{j+2} \left[\frac{p}{1-p+p\exp(-\lambda)}\right]^i S(j+1,i)\end{aligned} \tag{37}$$

where

$$S(n,k) = \left\{ \begin{matrix} n \\ k \end{matrix} \right\} = \frac{1}{k!} \sum_{i=0}^{k} (-1)^{k-i} \binom{k}{i} i^n \tag{38}$$

is the Stirling number of the second kind. Figure 4b shows the above modified combinants for the compound binomial distribution (CBD; a combination of a BD with a PD) with $K = 3$ and $\lambda = 10$ calculated for three different values of p in the BD: $p = 0.54, 0.62, 0.66$. We note that in general, the period of the oscillations is equal to 2λ, that is, in Figure 4b, where $\lambda = 10$ it is equal 20. The multiplicity distribution in this case is

$$P(0) = \left(1 - p + pe^{-\lambda}\right)^K \tag{39}$$

$$P(N) = \frac{1}{N!}\frac{d^N H(z)}{dz^N}\bigg|_{z=0} = \frac{1}{N!}\sum_{i=1}^{K} i!\binom{K}{i}\left(\lambda pe^{-\lambda}\right)^i\left(1 - p + pe^{-\lambda}\right)^{K-i} S(N,i) \tag{40}$$

The proper normalization comes from the fact that $H(1) = 1$. This shows that the choice of a BD as the basis of the CDs to be used seems to be crucial to obtain oscillatory C_j (e.g., a CD formed from a NBD and some other NBD provides smooth C_j).

Unfortunately, such a single component CBD (depending on three parameters: p, K and λ; $P(N) = h(N; p, K, \lambda)$ does not describe the experimental $P(N)$. We return therefore to the idea of using a multicomponent version of the CBD, for example, a three-component CBD defined as follows (with w_i being weights):

$$P(N) = \sum_{i=1,2,3} w_i h\left(N; p_i, K_i, \lambda_i\right); \qquad \sum_{i=1,2,3} w_i = 1 \tag{41}$$

As can be seen in Figure 5, in this case, the fit to $P(N)$ is quite good, and the modified combinants C_j follow an oscillatory pattern as far as the period of the oscillations is concerned, albeit their amplitudes still decaying too slowly.

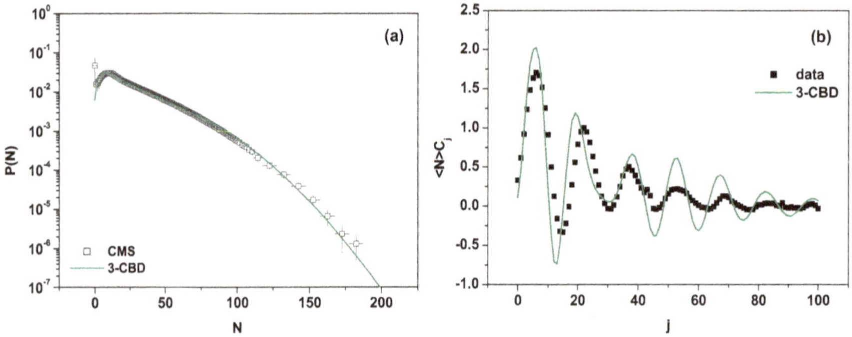

Figure 5. Results of using the compound binomial distribution (CBD) approach given by Equation (41) with parameters $\omega_1 = 0.34$, $\omega_2 = 0.4$, $\omega_3 = 0.26$; $p_1 = 0.22$, $p_2 = 0.22$, and $p_3 = 0.12$; $K_1 = 10$, $K_2 = 12$, and $K_2 = 30$; and $\lambda_1 = 4$, $\lambda_2 = 9$, and $\lambda_3 = 14$. (**a**) Charged hadron multiplicity distributions for $|\eta| < 2$ at $\sqrt{(s)} = 7$ TeV, as given by the CMS experiment [38] (points), compared with a three-component CBD (3-CBD; Equation (41)). (**b**) Coefficients C_j emerging from the CMS data used in panel (**a**) compared with the corresponding C_j obtained from the 3-CBD.

4. Summary

To summarize, we would like to mention that power law and quasi-power-law distributions are ubiquitous in many different, apparently very disparate, branches of science (such as, e.g., earthquakes, escape probabilities in chaotic maps close to crisis, biased diffusion of tracers in random systems, kinetic and dynamic processes in random quenched and fractal media, diffusion limited aggregates, growth models, or stock markets near financial crashes, to name only a few). Also ubiquitous is the fact that, in most cases, they are decorated with log-periodic oscillations of different kinds [23,24,26,27]. It is then natural to expect that oscillations of certain variables constitute a universal phenomenon, which should occur in a large class of stochastic processes, independently of the microscopic details. In this paper, we have concentrated on some specific oscillation phenomena seen at LHC energies in transverse momentum distributions. Their log-periodic characteristics suggest that either the exponent of the power-like behavior of these distributions is complex, or that there is a scale parameter that exhibits specific log-periodic oscillations. Whereas the most natural thing seems to be attributing the observed oscillations to some discrete scale invariance present in the system considered [23,24], it turns out that such scale-invariant functions also satisfy some specific wave equations showing a self-similarity property [21]. In both cases, these functions exhibit log-periodic behavior.

Concerning the second topic considered here, the presence of oscillations in counting statistics, one should realize that it is also a well-established phenomenon. The known examples include oscillations of the high-order cumulants of the probability distributions describing the transport through a double quantum dot, oscillations in quantum optics (in the photon distribution function in slightly squeezed states; see [47] for more information and references). In elementary particle physics, oscillations of the H_q moments, which represent ratios of the cumulants to factorial moments, also have a long history [11,35,36].

Our expectation that the oscillations discussed here could also be observed (and successfully measured) in multiparticle production processes is new. To see these, one must first deduce from the experimental data on the multiplicity distribution $P(N)$ the modified combinants C_j (which are defined by the recurrence relation presented in Equation (22); we note that, contrary to the H_q moments, the C_j are independent of the multiplicity distribution $P(N)$ for $N > j$). In the case for which these modified combinants show oscillatory behavior, they can be used to search for some underlying dynamical mechanism that could be responsible for it. The present situation is such that the measured multiplicity distributions $P(N)$ (for which, as we claim in [31], the corresponding modified combinants C_j oscillate), are most frequently described by NBDs (the modified combinants of which do not oscillate). Furthermore, with increasing collision energy and increasing multiplicity of produced secondaries, some systematic discrepancies between the data and the NBD form of the $P(N)$ used to fit them become more and more apparent. We propose therefore to use a novel phenomenological approach to the observed multiplicity distributions on the basis of the modified combinants C_j obtained from the measured multiplicity distributions. Together with the fitted multiplicity distributions $P(N)$, these would allow for a more detailed quantitative description of the complex structure of the multiparticle production process. We argue that the observed strong oscillations of the coefficients C_j in the pp data at LHC energies indicate the compound characteristics of the measured distributions $P(N)$ with a central role played by the BD, which provides the oscillatory characteristics of the C_j. This must be supplemented by some other distribution in such a way that the CD fits both the observed $P(N)$ and the C_j deduced from it. However, at the moment, we are not able to obtain fits to both $P(N)$ and C_j of acceptable quality. Therefore, these oscillations still await their physical justification, that is, the identification of some physical process (or combination of such processes) that would result in such phenomena.

We close by noting that both phenomena discussed here describe, in fact, different dynamical aspects of the multiparticle production process at high energies. The quasi-power-like distributions and the related log-periodic oscillations are related with events with rather small multiplicities of secondaries with large and very large momenta; these are called *hard collisions*, and they essentially probe the collision dynamics towards the edge of the phase space. The multiparticle distributions collect instead all produced particles, the majority of which come from the *soft collisions* concentrated in the middle of the phase space. In this sense, both the phenomena discussed provide us with complementary new information on these processes and, because of this, they should be considered, as much as possible, jointly. (Because of some similarities observed between hadronic, nuclear and e^+e^- collisions [11,48,49] (see also Chapter 20 of [50]), one might expect that the phenomena discussed above will also appear in these reactions. However, this is a separate problem, too extensive and not yet sufficiently discussed to be presented here).

Acknowledgments: This research was supported in part (GW) by the National Science Center (NCN) under contract 2016/22/M/ST2/00176. We would like to warmly thank Nicholas Keeley for reading the manuscript.

Author Contributions: Both authors analyzed experimental data, described the observed oscillation and equally contributed to the work on the manuscript. All authors have read and approved the final manuscript.

Conflicts of Interest: The authors declare no conflict of interest.

References

1. Tsallis, C. Possible generalization of Boltzman–Gibbs statistics. *J. Stat. Phys.* **1998**, *52*, 479–487.
2. Tsallis, C. *Introduction to Nonextensive Statistical Mechanics*; Springer: New York, NY, USA, 2009.
3. Wilk, G.; Włodarczyk, Z. Quasi-power law ensembles. *Acta Phys. Pol. B* **2015**, *46*, 1103, doi:10.5506/APhysPolB.46.1103.
4. Wilk, G.; Włodarczyk, Z. Self-similarity in jet events following from *pp* collisions at LHC. *Phys. Lett. B* **2013**, *727*, 163–167.
5. Wilk, G.; Włodarczyk, Z. Power laws in elementary and heavy-ion collisions. *Eur. Phys. J. A* **2009**, *40*, 299–312.
6. Hagedorn, R.; Ranft, R. Statistical thermodynamics of strong interactions at high energies. II–Momentum spectra of particles produced in *pp* collisions. *Suppl. Nuovo Cim.* **1968**, *6*, 169–310.
7. Hagedorn, R. Remarks of the thermodynamical model of strong interactions. *Nucl. Phys. B* **1970**, *24*, 93–139.
8. Deppman, A. Thermodynamics with fractal structure, tsallis statistics, and hadrons. *Phys. Rev. D* **2016**, *93*, 054001, doi:10.1103/PhysRevD.93.054001.
9. Calucci, G.; Treleani, D. Self-similarity of the negative binomial multiplicity distributions. *Phys. Rev. D* **1998**, *57*, 602–605.
10. De Wolf, E.A.; Dremin, I.M.; Kittel, W. Scaling laws for density correlations and fluctuations and fluctuations in multiparticle dynamics. *Phys. Rep.* **1996**, *270*, 1–141.
11. Kittel, W.; De Wolf, E.A. *Soft Multihadron Dynamics*; World Scientific: Singapore, 2005.
12. Aad, G.; Abbott, B.; Abdallah, J.; Abdelalim, A.A.; Abdesselam, A.; Abdinov, O.; Abi, B.; Abolins, M.; Abramowicz, H.; Abreu, H.; et al. Charged-particle multiplicities in *pp* interactions measured with the ATLAS detector at the LHC. *New J. Phys.* **2011**, *13*, 053033, doi:10.1088/1367-2630/13/5/053033.
13. Aad, G.; Abbott, B.; Abdallah, J.; Abdelalim, A.; Abdesselam, A.; Abdinov, O.; Abi, B.; Abolins, M.; Abramowicz, H.; Abreu, H.; et al. Properties of jets measured from tracks in proton-proton collisions at center-of-mass energy $\sqrt{s} = 7$ TeV with the ATLAS detector. *Phys. Rev. D* **2011**, *84*, 054001, doi:10.1103/PhysRevD.84.054001.
14. Khachatryan, V.; Sirunyan, A.M.; Tumasyan, A.; Adam, W.; Bergauer, T.; Dragicevic, M.; Ero, J.; Friedl, M.; Fruehwirth, R.; Ghete, V.M.; et al. Transverse-momentum and pseudorapidity distributions of charged hadrons in *pp* collisions at $\sqrt{s} = 0.9$ and 2.36 TeV. *J. High Energy Phys.* **2010**, *2*, 41, doi:10.1007/JHEP02(2010)041.
15. Khachatryan, V.; Sirunyan, A.M.; Tumasyan, A.; Adam, W.; Bergauer, T.; Dragicevic, M.; Ero, J.; Fabjan, C.; Friedl, M.; Fruehwirth, R.; et al. Transverse-momentum and pseudorapidity distributions of charged hadrons in *pp* collisions at $\sqrt{s} = 7$ TeV. *Phys. Rev. Lett.* **2010**, *105*, 022002, doi:10.1103/PhysRevLett.105.022002.
16. Wong, C.-T.; Wilk, G.; Cirto, L.J.L.; Tsallis, C. From QCD-based hard-scattering to nonextensive statistical mechanical descriptions of transverse momentum spectra in high-energy *pp* and $\bar{p}p$ collisions. *Phys. Rev. D* **2015**, *91*, 114027, doi:10.1103/PhysRevD.91.114027.
17. Aad, G.; Brad A.; Jalal, A.; Ahmed, A.; Abdelouahab, A.; Ovsat, A.; Babak, A.; Maris, A.; Halina, A.; Henso, A.; et al. Measurement of the jet fragmentation function and transverse profile in proton-proton collisions at a center-of-mass energy of 7 TeV with the ATLAS detector. *Eur. Phys. J. C* **2011**, *71*, 1795, doi:10.1140/epjc/s10052-011-1795-y.
18. Wróblewski, A. Multiplicity distributions in proton-proton collisions. *Acta Phys. Pol. B* **1973**, *4*, 857–884.
19. Geich-Gimbel, C. Particle production at collider energies. *Int. J. Mod. Phys. A* **1989**, *4*, 1527–1680.
20. Aad, G.; Abbott, B.; Abdallah, J.; Abdinov, O.; Abeloos, B.; Aben, R.; Abolins, M.; AbouZeid, O.S.; Abraham, N.L.; Abramowicz, H.; et al. Measurement of the charged particle multiplicity inside jets from $\sqrt{s} = 8$ TeV *pp* collisions with the ATLAS detector. *Eur. Phys. J. C* **2016**, *76*, 322, doi:10.1140/epjc/s10052-016-4126-5.
21. Wilk, G.; Włodarczyk, Z. Temperature oscillations and sound waves in hadronic matter. *Phys. A* **2017**, *486*, 579–586.

22. Chatrchyan, S.; Khachatryan, V.; Sirunyan, A.M.; Tumasyan, A.; Adam, W.; Bergauer, T.; Dragicevic, M.; Ero, J.; Fabjan, C.; Friedl, M.; et al. Charged particle transverse momentum spectra in pp collisions at $\sqrt{s} = 0.9$ and 7 TeV. *J. High Energy Phys.* **2011**, *8*, 86, doi:10.1007/JHEP08(2011)086.
23. Sornette, D. Discrete-scale invariance and complex dimensions. *Phys. Rep.* **1998**, *297*, 239–270.
24. Wilk, G.; Włodarczyk, Z. Tsallis distribution with complex nonextensivity parameter q. *Phys. A* **2014**, *413*, 53–58.
25. Rybczyński, M.; Wilk, G.; Włodarczyk, Z. System size dependence of the log-periodic oscillations of transverse momentum spectra. *EPJ Web Conf.* **2015**, *90*, 01002, doi:10.1051/epjconf/20159001002.
26. Wilk, G.; Włodarczyk, Z. Tsallis distribution decorated with log-periodic oscillation. *Entropy* **2015**, *17*, 384–400.
27. Wilk, G.; Włodarczyk, Z. Quasi-power laws in multiparticle production processes. *Chaos Solitons Fractals* **2015**, *81*, 487–496.
28. Wilk, G.; Włodarczyk, Z. Nonextensive information entropy for stochastic networks. *Acta Phys. Pol. B* **2004**, *35*, 871–879.
29. Landau, L.D.; Lifshitz, E.M. *Fluid Mechanics*; Pergamon Press: Oxford, UK, 1987.
30. Barenblatt, G.I. *Scaling, Self-Similarity, and Intermediate Asymptotics*; Cambridge University Press: Cambridge, UK, 1996.
31. Wilk, G.; Włodarczyk, Z. How to retrieve additional information from the multiplicity distributions. *J. Phys. G* **2017**, *44*, 015002, doi:10.1088/0954-3899/44/1/015002.
32. Giovannini, A.; Ugocciono, R. Signals of new physics in global event properties in pp collisions in the TeV energy domain. *Phys. Rev. D* **2003**, *68*, 034009, doi:10.1103/PhysRevD.68.034009.
33. Zborovsky, I.J. A three-component description of multiplicity distributions in pp collisions at the LHC. *J. Phys. G* **2013**, *40*, 055005, doi:10.1088/0954-3899/40/5/055005.
34. Dremin, I.M.; Nechitailo, V.A. Independent pair parton interactions model of hadron interactions. *Phys. Rev. D* **2004**, *70*, 034005, doi:10.1103/PhysRevD.70.034005.
35. Dremin, I.M.; Gary, J.W. Hadron multiplicities. *Phys. Rep.* **2001**, *349*, 301–393.
36. Grosse-Oetringhaus, J.F.; Teygers, K. Charged-particle multiplicity in proton–proton collisions. *J. Phys. G* **2010**, *37*, 083001, doi:10.1088/0954-3899/37/8/083001.
37. Saleh, B.E.A.; Teich, M.K. Multiplied-Poisson Noise in Pulse, Particle, and Photon Detection. *Proc. IEEE* **1982**, *70*, 229–245.
38. Khachatryan, V.; Sirunyan, A.M.; Tumasyan, A.; Adam, W.; Bergauer, T.; Dragicevic, M.; Ero, J.; Fabjan, C.; Friedl, M.; Fruehwirth, R.; et al. Charged particle multiplicities in pp interactions at $\sqrt{s} = 0.9$, 2.36, and 7 TeV. *J. High Energy Phys.* **2011**, *2011*, 79, doi:10.1007/JHEP01(2011)079.
39. Ghosh, P. Negative binomial multiplicity distribution in proton-proton collisions in limited pseudorapidity intervals at LHC up to $\sqrt{s} = 7$ TeV and the clan model. *Phys. Rev. D* **2012**, *85*, 054017, doi:10.1103/PhysRevD.85.054017.
40. Kauffmann, S.K.; Gyulassy, M. Multiplicity distributions of created bosons: The method of combinants. *J. Phys. A* **1978**, *11*, 1715–1727.
41. Balantekin, A.B.; Seger, J.E. Description of pion multiplicities using combinants. *Phys. Lett. B* **1991**, *266*, 231–235.
42. Hegyi, S. Correlation studies in quark jets using combinants. *Phys. Lett. B* **1999**, *463*, 126–131.
43. Biyajima, M.; Suzuki, N.; Wilk, G.; Włodarczyk, Z. Totally chaotic poissonian-like sources in multiparticle production processes? *Phys. Lett. B* **1996**, *386*, 297–303.
44. Giovannini, A.; Van Hove, L. Negative binomial multiplicity distributions in high energy hadron collisions. *Z. Phys. C* **1986**, *30*, 391–400.
45. Sundt, B.; Vernic, R. *Recursions for Convolutions and Compound Distributions with Insurance Applications*; Springer: Berlin/Heidelberg, Germany, 2009.
46. Panjer, H.H. Recursive evaluation of a family of compound distributions. *ASTIN Bull.* **1981**, *12*, 22–26.
47. Flindt, C.; Fricke, C.; Hohls, F.; Novotny, T.; Netocny, K.; Brandes, T.; Haug, R.J. Universal oscillations in counting statistics. *Proc. Natl. Acad. Sci. USA* **2009**, *106*, 10116–10119.
48. Sarkisyan, E.K.; Mishra, A.N.; Sahoo, R.; Alexander, S.; Sakharov, A.S. Multihadron production dynamics exploring the energy balance in hadronic and nuclear collisions. *Phys. Rev. D* **2016**, *93*, 054046, doi:10.1103/PhysRevD.93.054046.

49. Bzdak, A. Universality of multiplicity distribution in proton-proton and electron-positron collisions. *Phys. Rev. D* **2017**, *96*, 036007, doi:10.1103/PhysRevD.96.036007.
50. Patrignani, C.; Agashe, K.; Aielli, G.; Amsler, C.; Antonelli, M.; Asner, D.M.; Baer, H.; Banerjee, S.; Barnett, R.M.; Basaglia, T.; et al. Review of particle physics. *Chin. Phys. C* **2016**, *40*, 100001, doi:10.1088/1674-1137/40/10/100001.

entropy

MDPI

Article

Minimising the Kullback–Leibler Divergence for Model Selection in Distributed Nonlinear Systems

Oliver M. Cliff [1,2,*], Mikhail Prokopenko [2] and Robert Fitch [1,3]

1 Australian Centre for Field Robotics, The University of Sydney, Sydney NSW 2006, Australia; rfitch@uts.edu.au
2 Complex Systems Research Group, The University of Sydney, Sydney NSW 2006, Australia; mikhail.prokopenko@sydney.edu.au
3 Centre for Autonomous Systems, University of Technology Sydney, Ultimo NSW 2007, Australia
* Correspondence: o.cliff@acfr.usyd.edu.au; Tel.: +61-2-9351-3040

Received: 21 December 2017; Accepted: 18 January 2018; Published: 23 January 2018

Abstract: The Kullback–Leibler (KL) divergence is a fundamental measure of information geometry that is used in a variety of contexts in artificial intelligence. We show that, when system dynamics are given by distributed nonlinear systems, this measure can be decomposed as a function of two information-theoretic measures, transfer entropy and stochastic interaction. More specifically, these measures are applicable when selecting a candidate model for a distributed system, where individual subsystems are coupled via latent variables and observed through a filter. We represent this model as a directed acyclic graph (DAG) that characterises the unidirectional coupling between subsystems. Standard approaches to structure learning are not applicable in this framework due to the hidden variables; however, we can exploit the properties of certain dynamical systems to formulate exact methods based on differential topology. We approach the problem by using reconstruction theorems to derive an analytical expression for the KL divergence of a candidate DAG from the observed dataset. Using this result, we present a scoring function based on transfer entropy to be used as a subroutine in a structure learning algorithm. We then demonstrate its use in recovering the structure of coupled Lorenz and Rössler systems.

Keywords: Kullback–Leibler divergence; model selection; information theory; transfer entropy; stochastic interaction; nonlinear systems; complex networks; state space reconstruction

1. Introduction

Distributed information processing systems are commonly studied in complex systems and machine learning research. We are interested in inferring data-driven models of such systems, specifically in the case where each subsystem can be viewed as a nonlinear dynamical system. In this context, the Kullback–Leibler (KL) divergence is commonly used to measure the quality of a statistical model [1–3]. When a model is compared with fully observed data, computing the KL divergence can be straightforward. However, in the case of spatially distributed dynamical systems, where individual subsystems are coupled via latent variables and observed through a filter, the presence of hidden variables renders typical approaches unusable. We derive the KL divergence in such systems as a function of two information-theoretic measures using methods from differential topology.

The *model selection* problem has applications in a wide variety of areas due to its usefulness in performing efficient inference and understanding the underlying phenomena being studied. Dynamical systems are an expressive model characterised by a map that describes their evolution over time and a read-out function through which we observe the latent state. Our research focuses on the more general case of a multivariate system, where a set of these subsystems are distributed and unidirectionally coupled to one another. The problem of inferring this coupling is an important multidisciplinary study in fields such as ecology [4], neuroscience [5,6], multi-agent systems [7–9], and various others that focus on artificial and biological networks [10].

We represent such a spatially distributed system as a probabilistic graphical model termed a *synchronous graph dynamical system (GDS)* [11,12], whose structure is given by a directed acyclic graph (DAG). Model selection in this context is the problem of inferring directed relationships between hidden variables from an observed dataset, also known as *structure learning*. A main challenge in structure learning for DAGs is the case where variables are unobserved. Exact methods are known for fully observable systems (i.e., Bayesian networks (BNs)) [13]; however, these are not applicable in the more expressive case when the state variables in dynamical systems are latent. The main focus of this paper is to analytically derive a measure for comparing a candidate graph to the underlying graph that generated a measured dataset. Such a measure can then be used to solve the two subproblems that comprise structure learning, *evaluation* and *identification* [14], and hence find the optimal model that explains the data.

For the evaluation problem, it is desirable to select the *simplest* model that incorporates all statistical knowledge. This concept is commonly expressed via information theory, where an established technique is to evaluate the encoding length of the data, given the model [1,15,16]. The simplest model should aim to minimise code length [2], and therefore we can simplify our problem to that of minimising KL divergence for the synchronous GDS. Using this measure, we find a factorised distribution (given by the graph structure) that is closest to the complete (unfactorised) distribution. We first analytically derive an expression for this divergence, and build on this result to present a scoring function for evaluating candidate graphs based on a dataset.

The main result of this paper is an exact decomposition of the KL divergence for synchronous GDSs. We show that this measure can be decomposed as the difference between two well-known information-theoretic measures, stochastic interaction [17,18] and collective transfer entropy [19]. We establish this result by first representing discrete-time multivariate dynamical systems as dynamic Bayesian networks (DBNs) [20]. In this form, both the complete and factorised distributions cannot be directly computed due to the hidden system state. Thus, we draw on state space reconstruction methods from differential topology to reformulate the KL divergence in terms of computable distributions. Using this expression, we show that the maximum transfer entropy graph is the most likely to have generated the data. This is experimentally validated using toy examples of a Lorenz–Rössler system and a network of coupled Lorenz attractors (Figure 1) of up to four nodes. These results support the conjecture that transfer entropy can be used to infer effective connectivity in complex networks.

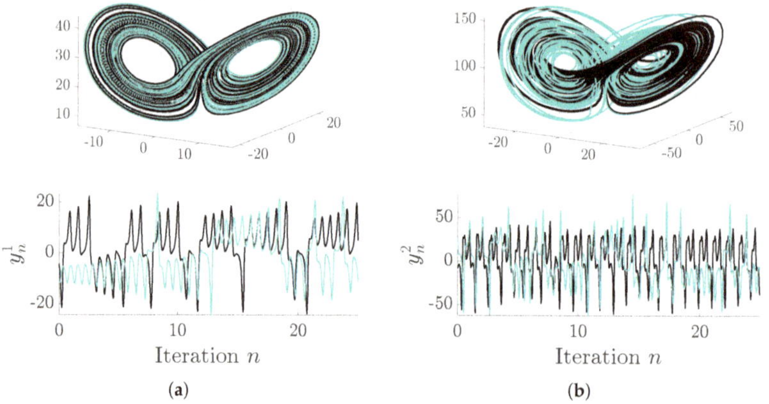

(a) (b)

Figure 1. Trajectory of a pair of coupled Lorenz systems. *Top row*: original state of the subsystems. *Bottom row:* time-series measurements of the subsystems. In each figure, the black lines represent an uncoupled simulation ($\lambda = 0$), and teal lines illustrate a simulation where the first (leftmost) subsystem was coupled to the second ($\lambda = 10$). (**a**) $\sigma = 10, \beta = 8/3, \rho = 28$; (**b**) $\sigma = 10, \beta = 8/3, \rho = 90$.

2. Related Work

Networks of coupled dynamical systems have been introduced under a variety of terms, such as complex networks [10], distributed dynamical systems [6] and master–slave configurations [21]. The defining feature of these networks is that the dynamics of each subsystem are given by a set of either discrete-time maps or first-order ordinary differential equations (ODEs). In this paper, we use the discrete-time formulation, where a map can be obtained numerically by integrating ODEs or recording observations at discrete-time intervals [22].

An important precursor to network reconstruction is inferring causality and coupling strength between complex nonlinear systems. Causal inference is intractable when the experimenter can not intervene with the dataset [23], and so we focus our attention on methods that determine conditional independence (coupling) rather than causality. In seminal work, Granger [24] proposed *Granger causality* for quantifying the predictability of one variable from another; however, a key requirement of this measure is linearity of the system, implying subsystems are separable [4]. Schreiber [25] extended these ideas and introduced *transfer entropy* using the concept of finite-order Markov processes to quantify the information transfer between coupled nonlinear systems. Transfer entropy and Granger causality are equivalent for linearly-coupled Gaussian systems (e.g., Kalman models) [26]; however, there are clear distinctions between the concepts of information transfer and causal effect [27]. Although transfer entropy has received criticism over spuriously identifying causality [28–30], we are concerned with statistical modelling and not causality of the underlying process.

Recently, a number of measures have been proposed to infer coupling between distributed dynamical systems based on reconstruction theorems. Sugihara et al. [4] proposed convergent cross-mapping that involves collecting a history of observed data from one subsystem and uses this to predict the outcome of another subsystem. This history is the delay reconstruction map described by Takens' Delay Embedding Theorem [31]. Similarly, Schumacher et al. [6] used the Bundle Delay Embedding Theorem [32,33] to infer causality and perform inference via Gaussian processes. Although the algorithms presented in these papers can infer driving subsystems in a spatially distributed dynamical system, the results obtained differ from ours as inference is not considered for an entire network structure, nor is a formal derivation presented. Contrasting this, we recently derived an information criterion for learning the structure of distributed dynamical systems [12]. However, the criterion proposed required parametric modelling of the probability distributions, and thus a detailed understanding of the physical phenomena being studied. In this paper, we extend this framework by first showing that KL divergence can be decomposed as information-theoretically useful measures, and then arriving at a similar result but employing non-parametric density estimation techniques to allow for no assumptions about the underlying distributions.

It is important to distinguish our approach from dynamic causal modelling (DCM), which attempts to infer the parameters of explicit dynamic models that cause (generate) data. In DCM, the set of potential models is specified a priori (typically in the form of ODEs) and then scored via marginal likelihood or evidence. The parameters of these models include *effective connectivity* such that their posterior estimates can be used to infer coupling among distributed dynamical systems [34]. As a consequence, these approaches can be used to recover networks that reveal the effective structure of observed systems [35,36]. In contrast, our approach does not require an explicitly specified model because the scoring function can be computed directly from the data. However, it does assume an implicit model in the form of a DAG where the subsystem processes are generated by generic functions.

Unlike effective connectivity, which is defined in relation to a (dynamic causal) model, the concept of *functional connectivity* refers to recovering statistical dependencies [37]. Consequently, statistical measures such as Granger causality and transfer entropy are typically used to identify functional, rather than effective structure. For example, transfer entropy has been used previously to infer networks in numerous fields, e.g., computational neuroscience [5,38], multi-agent systems [8], financial markets [39], supply-chain networks [40], and biology [41]. However, most of these results build on the work of Schreiber [25] by assuming the system is composed of finite-order Markov chains and thus there is a dearth of work that provides formal derivations for the use of this measure for inferring effective connectivity. Our work allows us to compute scoring functions directly from multivariate time series (as in functional connectivity), yet still assumes an implicit model (albeit with weaker assumptions on the model than those considered in inferring effective connectivity).

3. Background

3.1. Notation

We use the convention that $(\cdot)$ denotes a sequence, $\{\cdot\}$ a set, and $\langle\cdot\rangle$ a vector. In this work, we consider a collection of stationary stochastic temporal processes $\mathbf{Z}$. Each process Z^i comprises a sequence of random variables $(Z^i_1, \ldots, Z^i_N)$ with realisation $(z^i_1, \ldots, z^i_N)$ for countable time indices $n \in \mathbb{N}$. Given these processes, we can compute probability distributions of each variable by counting relative frequencies or by density estimation techniques [42,43]. We use bold to denote the set of all variables, e.g., $\boldsymbol{z}_n = \{z^1_n, \ldots, z^M_n\}$ is the collection of M realisations at index n. Furthermore, unless otherwise stated, X^i_n is a latent (hidden) variable, Y^i_n is an observed variable, and Z^i_n is an arbitrary variable; thus, $\boldsymbol{Z}_n = \{\boldsymbol{X}_n, \boldsymbol{Y}_n\}$ is the set of all hidden and observed variables at temporal index n. Given a graphical model G, the p^i parents of variable Z^i_{n+1} are given by the parent set $\Pi_G(Z^i_{n+1}) = \{Z^{ij}_n\}_j = \{Z^{i1}_n, \ldots, Z^{ip^i}_n\}$. Finally, let the superscript $z^{i,(k)}_n = \langle z^i_n, z^i_{n-1}, \ldots, z^i_{n-k+1}\rangle$ denote the vector of k previous values taken by variable Z^i_n.

3.2. Representing Distributed Dynamical Systems as Probabilistic Graphical Models

We are interested in modelling discrete-time multivariate dynamical systems, where the state is a vector of real numbers given by a point $\boldsymbol{x}_n$ lying on a compact d-dimensional manifold $\mathcal{M}$. A map $f : \mathcal{M} \to \mathcal{M}$ describes the temporal evolution of the state at any given time, such that the state at the next time index $\boldsymbol{x}_{n+1} = f(\boldsymbol{x}_n)$. Furthermore, in many practical scenarios, we do not have access to $\boldsymbol{x}_n$ directly, and can instead observe it through a *measurement function* $\psi : \mathcal{M} \to \mathbb{R}^M$ that yields a scalar representation $\boldsymbol{y}_n = \psi(\boldsymbol{x}_n)$ of the latent state [22,44]. We assume the multivariate system can be factorised and modelled as a DAG with spatially distributed dynamical subsystems, termed a synchronous GDS (see Figure 2a). This definition is restated from [12] as follows.

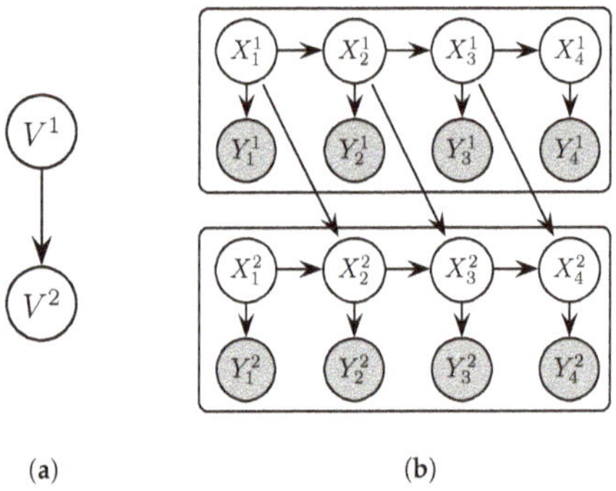

Figure 2. Representation of (**a**) the synchronous GDS with two vertices (V^1 and V^2), and (**b**) the rolled-out DBN of the equivalent structure. Subsystems V^1 and V^2 are coupled by virtue of the edge $X^1_n \to X^2_{n+1}$.

Definition 1 (Synchronous GDS). *A synchronous GDS* $(G, \boldsymbol{x}_n, \boldsymbol{y}_n, \{f^i\}, \{\psi^i\})$ *is a tuple that consists of: a finite, directed graph* $G = (\mathcal{V}, \mathcal{E})$ *with edge-set* $\mathcal{E} = \{E^i\}$ *and M vertices comprising the vertex set* $\mathcal{V} = \{V^i\}$*; a multivariate state* $\boldsymbol{x}_n = \langle x^i_n\rangle$*, composed of states for each vertex* V^i *confined to a* d^i*-dimensional manifold* $x^i_n \in \mathcal{M}^i$*; an M-variate observation* $\boldsymbol{y}_n = \langle y^i_n\rangle$*, composed of scalar observations for each vertex* $y^i_n \in \mathbb{R}$*; a set of* *local maps* $\{f^i\}$ *of the form* $f^i : \mathcal{M} \to \mathcal{M}^i$*, which update synchronously and induce a global map* $f : \mathcal{M} \to \mathcal{M}$*; and a set of local observation functions* $\{\psi^1, \psi^2, \ldots, \psi^M\}$ *of the form* $\psi^i : \mathcal{M}^i \to \mathbb{R}$.

The global dynamics and observations can therefore be described by the set of local functions [12]:

$$x^i_{n+1} = f^i(x^i_n, \langle x^{ij}_n \rangle_j) + v_{fi}, \tag{1}$$

$$y^i_{n+1} = \psi^i(x^i_{n+1}) + v_{\psi^i}, \tag{2}$$

where v_{fi} and v_{ψ^i} are additive noise terms. The subsystem dynamics (1) are a function of the subsystem state x^i_n and the subsystem parents' state $\langle x^{ij}_n \rangle_j$ at the previous time index, i.e., $f^i : (\mathcal{M}^i \times_j \mathcal{M}^{ij}) \to \mathcal{M}^i$. However, the observation y^i_{n+1} is a function of the subsystem state alone, i.e., $\psi^i : \mathcal{M}^i \to \mathbb{R}$. We assume that the maps $\{f^i\}$ and $\{\psi^i\}$, as well as the graph G, are time-invariant.

The discrete-time mapping for the dynamics (1) and measurement functions (2) can be modelled as a DBN in order to facilitate structure learning of the graph [12] (see Figure 2b). DBNs are a probabilistic graphical model that represent probability distributions over trajectories of random variables $(\mathbf{Z}_1, \mathbf{Z}_2, \ldots)$ using a prior BN and a *two-time-slice BN (2TBN)* [45]. To model the maps, however, we need only to consider the 2TBN $B = (G, \Theta_G)$, which can model a first-order Markov process $p_B(\mathbf{z}_{n+1} \mid \mathbf{z}_n)$ graphically via a DAG G and a set of conditional probability distribution (CPD) parameters Θ_G [45]. Given a set of stochastic processes $(\mathbf{Z}_1, \mathbf{Z}_2, \ldots, \mathbf{Z}_N)$, the realisation of which constitutes the sample path $(\mathbf{z}_1, \mathbf{z}_2, \ldots, \mathbf{z}_N)$, the 2TBN distribution is given by $p_B(\mathbf{z}_{n+1} \mid \mathbf{z}_n) = \prod_i \Pr(z^i_{n+1} \mid \pi_G(Z^i_{n+1}))$, where $\pi_{\mathcal{G}}(Z^i_{n+1})$ denotes the (index-ordered) set of realisations $\{z^j_o : Z^j_o \in \Pi_{\mathcal{G}}(Z^i_{n+1})\}$.

To model the synchronous GDS as a DBN, we associate each subsystem vertex V^i with a state variable X^i_n and an observation variable Y^i_n. The parents of subsystem V^i are denoted $\Pi_G(V^i)$ [12]. From the dynamics (1), variables in the set $\Pi_G(X^i_{n+1})$ come strictly from the preceding time slice, and additionally, from the measurement function (2), $\Pi_G(Y^i_{n+1}) = X^i_{n+1}$. Thus, we can build the edge set $\mathcal{E}$ in the GDS by means of the edges in the DBN [12], i.e., given an edge $X^i_n \to X^j_{n+1}$ of the DBN, the equivalent edge $V^i \to V^j$ exists for the GDS. The distributions for the dynamics (1) and observation (2) maps of M arbitrary subsystems can therefore be factorised according to the DBN structure such that [12]

$$p_B(\mathbf{z}_{n+1} \mid \mathbf{z}_n) = \prod_{i=1}^{M} \Pr(x^i_{n+1} \mid x^i_n, \langle x^{ij}_n \rangle_j) \cdot \Pr(y^i_{n+1} \mid x^i_{n+1}). \tag{3}$$

The goal of learning nonlinear dynamical networks thus becomes that of inferring the parent set $\Pi_G(X^i_n)$ for each latent variable X^i_n.

Finally, recall that the parents of each observation are constrained such that $\Pi_G(Y^i_{n+1}) = X^i_{n+1}$. As a consequence, we use the shorthand notation y^{ij}_n to denote the observation of the j-th parent of the i-th subsystem at time n (and the same for x^{ij}_n).

3.3. Network Scoring Functions

A number of exact and approximate DBN structure learning algorithms exist that are based on Bayesian statistics and information theory. We have shown in prior work how to compute the log-likelihood function for synchronous GDSs. In this section, we will briefly summarise the problem of structure learning for DBNs, focusing on the factorised distribution (3).

The *score and search* paradigm [46] is a common method for recovering graphical models from data. Given a dataset $D = (\mathbf{y}_1, \mathbf{y}_2, \ldots, \mathbf{y}_N)$, the objective is to find a DAG G^* such that

$$G^* = \arg\max_{G \in \mathcal{G}} g(B : D), \tag{4}$$

where $g(B{:}D)$ is a scoring function measuring the degree of fitness of a candidate DAG G to the data set D, and $\mathcal{G}$ is the set of all DAGs. Finding the optimal graph G^* in Equation (4) requires solutions to the two subproblems that comprise structure learning: the *evaluation* problem and the *identification* problem [14]. The main problem we focus on in this paper is the evaluation problem, i.e., determining a score that quantifies the quality of a graph, given data. Later, we will address the identification problem by discussing the attributes of this scoring function in efficiently finding the optimal graph structure.

In prior work, we developed a score based on the posterior probability of the network structure G, given data D. That is, we considered maximising the expected log-likelihood [12]

$$\ell(\hat{\Theta}_G : D) = \mathbf{E}\left[\log \Pr(D \mid G, \hat{\Theta}_G)\right] = \mathbf{E}\left[\log\left(p_B(z_{n+1} \mid z_n)\right)\right], \tag{5}$$

where the expectation $\mathbf{E}[Z] = \int_{-\infty}^{\infty} z \Pr(z) \mathrm{d}z$. It was shown that state space reconstruction techniques (see Appendix A) can be used to compute the log-likelihood of Equation (3) as a difference of conditional entropy terms [12]. In the same work, we illustrated that the log-likelihood ratio of a candidate DAG G to the empty network $G_\emptyset$ is given by collective transfer entropy (see Appendix B), i.e.,

$$\ell(\hat{\Theta}_G : D) - \ell(\hat{\Theta}_{G_\emptyset} : D) = N \cdot \sum_{i=1}^{M} T_{\langle Y^{ij}\rangle_j \to Y^i}. \tag{6}$$

For the nested log-likelihoods above, the statistics of $2(\ell(\hat{\Theta}_G : D) - \ell(\hat{\Theta}_{G_\emptyset} : D))$ asymptotically follow the χ^2_q-distribution, where q is the difference between the number of parameters of each model [47,48]. We will draw on this log-likelihood decomposition in later sections for statistical significance testing.

4. Computing Conditional KL Divergence

In this section, we present our main result, which is an analytical expression of KL divergence that facilitates structure learning in distributed nonlinear systems. We begin by considering the problem of finding an optimal DBN structure as searching for a parsimonious *factorised distribution* p_B that best represents the complete digraph distribution p_{K_M}. That is, p_{K_M} is the joint distribution yielded by assuming no factorisation (the complete graph K_M) and thus no information loss. The distribution is expressed as:

$$p_{K_M}(z_{n+1} \mid z_n^{(n)}) = \Pr\left(\{z^1_{n+1}, \ldots, z^M_{n+1}\} \mid \{z^1_n, \ldots, z^M_n\}, \{z^1_{n-1}, \ldots, z^M_{n-1}\}, \{z^1_1, \ldots, z^M_1\}\right). \tag{7}$$

We quantify the similarity of the factorised distribution p_B to this joint distribution via KL divergence. In prior work, De Campos [3] derived the *MIT* scoring function for BNs by this approach and it was later used for DBN structure learning with complete data [49]. We extend the analysis to DBNs with latent variables, i.e., we compare the joint and factorised distributions of time slices, given the entire history,

$$\begin{aligned} D_{\mathrm{KL}}\left[p_{K_M} \parallel p_B\right] &= D_{\mathrm{KL}}\left[p_{K_M}(z_{n+1} \mid z_n^{(n)}) \parallel p_B(z_{n+1} \mid z_n^{(n)})\right] \\ &= \sum_{z_n^{(n)}} \Pr(z_n^{(n)}) \sum_{z_{n+1}} \Pr(z_{n+1} \mid z_n^{(n)}) \log \frac{\Pr(z_{n+1} \mid z_n^{(n)})}{p_B(z_{n+1} \mid z_n^{(n)})} \\ &= \mathbf{E}\left[\log \frac{\Pr(z_{n+1} \mid z_n^{(n)})}{p_B(z_{n+1} \mid z_n)}\right]. \end{aligned} \tag{8}$$

Substituting the synchronous GDS model (3) into Equation (8), we get

$$D_{\mathrm{KL}}\left[p_{K_M} \parallel p_B\right] = \mathbf{E}\left[\log \frac{\Pr(z_{n+1} \mid z_n^{(n)})}{\prod_{i=1}^{M} \Pr(x_{n+1}^i \mid x_n^i, \langle x_n^{ij} \rangle_j) \cdot \Pr(y_{n+1}^i \mid x_{n+1}^i)}\right]. \tag{9}$$

However, Equation (9) comprises maximum likelihood distributions with unobserved (latent) states $\boldsymbol{x}_n$. It is common in model selection to decompose the KL divergence as

$$D_{\mathrm{KL}}\left[p_{K_M} \parallel p_B\right] = \mathbf{E}\left[\log\left(\Pr(z_{n+1} \mid z_n^{(n)})\right)\right] - \mathbf{E}\left[\log\left(p_B(z_{n+1} \mid z_n)\right)\right], \tag{10}$$

where the second term is simply the log-likelihood (5). In this form, p_{K_M} is often identical for all models considered and, in practice, it suffices to ignore this term and thus avoid the problem of computing distributions of latent variables. The resulting simpler expression can be viewed as log-likelihood maximisation (as in our previous work outlined in Section 3.3). However, as we show in this section, p_{K_M} is not equivalent for all models unless certain parameters of the dynamical systems are known. Hence, for now, we cannot ignore the first term of Equation (10) and we instead propose an alternative decomposition of KL divergence that comprises only observed variables.

4.1. A Tractable Expression via Embedding Theory

In order to compute the distributions in (9), we use the Bundle Delay Embedding Theorem [32,33] to reformulate the factorised distribution (denominator), and the Delay Embedding Theorem for Multivariate Observation Functions [50] for the joint distribution (numerator). We describe these theorems in detail in Appendix A, along with the technical assumptions required for (f, ψ). Although the following theorems assume a diffeomorphism, we also discuss application of the theory towards inferring the structure of endomorphisms (e.g., coupled map lattices [51]) in the same appendix.

The first step is to reproduce a prior result for computing the factorised distribution (denominator) in Equation (9). First, the embedding

$$y_n^{i,(\kappa^i)} = \langle y_n^i, y_{n-\tau^i}^i, \ldots, y_{n-(\kappa^i-1)\tau^i}^i \rangle, \tag{11}$$

where τ^i is the (strictly positive) lag, and κ^i is the embedding dimension of the i-th subsystem (the *embedding parameters*). Note that, although we can take either the future or past delay embedding (11) for diffeomorphisms, we explicitly consider a *history* of values to account for both endomorphisms and diffeomorphisms. Moreover, an important assumption of our approach is that the the structure (enforced by coupling between subsystems) is a DAG; this comes from the Bundle Delay Embedding Theorem [32,33] (see Lemma 1 of [12] for more detail). Our previous result is expressed as follows.

Lemma 1 (Cliff et al. [12]). *Given an observed dataset D, where $\boldsymbol{y}_n \in \mathbb{R}^M$, generated by a directed and acyclic synchronous GDS $(G, \boldsymbol{x}_n, \boldsymbol{y}_n, \{f^i\}, \{\psi^i\})$, the 2TBN distribution can be written as*

$$\prod_{i=1}^{M} \Pr(x_{n+1}^i \mid x_n^i, \langle x_n^{ij} \rangle_j) \cdot \Pr(y_{n+1}^i \mid x_{n+1}^i) = \frac{\prod_{i=1}^{M} \Pr(y_{n+1}^i \mid y_n^{i,(\kappa^i)}, \langle y_n^{ij,(\kappa^{ij})} \rangle_j)}{\Pr(\boldsymbol{x}_n \mid \langle y_n^{i,(\kappa^i)} \rangle)}. \tag{12}$$

Next, we present a method for computing the joint distribution (numerator) in Lemma 3. For convenience, Lemma 2 restates part of the delay embedding theorem in [50] in terms of subsystems of a synchronous GDS and establishes existence of a map $\mathbf{G}$ for predicting future observations from a history of observations.

Lemma 2. *Consider a diffeomorphism $f : \mathcal{M} \to \mathcal{M}$ on a d-dimensional manifold $\mathcal{M}$, where the multivariate state $\boldsymbol{x}_n$ consists of M subsystem states $\langle x_n^1, \ldots, x_n^M \rangle$. Each subsystem state x_n^i is confined to a submanifold*

$\mathcal{M}^i \subseteq \mathcal{M}$ of dimension $d^i \leq d$, where $\sum_i d^i = d$. The multivariate observation is given, for some map $\mathbf{G}$, by $\boldsymbol{y}_{n+1} = \mathbf{G}(\langle y_n^{i,(\kappa^i)} \rangle)$.

Proof. The proof restates part of the proof of Theorem 2 of Deyle and Sugihara [50] in terms of subsystems.

Given M inhomogeneous observation functions $\{\psi^i\}$, the following map

$$\boldsymbol{\Phi}_{f,\psi}(\boldsymbol{x}) = \langle \boldsymbol{\Phi}_{f^1,\psi^1}(\boldsymbol{x}), \boldsymbol{\Phi}_{f^2,\psi^2}(\boldsymbol{x}), \ldots, \boldsymbol{\Phi}_{f^M,\psi^M}(\boldsymbol{x}) \rangle \tag{13}$$

is an embedding where each subsystem (local) map $\boldsymbol{\Phi}_{f^i,\psi^i} : \mathcal{M} \to \mathbb{R}^{\kappa^i}$, smoothly (at least $\mathbb{C}^2$), and, at time index n is described by

$$\begin{aligned} \boldsymbol{\Phi}_{f^i,\psi^i}(\boldsymbol{x}_n) &= \langle \psi^i(\boldsymbol{x}_n), \psi^i(\boldsymbol{x}_{n-\tau}), \ldots, \psi^i(\boldsymbol{x}_{n-(k-1)\tau}) \rangle \\ &= y_n^{i,(\kappa^i)}, \end{aligned} \tag{14}$$

where $\sum_i \kappa^i = 2d + 1$ [50]. Note that, from (13) and (14), we have the global map

$$\boldsymbol{\Phi}_{f,\psi}(\boldsymbol{x}_n) = \langle y_n^{i,(\kappa^i)} \rangle = \langle y_n^{1,(\kappa^1)}, \ldots, y_n^{m,(\kappa^M)} \rangle.$$

Now, since $\boldsymbol{\Phi}_{f,\psi}$ is an embedding, it follows that the map $\mathbf{F} = \boldsymbol{\Phi}_{f,\psi} \circ f \circ \boldsymbol{\Phi}_{f,\psi}^{-1}$ is well defined and a diffeomorphism between two observation sequences $\mathbf{F} : \mathbb{R}^{2d+1} \to \mathbb{R}^{2d+1}$, i.e.,

$$\begin{aligned} \langle y_{n+1}^{i,(\kappa^i)} \rangle &= \boldsymbol{\Phi}_{f,\psi}(\boldsymbol{x}_{n+1}) = \boldsymbol{\Phi}_{f,\psi}(f(\boldsymbol{x}_n)) \\ &= \boldsymbol{\Phi}_{f,\psi}\left(f\left(\boldsymbol{\Phi}_{f,\psi}^{-1}\left(\langle y_n^{i,(\kappa^i)} \rangle\right)\right)\right) = \mathbf{F}(\langle y_n^{i,(\kappa^i)} \rangle). \end{aligned}$$

The last $2d + 1$ components of $\mathbf{F}$ are trivial, i.e., the set $\langle y_n^{i,(\kappa^i)} \rangle$ is observed; denote the first M components by $\mathbf{G} : \boldsymbol{\Phi}_{f,\psi} \to \mathbb{R}^M$, and then we have $\boldsymbol{y}_{n+1} = \mathbf{G}(\langle y_n^{i,(\kappa^i)} \rangle)$. □

We now use the result of Lemma 2 to obtain a computable form of the KL divergence.

Lemma 3. *Consider a discrete-time multivariate dynamical system with generic (f, ψ) modelled as a directed and acyclic synchronous GDS $(G, \boldsymbol{x}_n, \boldsymbol{y}_n, \{f^i\}, \{\psi^i\})$ with M subsystems. The KL divergence of a candidate graph G from the observed dataset D can be computed from tractable probability distributions:*

$$D_{\text{KL}}\left[p_{K_M} \,\|\, p_B\right] = \mathbf{E}\left[\log \frac{\Pr(\boldsymbol{y}_{n+1} \mid \langle y_n^{i,(\kappa^i)} \rangle)}{\prod_{i=1}^M \Pr(y_{n+1}^i \mid y_n^{i,(\kappa^i)}, \langle y_n^{ij,(\kappa^{ij})} \rangle_j)}\right]. \tag{15}$$

Proof. Lemma 1, we can substitute (12) into (9), and express the KL divergence $D_{\text{KL}}\left[p_{K_M} \,\|\, p_B\right]$ as

$$D_{\text{KL}}\left[p_{K_M} \,\|\, p_B\right] = \mathbf{E}\left[\log\left(\Pr(\boldsymbol{z}_{n+1} \mid \boldsymbol{z}_n^{(n)}) \cdot \frac{\Pr(\boldsymbol{x}_n \mid \langle y_n^{i,(\kappa^i)} \rangle)}{\prod_{i=1}^M \Pr(y_{n+1}^i \mid y_n^{i,(\kappa^i)}, \langle y_n^{ij,(\kappa^{ij})} \rangle_j)}\right)\right]. \tag{16}$$

We now focus on $p_{K_M}(\boldsymbol{z}_{n+1} | \boldsymbol{z}_n^{(n)})$. Using the chain rule,

$$p_{K_M}(\boldsymbol{z}_{n+1} \mid \boldsymbol{z}_n^{(n)}) = \Pr(\boldsymbol{x}_{n+1} \mid \boldsymbol{z}_n^{(n)}) \cdot \Pr(\boldsymbol{y}_{n+1} \mid \boldsymbol{x}_{n+1}, \boldsymbol{z}_n^{(n)}).$$

Given the Markov property of the dynamics (1) and observation (2) maps, we get

$$p_{K_M}(\boldsymbol{z}_{n+1} \mid \boldsymbol{z}_n^{(n)}) = \Pr(\boldsymbol{X}_{n+1} = f(\boldsymbol{x}_n) \mid \boldsymbol{x}_n) \cdot \Pr(\boldsymbol{Y}_{n+1} = \psi(\boldsymbol{x}_{n+1}) \mid \boldsymbol{x}_{n+1}). \tag{17}$$

Now, recall fom Lemma 2 that global equations for the entire system state x_n and observation y_n are

$$x_{n+1} = f(x_n) + v_f = f\left(\Phi^{-1}_{f,\psi}(\langle y_n^{i,(\kappa^i)}\rangle)\right) + v_f, \tag{18}$$

$$y_{n+1} = \psi(x_{n+1}) + v_\psi = \mathbf{G}(\langle y_n^{i,(\kappa^i)}\rangle) + v_\psi. \tag{19}$$

Given the assumption of i.i.d noise on the function f, from (18), we express the probability of the dynamics x_{n+1}, given by the embedding, as

$$\begin{aligned}\Pr\left(x_{n+1} \mid \langle y_n^{i,(\kappa^i)}\rangle\right) &= \Pr\left(X_{n+1} = f\left(\Phi^{-1}_{f,\psi}\left(\langle y_n^{i,(\kappa^i)}\rangle\right)\right) \mid \langle y_n^{i,(\kappa^i)}\rangle\right) \\ &= \Pr\left(X_n = \Phi^{-1}_{f,\psi}\left(\langle y_n^{i,(\kappa^i)}\rangle\right) \mid \langle y_n^{i,(\kappa^i)}\rangle\right) \cdot \Pr\left(X_{n+1} = f(x_n) \mid x_n\right).\end{aligned} \tag{20}$$

By assumption, the observation noise is i.i.d or dependent only on the state x_{n+1}, and thus the probability of observing y_{n+1}, from (19) is

$$\begin{aligned}\Pr\left(y_{n+1} \mid \langle y_n^{i,(\kappa^i)}\rangle\right) &= \Pr\left(Y_{n+1} = \mathbf{G}(\langle y_n^{i,(\kappa^i)}\rangle) \mid \langle y_n^{i,(\kappa^i)}\rangle\right) \\ &= \Pr\left(X_{n+1} = f\left(\Phi^{-1}_{f,\psi}\left(\langle y_n^{i,(\kappa^i)}\rangle\right)\right) \mid \langle y_n^{i,(\kappa^i)}\rangle\right) \\ &\quad \times \Pr\left(Y_{n+1} = \psi(x_{n+1}) \mid x_{n+1}\right).\end{aligned} \tag{21}$$

By (20) and (21), we have that

$$\Pr(x_{n+1} \mid x_n) \cdot \Pr(y_{n+1} \mid x_{n+1}) = \frac{\Pr(y_{n+1} \mid \langle y_n^{i,(\kappa^i)}\rangle)}{\Pr(x_n \mid \langle y_n^{i,(\kappa^i)}\rangle)}. \tag{22}$$

Substituting Equation (22) into (17) gives

$$p_{K_M}(z_{n+1} \mid z_n^{(n)}) = \frac{\Pr(y_{n+1} \mid \langle y_n^{i,(\kappa^i)}\rangle)}{\Pr(x_n \mid \langle y_n^{i,(\kappa^i)}\rangle)}. \tag{23}$$

Finally, substituting (23) back into (16) yields the statement of the theorem. □

Given all variables in (15) are observed, it is now straightforward to compute KL divergence; however, as we will see, it is more convenient to express (15) as a function of known information-theoretic measures.

4.2. Information-Theoretic Interpretation

The main theorem of this paper states KL divergence in terms of transfer entropy and stochastic interaction. These information-theoretic concepts are defined in Appendix B for convenience.

Theorem 4. *Consider a discrete-time multivariate dynamical system with generic (f, ψ) represented as a directed and acyclic synchronous GDS $(G, x_n, y_n, \{f^i\}, \{\psi^i\})$ with M subsystems. The KL divergence $D_{\mathrm{KL}}\left[p_{K_M} \parallel p_B\right]$ of a candidate graph G from the observed dataset D can be expressed as the difference between stochastic interaction (A9) and collective transfer entropy (A8), i.e.,*

$$D_{\mathrm{KL}}\left[p_{K_M} \parallel p_B\right] = S_Y - \sum_{i=1}^{M} T_{\{Y^{ij}\}_j \to Y^i}. \tag{24}$$

Proof. We can reformulate the KL divergence in (15) as

$$
\begin{aligned}
D_{\mathrm{KL}}\left[p_{K_M} \parallel p_B\right] &= \mathbf{E}\left[\log\left(\Pr(\boldsymbol{y}_{n+1} \mid \langle y_n^{i,(\kappa^i)}\rangle)\right)\right] - \mathbf{E}\left[\log\left(\prod_{i=1}^{M}\Pr(y_{n+1}^{i} \mid y_n^{i,(\kappa^i)}, \langle y_n^{ij,(\kappa^{ij})}\rangle_j)\right)\right] \\
&= -H(\boldsymbol{Y}_{n+1} \mid \{Y_n^{(\kappa^i)}\}) + \sum_{i=1}^{M} H(Y_{n+1}^{i} \mid Y_n^{i,(\kappa^i)}, \{Y_n^{ij,(\kappa^{ij})}\}_j) \\
&= -H(\boldsymbol{Y}_{n+1} \mid \{Y_n^{(\kappa^i)}\}) + \sum_{i=1}^{M} H(Y_{n+1}^{i} \mid Y_n^{i,(\kappa^i)}) \\
&\quad + \sum_{i=1}^{M}\left(H(Y_{n+1}^{i} \mid Y_n^{i,(\kappa^i)}, \{Y_n^{ij,(\kappa^{ij})}\}_j) - H(Y_{n+1}^{i} \mid Y_n^{i,(\kappa^i)})\right).
\end{aligned}
\tag{25}
$$

Substituting in the definitions of transfer entropy (A8) and stochastic interaction (A9) completes the proof. □

To conclude this section, we present the following corollary showing that, when we assume a maximum or fixed embedding dimension κ^i and time delay τ^i, it suffices to maximise the collective transfer entropy alone in order to minimise KL divergence for a synchronous GDS.

Corollary 1. *Fix an embedding dimension κ^i and time delay τ^i for each subsystem $V^i \in \mathcal{V}$. Then, the graph G that minimises the KL divergence $D_{\mathrm{KL}}\left[p_{K_M} \parallel p_B\right]$ is equivalent to the graph that maximises transfer entropy, i.e.,*

$$
\underset{G \in \mathcal{G}}{\arg\min}\, D_{\mathrm{KL}}\left[p_{K_M} \parallel p_B\right] = \underset{G \in \mathcal{G}}{\arg\max} \sum_{i=1}^{M} T_{\{Y^{ij}\}_j \to Y^i}. \tag{26}
$$

Proof. The first term of (24) is constant, given a constant vertex set $\mathcal{V}$, time delay τ and embedding dimension κ and is thus unaffected by the parent set $\Pi_G(V^i)$ of a variable. As a result, S_Y does not depend on the graph G being considered, and, therefore, we only need to consider transfer entropy when optimising KL divergence (24). □

As mentioned above, Corollary 1 is, in practice, equivalent to the maximum log-likelihood (5) and log-likelihood ratio (6) approaches. However, the statement only holds for constant embedding parameters. In the general case, where these parameters are unknown, one requires Theorem 4 to perform structure learning. Given this result, we can now confidently derive scoring functions from Corollary 1.

5. Application to Structure Learning

We now employ the results above in selecting a synchronous GDS that best fits data generated by a multivariate dynamical system. The most natural way to find an optimal model based on Theorem 4 is to minimise KL divergence. Here, we assume constant embedding parameters and use Corollary 1 to present the *transfer entropy score* and discuss some attributes of this score. We then use this scoring function as a subroutine for learning the structure of coupled Lorenz and Rössler attractors.

From Corollary 1, a naive scoring function can be defined as

$$
g_{\mathrm{TE}}(B : D) = \sum_{i=1}^{M} T_{\{Y^{ij}\}_j \to Y^i}. \tag{27}
$$

Given parameterised probability distributions, this score is insufficient, since the sum of transfer entropy in (27) is non-decreasing when including more parents in the graph [38]. Thus, we use statistical significance tests in our scoring functions to mitigate this issue.

5.1. Penalising Transfer Entropy by Independence Tests

Building on the maximum likelihood score (27), we propose using independence tests to define two new scores of practical value. Here, we draw on the result of de Campos [3], who derived a scoring function for BN structure learning based on conditional mutual information and statistical significance tests, called *MIT*. The central idea is to use collective transfer entropy $T_{\langle Y^{ij}\rangle_j \to Y^i}$ to measure the degree of interaction between each subsystem V^i and its parent subsystems $\Pi_G(V^i)$, but also to penalise this term with a value based on significance testing. As with the *MIT* score, this gives a principled way to re-scale the transfer entropy when including more edges in the graph.

To develop our scores, we form a *null hypothesis* H_0 that there is no interaction $T_{\langle Y^{ij}\rangle_j \to Y^i}$, and then compute a test statistic to penalise the measured transfer entropy. To compute the test statistic, it is necessary to consider the measurement distribution in the case where the hypothesis is true. Unfortunately, this distribution is only analytically tractable in the case of discrete and linear-Gaussian systems, where $2NT_{\langle Y^{ij}\rangle_j \to Y^i}$ is known to asymptotically approach the χ^2-distribution [48]. Since this distribution is a function of the parents of Y^i, we let it be described by the function $\chi^2(\{l^{ij}\}_j)$. Now, given this distribution, we can fix some *confidence level* α and determine the value $\chi_{\alpha,\{l^{ij}\}_j}$ such that $p(\chi^2(\{l^{ij}\}_j) \leq \chi_{\alpha,\{l^{ij}\}_j})$. This represents a conditional independence test: if $2NT_{\langle Y^{ij}\rangle_j \to Y^i} \leq \chi_{\alpha,\{l^{ij}\}_j}$, then we accept the hypothesis of conditional independence between Y^i and $\langle Y^{ij}\rangle_j$; otherwise, we reject it. We express this idea as the *TEA* score:

$$g_{TEA}(B : D) = \sum_{i=1}^{M} \left(2NT_{\{Y^{ij}\}_j \to Y^i} - \chi_{\alpha,\{l^{ij}\}_j} \right). \tag{28}$$

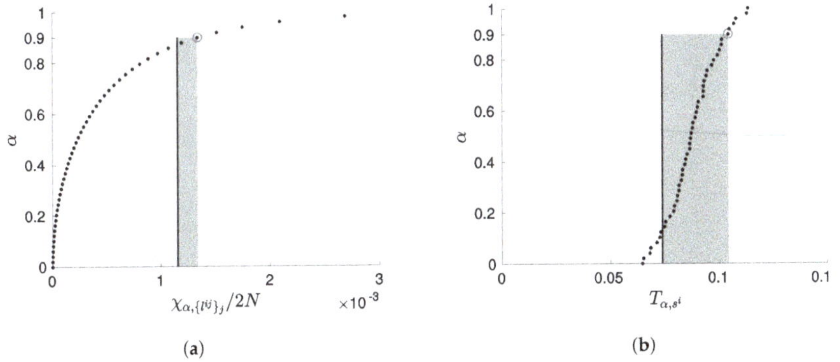

Figure 3. Distributions of the (**a**) *TEA* penalty function (28) and the (**b**) *TEE* penalty function (28). Both distributions were generated by observing the outcome of 1000 samples from two Gaussian variables with a correlation of 0.05. The figures illustrate: the distribution as a set of 100 sampled points (black dots); the area considered independent (grey regions); the measured transfer entropy (black line); and the difference between measurement and penalty term (dark grey region). Both tests use a value of $\alpha = 0.9$ (a p-value of 0.1). The distribution in (**a**) was estimated by assuming variables were linearly-coupled Gaussians, and the distribution in (**b**) was computed via a kernal box method (computed by the Java Information Dynamics Toolkit (JIDT), see [52] for details).

In general, we only have access to *continuous* measurements of dynamical systems, and so are limited by the discrete or linear-Gaussian assumption. We can, however, use *surrogate* measurements $T_{\langle Y^{ij}\rangle_j^s \to Y^i}$ to empirically compute the distribution under the assumption of H_0 [52]. This same technique has been used by [38] to derive a greedy structure learning algorithm for effective network analysis. Here, $\langle Y^{ij}\rangle_j^s$ are surrogate sets of variables for $\langle Y^{ij}\rangle_j$, which have the same statistical properties as $\langle Y^{ij}\rangle_j$, but the correlation between $\langle Y^{ij}\rangle_j^s$ and Y^i is removed. Let the distribution of these surrogate measurements be represented by some general function $T(s^i)$ where, for the discrete and linear-Gaussian systems, we could compute $T(s^i)$ analytically as an independent set of χ^2-distributions $\chi^2(\{l^{ij}\}_j)$. When no analytic distribution is known, we use a resampling method (i.e., permutation or bootstrapping), creating a large number of surrogate time-series pairs $\{\langle Y^{ij}\rangle_j^s, Y^i\}$ by shuffling (for permutations, or redrawing for bootstrapping) the samples of Y^i and computing a population of $T_{\langle Y^{ij}\rangle_j^s \to Y^i}$. As with the *TEA* score, we fix some confidence level α and determine the value T_{α,s^i}, such that $p(T(s^i) \leq T_{\alpha,s^i}) = \alpha$. This results in the *TEE* scoring function as

$$g_{TEE}(B : D) = \sum_{i=1}^{M} \left(T_{\{Y^{ij}\}_j \to Y^i} - T_{\alpha,s^i} \right). \tag{29}$$

We can obtain the value T_{α,s^i} by (1) drawing S samples $T_{\langle Y^{ij}\rangle_j^s \to Y^i}$ from the distribution $T(s^i)$ (by permutation or bootstrapping), (2) fixing $\alpha \in \{0, 1/S, 2/S, \ldots, 1\}$, and then (3) taking T_{α,s^i} such that

$$\alpha = \frac{1}{S} \sum_{T_{\{Y^{ij}\}_j \to Y^i}} \mathbb{1}_{T_{\{Y^{ij}\}_j^s \to Y^i} \leq T_{\alpha,s^i}}.$$

We can alternatively limit the number of surrogates S to $\lceil \alpha/(1-\alpha) \rceil$ and take the maximum as T_{α,s^i} [22]; however, taking a larger number of surrogates will improve the validity of the distribution $T(s^i)$.

Both the analytical (*TEA*) and empirical (*TEE*) scoring functions are illustrated in Figure 3. Note that the approach of significance testing is functionally equivalent to considering the log-likelihood ratio in (6), where, as stated, nested log-likelihoods (and thus transfer entropy) follows the above χ^2-distribution [48].

5.2. Implementation Details and Algorithm Analysis

The two main implementation challenges that arise when performing structure learning are: (1) computing the score for every candidate network and (2) obtaining a sufficient number of samples to recover the network. The main contributions of this work are theoretical justifications for measures already in use and, fortunately, algorithmic performance has already been addressed extensively using various heuristics. Here, we present an exact, exhaustive implementation for the purpose of validating our theoretical contributions.

First, for computing collective transfer entropy for the score (29), we require CPDs to be estimated from data. Given these CPDs, collective transfer entropy (A8) decomposes as a sum of p conditional transfer entropy (A7) terms, where $p = |\{Y^{ij}\}_j|$ is the size of the parent set (see Appendix B for details). Since most observations of dynamical systems are expected to be continuous, we employ a non-parametric, nearest-neighbour based approach to density estimation called the Kraskov–Stögbauer–Grassberger (KSG) estimator [43]. For any arbitrary decomposition of collective transfer entropy (i.e., any ordering of the parent set), this density estimation can be computed in time $O(\kappa(p+1)KN^{\kappa(p+1)}\log(N))$, where K is the number of nearest neighbours for each observation in a dataset of size N, and κ is the embedding dimension [52]. We upper bound this as $O(\kappa MKN^{\kappa M}\log(N))$ since the maximum p is $M-1$.

Now, the above density estimation was described for an arbitrary ordering of the parent set. In the case of parametric (discrete or linear-Gaussian) density estimation, every permutation of the parent set yields equivalent results, with potentially different $\chi_{\alpha,\{l^{ij}\}_j}$ values for each permutation [3]; however, this is not the case for non-parametric density estimation techniques, e.g., the KSG estimator. Hence, as a conservative estimate of the score, we compute all $p!$ permutations of the parent set and take the minimum collective transfer entropy. In order to obtain the surrogate distribution, we require S

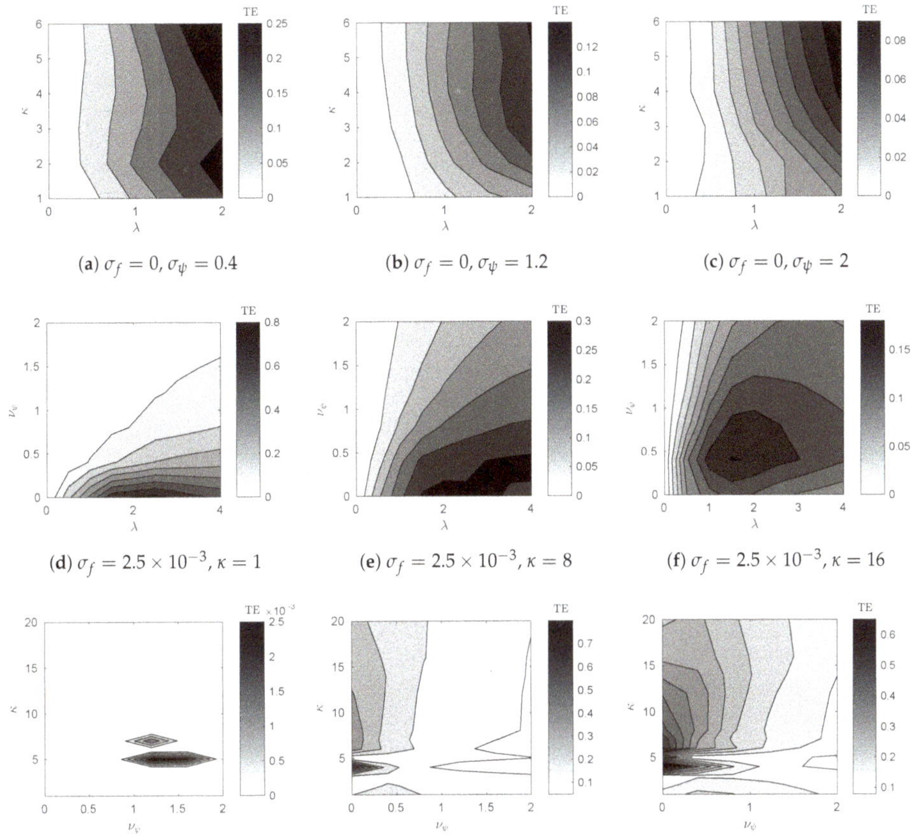

(**a**) $\sigma_f = 0, \sigma_\psi = 0.4$ (**b**) $\sigma_f = 0, \sigma_\psi = 1.2$ (**c**) $\sigma_f = 0, \sigma_\psi = 2$

(**d**) $\sigma_f = 2.5 \times 10^{-3}, \kappa = 1$ (**e**) $\sigma_f = 2.5 \times 10^{-3}, \kappa = 8$ (**f**) $\sigma_f = 2.5 \times 10^{-3}, \kappa = 16$

(**g**) $\sigma_f = 2.5 \times 10^{-3}, \lambda = 0$ (**h**) $\sigma_f = 2.5 \times 10^{-3}, \lambda = 2$ (**i**) $\sigma_f = 2.5 \times 10^{-3}, \lambda = 4$

Figure 5. Transfer entropy as a function of the parameters of a coupled Lorenz–Rössler system. These components are: coupling strength λ and embedding dimension κ in the top row (**a–c**); coupling strength λ and observation noise σ_ψ in the middle row (**d–f**); and observation noise σ_ψ and embedding dimension κ in the bottom row (**g–i**).

There are two interesting features in Figure 5 due to the dynamical systems studied. First, in the bottom row (Figure 5g–i), there is a bifurcation around $\kappa = 6$. The theoretical embedding dimension for this system is $\kappa = 2(d^1 + d^2) + 1 = 7$, and, in this case, for $\kappa < 6$, the embedding does not suffice to reconstruct the dynamics. Second, in Figure 5i, the transfer entropy decreases after about $\lambda = 2$. This appears to be the case of synchrony due to strong coupling, where the dynamics of the forced variable become subordinate to the forcing [4], thus reducing the information transferred between the two subsystems.

6.3. Case Study: Network of Lorenz Attractors

In this section, we evaluate the score (27) in learning the structure of distributed dynamical systems. We will look at systems of three and four nodes of coupled Lorenz subsystems with arbitrary topologies. Unfortunately, significantly higher number of nodes become computationally expensive due to an increased embedding dimension κ, number of data points N, and number of permutations required to calculate the collective transfer entropy. To evaluate the performance of the score (27), the dynamics noise is constant $\sigma_f = 0.01$, whereas the observation noise σ_ψ and the number of observations taken N are varied. We selected the theoretical maximum embedding dimension $\kappa = 2d + 1$ and $\tau = 1$ as is common given discrete-time measurements [22]. It should be noted that from the results from Section 6.2 that transfer entropy is sensitive to the numerous parameters used to generate the data, and thus depending on the scenario, a significant sample size can be required for recovering the underlying graph structure. We do not make an effort to reduce this sample size and instead show the effect of using a different number of samples on the accuracy of the structure learning procedure.

In order to evaluate the scoring function, we compute the recall (R, or true positive rate), fallout (F, or false positive rate), and precision (P, or positive predictive value) of the recovered graph. Let TP denote the number of true positives (correct edges); TN denote the number of true negatives (correctly rejected edges); FP denote the number of false positives (incorrect edges); and FN denote the number of false negatives (incorrectly rejected edges). Then, $\mathrm{R} = \mathrm{TP}/(\mathrm{TP}+\mathrm{FN})$, $\mathrm{F} = \mathrm{FP}/(\mathrm{FP}+\mathrm{TN})$, and $\mathrm{P} = \mathrm{TP}/(\mathrm{TP}+\mathrm{FP})$. Finally, the F_1-score gives the harmonic mean of precision and recall to give a measure of the tests accuracy, i.e., $F_1 = 2 \cdot \mathrm{R} \cdot \mathrm{P}/(\mathrm{R}+\mathrm{P})$. Note that the ideal recall, precision and F_1-score is 1, and ideal fallout is 0. Furthermore, a ratio of $\mathrm{R}/\mathrm{F} > 1$ suggests the classifier is better than random. As a summary statistic, Tables 1 and 2 presents the F_1-scores for all networks illustrated in Figure 4, and the full classification results (e.g., precision, recall, and fallout) are given in Appendix C. The F_1-scores are thus a measure of how relevant the recovered network is to the original (generating) network from our data-driven approach.

In general, the results of Tables 1 and 2 show that the scoring function is capable of recovering the network with high precision and recall, as well as low fallout. In the table, the cell colours are shaded to indicate higher (white) to lower (black) F_1 scores. The best performing score is that with a p-value of 0.01 and no penalisation (a p-value of ∞) has the second highest classification results. As expected, the graphs recovered from data with low observational noise ($\sigma_\psi = 1$) are more accurate than those inferred from noisier data ($\sigma_\psi = 10$). The results for three-node networks (shown in Table 1) yields mostly full recovery of the structure for a higher number of observations $N \geq 75$ K, whereas, the four-node networks (shown in Table 2) are more difficult to classify.

Table 1. F_1-scores for three-node ($M = 3$) networks. We present the classification summary for the three arbitrary topologies of coupled Lorenz systems represented by Figure 4b–d (network G^1 has no edges and thus an undefined F_1-score). The p-value of the *TEE* score is given in the top row of each table, with ∞ signifying using no significance testing, i.e., score (27).

		$p = \infty$		$p = 0.01$		$p = 0.001$		$p = 0.0001$	
Graph	N	$\sigma_\psi = 1$	$\sigma_\psi = 10$	$\sigma_\psi = 1$	$\sigma_\psi = 10$	$\sigma_\psi = 1$	$\sigma_\psi = 10$	$\sigma_\psi = 1$	$\sigma_\psi = 10$
G^2	5 K	0.8	0.5	0.8	0.5	0.8	0.5	0.8	0.5
	25 K	1	0.8	1	0.5	1	0.5	1	0.8
	100 K	1	0.5	1	1	1	1	1	0.8
G^3	5 K	1	0.67	1	1	1	1	1	0.67
	25 K	1	1	1	0.5	1	1	1	1
	100 K	1	1	1	1	1	1	1	1
G^4	5 K	0.8	-	0.8	0.8	0.8	0.5	0.8	-
	25 K	1	1	1	1	1	0.5	1	1
	100 K	1	1	1	1	1	1	1	1

Table 2. F_1-scores for four-node ($M = 4$) networks. We present the classification summary for the three arbitrary topologies of coupled Lorenz systems represented by Figure 4f–h (network G^5 has no edges and thus an undefined F_1-score). The p-value of the *TEE* score is given in the top row of each table, with ∞ signifying using no significance testing, i.e., score (27).

		$p = \infty$		$p = 0.01$		$p = 0.001$		$p = 0.0001$	
Graph	N	$\sigma_\psi = 1$	$\sigma_\psi = 10$	$\sigma_\psi = 1$	$\sigma_\psi = 10$	$\sigma_\psi = 1$	$\sigma_\psi = 10$	$\sigma_\psi = 1$	$\sigma_\psi = 10$
	5 K	0.57	0.5	0.57	0.29	0.57	0.29	0.57	-
G^6	25 K	0.75	0.33	0.75	0.33	0.75	0.29	0.75	0.33
	100 K	1	0.33	1	0.57	1	0.4	1	0.33
	5 K	1	0.25	1	0.29	0.75	0.25	0.75	0.57
G^7	25 K	1	0.5	1	0.86	1	0.86	1	0.5
	100 K	1	0.86	1	0.86	1	0.86	1	0.86
	5 K	1	0.25	1	0.57	1	0.75	1	0.25
G^8	25 K	1	0.86	1	0.86	1	0.86	1	0.86
	100 K	1	0.86	1	0.86	1	0.57	1	0.86

Interestingly, the statistical significance testing does not have a strong effect on the results. It is unclear if this is due to the use of the non-parametric density estimators, which, in effect, are parsimonious in nature since transfer entropy will likely reduce when conditioning on more variables with a fixed samples size. One challenging case is the empty networks G^1 and G^5; this is shown in Appendix C, where the fallout is rarely 0 for any of the p-values or sample sizes (although a large number of observations $N = 100$ K appears to reduce spurious edges). It would be expected that significance testing on these networks would outperform the naive score (27) given that a non-zero bias is introduced for a finite number of observations. Further investigation is required to understand why the null case fails.

7. Discussion and Future Work

We have presented a principled method to compute the KL divergence for model selection in distributed dynamical systems based on concepts from differential topology. The results presented in Figure 5 and Tables 1 and 2 illustrate that this approach is suitable for recovering synchronous GDSs from data. Further, KL divergence is related to model encoding, which is a fundamental measure used in complex systems analysis. Our result, therefore, has potential implications for other areas of research. For example, the notion of equivalence classes in BN structure learning [63] should lend insight into the area of effective network analysis [35,36].

More specifically, the approach proposed here complements explicit Bayesian identification and comparison of state space models. In DCM, and more generally in approximate Bayesian inference, models are identified in terms of their parameters via an optimisation of an approximate posterior density over model parameters with respect to a variational (free energy) bound on log evidence [64]. After these parameters have been identified, this bound can be used directly for model comparison and selection. Interestingly, free energy is derived from the KL divergence between the approximate and true posterior and thus automatically penalises more complex models; however, in Equation (8), these distributions are inverted. In future work, it would be interesting to explore the relationship between transfer entropy and the variational free energy bound. Specifically, computing an evidence bound directly from the transfer entropy may allow us to avoid the significance testing described in Section 5 and instead use an approximation to evidence for structure learning.

Multivariate extensions to transfer entropy are known to eliminate redundant pairwise relationships and take into account the influence of confounding relationships in a network (i.e., synergistic effects) [65,66]. In this work, we have shown that this intuition holds for distributed dynamical systems when confined to a DAG topology. We conjecture that these methods are also applicable when cyclic dependencies exist within a graph, given any generic observation can be used in reconstructing the dynamics [50]; however, the methods presented are more likely to reveal *one* source in the cycle, rather than all information sources due to redundancy.

There are a number of extensions that should be considered for further practical implementations of this algorithm. Currently, we assume that the dimensionality of each subsystem is known, and thus we can bound the embedding dimension κ for recovering the hidden structure. However, this is generally infeasible in practice and a more general algorithm would infer the embedding dimension and time delay for an unknown system. Fortunately, there are numerous techniques to recover these parameters [54,55]. Furthermore, evaluating the quality of large graphs is infeasible with our current approach. However, our exact algorithm illustrates the feasibility of state space reconstruction in recovering a graph in practice. In the future, we aim to leverage the structure learning literature on reducing the search space and approximating scoring functions to produce more efficient algorithms.

Finally, the theoretical results of this work supplements understanding in fields where transfer entropy is commonly employed. Point processes are being increasingly viewed as models for a variety of information processing systems, e.g., as spiking neural trains [67] and adversaries in robotic patrolling models [68]. It was recently shown how transfer entropy can be computed for continuous time point processes such as these [67], allowing for efficient use of our analytical scoring function g_{TEA} in a number of contexts. Another intriguing line of research is the physical and thermodynamic interpretation of transfer entropy [69], particularly its relationship to the arrow of time [70]; this relationship between endomorphisms as discussed here and time asymmetry of thermodynamics should be explored further.

Acknowledgments: This work was supported in part by the Australian Centre for Field Robotics; the New South Wales Government; and the Faculty of Engineering & Information Technologies, The University of Sydney, under the Faculty Research Cluster Program. Special thanks go to Jürgen Jost, Michael Small, Joseph Lizier, and Wolfram Martens for their useful discussions.

Author Contributions: O.C, M.P. and R.F. conceived and designed the experiments; O.C. performed the experiments; O.C. and M.P. analyzed the data; O.C., M.P., and R.F. wrote the paper. All authors have read and approved the final manuscript.

Conflicts of Interest: The authors declare no conflict of interest.

Appendix A. Embedding Theory

We refer here to embedding theory as the study of inferring the (hidden) state $x_n \in \mathcal{M}$ of a dynamical system from a sequence of observations $y_n \in \mathbb{R}$. This section will cover reconstruction theorems that define the conditions under which we can use delay embeddings for recovering the original dynamics f from this observed time series.

In differential topology, an *embedding* refers to a smooth map $\mathbf{\Phi} : \mathcal{M} \to \mathcal{N}$ between manifolds $\mathcal{M}$ and $\mathcal{N}$ if it maps $\mathcal{M}$ diffeomorphically onto its image. In Takens' seminal work on turbulent flow [31], he proposed a map $\mathbf{\Phi}_{f,\psi} : \mathcal{M} \to \mathbb{R}^\kappa$, that is composed of delayed observations, can be used to reconstruct the dynamics for typical (f, ψ). That is, fix some κ (the *embedding dimension*) and τ (the *time delay*), the *delay embedding map*, given by

$$\mathbf{\Phi}_{f,\psi}(x_n) = y_n^{(\kappa)} = \langle y_n, y_{n+\tau}, y_{n+2\tau}, \ldots, y_{n+(\kappa-1)\tau} \rangle, \tag{A1}$$

is an embedding. More formally, denote $\mathbf{\Phi}_{f,\psi}$, $\mathcal{D}^r(\mathcal{M}, \mathcal{M})$ as the space of C^r-diffeomorphisms on $\mathcal{M}$ and $C^r(\mathcal{M}, \mathbb{R})$ as the space of C^r-functions on $\mathcal{M}$, then the theorem can be expressed as follows.

Theorem A1 (Delay Embedding Theorem for Diffeomorphisms [31])**.** *Let $\mathcal{M}$ be a compact manifold of dimension $d \geq 1$. If $\kappa \geq 2d + 1$ and $r \geq 1$, then there exists an open and dense set $(f, \psi) \in \mathcal{D}^r(\mathcal{M}, \mathcal{M}) \times C^r(\mathcal{M}, \mathbb{R})$ for which the map $\mathbf{\Phi}_{f,\psi}$ is an embedding of $\mathcal{M}$ into $\mathbb{R}^\kappa$.*

The implication of Theorem A1 is that, for typical (f, ψ), the image $\mathbf{\Phi}_{f,\psi}(\mathcal{M})$ of $\mathcal{M}$ under the delay embedding map $\mathbf{\Phi}_{f,\psi}$ is completely equivalent to $\mathcal{M}$ itself, apart from the smooth invertible change of coordinates given by the mapping $\mathbf{\Phi}_{f,\psi}$. An important consequence of this result is that we can define a map $\mathbf{F} = \mathbf{\Phi}_{f,\psi} \circ f \circ \mathbf{\Phi}_{f,\psi}^{-1}$ on $\mathbf{\Phi}_{f,\psi}$, such that $y_{n+1}^{(\kappa)} = \mathbf{F}(y_n^{(\kappa)})$ [44]. The bound for the open and dense set referred to in Theorem A1 is given by a number of technical assumptions. Denote $(Df)_x$ as the derivative of function f at a point x in the domain of f. The set of periodic points A of f with period less than τ has finitely many points. In addition, the eigenvalues of $(Df)_x$ at each x in a compact neighbourhood A are distinct and not equal to 1.

Theorem A1 was established for diffeomorphisms $\mathcal{D}^r$; by definition, the dynamics are thus invertible in time. Thus, the time delay τ in (A1) can be either positive (delay lags) or negative (delay leads). Takens later proved a similar result for endomorphisms, i.e., non-invertible maps that restricts the time delay to a negative integer. Denote by $\mathcal{E}(\mathcal{M}, \mathcal{M})$ the set of the space of C^r-endomorphisms on $\mathcal{M}$, then the reconstruction theorem for endomorphisms can be expressed as the following.

Theorem A2 (Delay Embedding Theorem for Endomorphisms [71]). *Let $\mathcal{M}$ be a compact m dimensional manifold. If $\kappa \geq 2d + 1$ and $r \geq 1$, then there exists an open and dense set $(f, \psi) \in \mathcal{D}^r(\mathcal{M}, \mathcal{M}) \times C^r(\mathcal{M}, \mathbb{R})$ for which there is a map $\pi_\kappa : \mathcal{X}_\kappa \to \mathcal{M}$ with $\pi_\kappa \mathbf{\Phi}_{f,\psi} = f^{\kappa-1}$. Moreover, the map π_κ has bounded expansion or is Lipschitz continuous.*

As a result of Theorem A2, a sequence of κ successive measurements from a system determines the system state *at the end* of the sequence of measurements [71]. That is, there exists an endomorphism $\mathbf{F} = \mathbf{\Phi}_{f,\psi} \circ f \circ \mathbf{\Phi}_{f,\psi}^{-1}$ to predict the next observation if one takes a negative time (lead) delay τ in (A1).

In this work, we consider two important generalisations of the Delay Embedding Theorem A1. Both of these theorems follow similar proofs to the original and have thus been derived for diffeomorphisms, not endomorphisms. However, encouraging empirical results in [6] support the conjecture that they can both be generalised to the case of endomorphisms by taking a negative time delay, as is done in Theorem A2 above. This would allow for not only distributed flows that are used in our work, but endomorphic maps, e.g., the well-studied coupled map lattice structure [51].

The first generalisation is by Stark et al. [44] and deals with a skew-product system. That is, f is now forced by some second, independent system $g : \mathcal{N} \to \mathcal{N}$. The dynamical system on $\mathcal{M} \times \mathcal{N}$ is thus given by the set of equations

$$x_{n+1} = f(x_n, \omega_n), \qquad \omega_{n+1} = g(\omega_n). \tag{A2}$$

In this case, the delay map is written as

$$\mathbf{\Phi}_{f,g,\psi}(x, \omega) = \langle y_n, y_{n+\tau}, y_{n+2\tau}, \ldots, y_{n+(\kappa-1)\tau} \rangle, \tag{A3}$$

and the theorem can be expressed as follows.

Theorem A3 (Bundle Delay Embedding Theorem [44]). *Let $\mathcal{M}$ and $\mathcal{N}$ be compact manifolds of dimension $d \geq 1$ and e, respectively. Suppose that $\kappa \geq 2(d+e)+1$ and the periodic orbits of period $\leq d$ of $g \in \mathcal{D}^r(\mathcal{N})$ are isolated and have distinct eigenvalues. Then, for $r \geq 1$, there exists an open and dense set of $(f, \psi) \subset \mathcal{D}^r(\mathcal{M} \times \mathcal{N}, \mathcal{M}) \times C^r(\mathcal{M}, \mathbb{R})$ for which the map $\mathbf{\Phi}_{f,g,\psi}$ is an embedding of $\mathcal{M} \times \mathcal{N}$ into $\mathbb{R}^\kappa$.*

Finally, all theorems up until now have assumed a single read-out function for the system in question. Recently, Sugihara et al. [4] showed that multivariate mappings also form an embedding, with minor changes to the technical assumptions underlying Takens' original theorem. That is, given $M \leq 2d + 1$ different observation functions, the delay map can be written as

$$\mathbf{\Phi}_{f,\langle\psi^i\rangle}(x) = \langle \mathbf{\Phi}_{f,\psi^1}(x), \mathbf{\Phi}_{f,\psi^2}(x), \ldots, \mathbf{\Phi}_{f,\psi^M}(x) \rangle, \tag{A4}$$

where each delay map $\mathbf{\Phi}_{f,\psi^i}$ is as per (A1) for individual embedding dimension $\kappa^i \leq \kappa$. The theorem can then be stated as follows.

Theorem A4 (Delay Embedding Theorem for Multivariate Observation Functions [50]). *Let $\mathcal{M}$ be a compact manifold of dimension $d \geq 1$. Consider a diffeomorphism $f \in \mathcal{D}^r(\mathcal{M}, \mathcal{M})$ and a set of at most $2d+1$ observation functions $\langle \psi^i \rangle$ where each $\psi^i \in C^r(\mathcal{M}, \mathbb{R})$ and $r \geq 2$. If $\sum_i \kappa^i \geq 2d+1$, then, for generic $(f, \langle \psi^i \rangle)$, the map $\mathbf{\Phi}_{f,\langle \psi^i \rangle}$ is an embedding.*

Appendix B. Information Theory

In this section, we introduce some key concepts of information theory: conditional entropy; conditional and collective transfer entropy; and stochastic interaction.

Consider two arbitrary random variables X and Y; the *conditional entropy* $H(X \mid Y)$ represents the uncertainty of X after taking into account the outcomes of another random variable Y by the equation

$$H(X \mid Y) = -\sum_{x,y} \Pr(x,y) \log \Pr(x \mid y) = \mathbf{E}\left[\Pr(x \mid y)\right]. \tag{A5}$$

Transfer entropy detects the directed exchange of information between random processes by marginalising out common history and static correlations between variables; it is thus considered a measure of information transfer within a system [25]. Let the processes X and Y have associated embedding dimensions κ^X and κ^Y. The *transfer entropy* of X to Y is given in terms of conditional entropy:

$$T_{X \to Y} = H(Y_{n+1} \mid Y_n^{(\kappa^Y)}) - H(Y_{n+1} \mid X_n^{i,(\kappa^X)}, Y_n^{(\kappa^Y)}). \tag{A6}$$

Now, given a third process Z with embedding dimension κ^Z, we can compute the information transfer of X to Y in the context of Z as:

$$T_{X \to Y \mid Z} = H(Y_{n+1} \mid Y_n^{(\kappa^Y)}, Z_n^{i,(\kappa^Z)}) - H(Y_{n+1} \mid X_n^{i,(\kappa^X)}, Y_n^{(\kappa^Y)}, Z_n^{i,(\kappa^Z)}). \tag{A7}$$

The collective transfer entropy computes the information transfer between a set of M source processes and a single destination process [19]. Consider the set $\mathbf{Y} = \{Y^i\}$ of source processes. We can compute the collective transfer entropy from $\mathbf{Y}$ to the destination process X as a function of conditional entropy (A5) terms:

$$T_{\mathbf{Y} \to X} = T_{Y^1 \to X} + \sum_{i=1}^{M} T_{Y^i \to X \mid \{Y^1, \ldots, Y^{i-1}\}}, \tag{A8}$$

where the ordering of the source processes are arbitrary.

Stochastic interaction measures the complexity of dynamical systems by quantifying the excess of information processed, in time, by the system beyond the information processed by each of the nodes [17,18,72,73]. Using the same notation, stochastic interaction of the collection of processes $\mathbf{Y}$ is

$$S_{\mathbf{Y}} = -H(\mathbf{Y}_{n+1} \mid \{Y_n^{i,(\kappa^i)}\}) + \sum_{i=1}^{M} H(Y_{n+1}^i \mid Y_n^{i,(\kappa^i)}). \tag{A9}$$

The standard definition assumes a first-order Markov process [17,18]; In (A9), we generalise stochastic interaction to arbitrary κ-order Markov chains.

Appendix C. Extended Results

Here, we present the extended results of Tables 1 and 2. That is, we give the precision, recall, fallout, and F_1-scores for the eight networks of Lorenz attractors shown in Figure 4. These results are given for a number of different sample sizes to illustrate the sample complexity of this problem:

$N = 5000$ (Tables A1 and A2), $N = 10{,}000$ (Tables A3 and A4), $N = 25{,}000$ (Tables A5 and A6), $N = 50{,}000$ (Tables A7 and A8), and $N = 100{,}000$ (Tables A9 and A10). Each table has results for various p-values (with a p-value of ∞ denoting the maximum likelihood score (27)), as well as two different observation noise variances, $\sigma_\psi = 1$ and $\sigma_\psi = 10$.

Table A1. Classification results for three-node ($M = 3$) networks for $N = 5000$ samples. We present the precision (P), recall (R), fallout (F), and F_1-score for the eight arbitrary topologies of coupled Lorenz systems represented by Figure 4.

Graph	p-Value	∞		0.01		0.001		0.0001	
	σ_ψ	1	10	1	10	1	10	1	10
G^1	R	-	-	-	-	-	-	-	-
	F	0.33	0.22	0.33	0.22	0.22	0.33	0.33	0.22
	P	0	0	0	0	0	0	0	0
	F_1	-	-	-	-	-	-	-	-
G^2	R	1	0.5	1	0.5	1	0.5	1	0.5
	F	0.14	0.14	0.14	0.14	0.14	0.14	0.14	0.14
	P	0.67	0.5	0.67	0.5	0.67	0.5	0.67	0.5
	F_1	0.8	0.5	0.8	0.5	0.8	0.5	0.8	0.5
G^3	R	1	0.5	1	1	1	1	1	0.5
	F	0	0	0	0	0	0	0	0
	P	1	1	1	1	1	1	1	1
	F_1	1	0.67	1	1	1	1	1	0.67
G^4	R	1	0	1	1	1	0.5	1	0
	F	0.14	0.43	0.14	0.14	0.14	0.14	0.14	0.43
	P	0.67	0	0.67	0.67	0.67	0.5	0.67	0
	F_1	0.8	-	0.8	0.8	0.8	0.5	0.8	-

Table A2. Classification results for four-node ($M = 4$) networks for $N = 5000$ samples. We present the precision (P), recall (R), fallout (F), and F_1-score for the eight arbitrary topologies of coupled Lorenz systems represented by Figure 4.

Graph	p-Value	∞		0.01		0.001		0.0001	
	σ_ψ	1	10	1	10	1	10	1	10
G^5	R	-	-	-	-	-	-	-	-
	F	0.31	0.25	0.31	0.19	0.31	0.25	0.31	0.19
	P	0	0	0	0	0	0	0	0
	F_1	-	-	-	-	-	-	-	-
G^6	R	0.67	0.67	0.67	0.33	0.67	0.33	0.67	0
	F	0.15	0.23	0.15	0.23	0.15	0.23	0.15	0.31
	P	0.5	0.4	0.5	0.25	0.5	0.25	0.5	0
	F_1	0.57	0.5	0.57	0.29	0.57	0.29	0.57	-
G^7	R	1	0.25	1	0.25	0.75	0.25	0.75	0.5
	F	0	0.25	0	0.17	0.083	0.25	0.083	0.083
	P	1	0.25	1	0.33	0.75	0.25	0.75	0.67
	F_1	1	0.25	1	0.29	0.75	0.25	0.75	0.57
G^8	R	1	0.25	1	0.5	1	0.75	1	0.25
	F	0	0.25	0	0.083	0	0.083	0	0.25
	P	1	0.25	1	0.67	1	0.75	1	0.25
	F_1	1	0.25	1	0.57	1	0.75	1	0.25

Table A3. Classification results for three-node ($M = 3$) networks for $N = 10{,}000$ samples. We present the precision (P), recall (R), fallout (F), and F_1-score for the eight arbitrary topologies of coupled Lorenz systems represented by Figure 4.

Graph	*p*-Value	∞		0.01		0.001		0.0001	
	σ_ψ	1	10	1	10	1	10	1	10
G^1	R	-	-	-	-	-	-	-	-
	F	0.22	0.11	0.22	0.11	0.22	0.22	0.22	0.11
	P	0	0	0	0	0	0	0	0
	F_1	-	-	-	-	-	-	-	-
G^2	R	1	0.5	1	0.5	1	0.5	1	0.5
	F	0	0.14	0	0.14	0	0.14	0	0.14
	P	1	0.5	1	0.5	1	0.5	1	0.5
	F_1	1	0.5	1	0.5	1	0.5	1	0.5
G^3	R	1	0.5	1	1	1	0	1	0.5
	F	0	0.14	0	0	0	0.29	0	0.14
	P	1	0.5	1	1	1	0	1	0.5
	F_1	1	0.5	1	1	1	-	1	0.5
G^4	R	1	1	1	0.5	1	0.5	1	1
	F	0.14	0.14	0	0	0.14	0.14	0.14	0.14
	P	0.67	0.67	1	1	0.67	0.5	0.67	0.67
	F_1	0.8	0.8	1	0.67	0.8	0.5	0.8	0.8

Table A4. Classification results for four-node ($M = 4$) networks for $N = 10{,}000$ samples. We present the precision (P), recall (R), fallout (F), and F_1-score for the eight arbitrary topologies of coupled Lorenz systems represented by Figure 4.

Graph	*p*-Value	∞		0.01		0.001		0.0001	
	σ_ψ	1	10	1	10	1	10	1	10
G^5	R	-	-	-	-	-	-	-	-
	F	0.31	0.25	0.31	0.19	0.31	0.19	0.31	0.25
	P	0	0	0	0	0	0	0	0
	F_1	-	-	-	-	-	-	-	-
G^6	R	0.67	0.33	0.67	0	1	1	0.67	0.33
	F	0.15	0.15	0.15	0.15	0.15	0.15	0.15	0.15
	P	0.5	0.33	0.5	0	0.6	0.6	0.5	0.33
	F_1	0.57	0.33	0.57	-	0.75	0.75	0.57	0.33
G^7	R	0.75	0.5	1	0.5	1	0.25	0.75	0.5
	F	0.083	0.083	0	0.083	0	0.17	0.083	0.083
	P	0.75	0.67	1	0.67	1	0.33	0.75	0.67
	F_1	0.75	0.57	1	0.57	1	0.29	0.75	0.57
G^8	R	1	0.25	1	0.25	1	0	1	0.25
	F	0	0.17	0	0.17	0	0.25	0	0.17
	P	1	0.33	1	0.33	1	0	1	0.33
	F_1	1	0.29	1	0.29	1	-	1	0.29

Table A5. Classification results for three-node ($M = 3$) networks for $N = 25{,}000$ samples. We present the precision (P), recall (R), fallout (F), and F_1-score for the eight arbitrary topologies of coupled Lorenz systems represented by Figure 4.

Graph	p-Value	∞		0.01		0.001		0.0001	
	σ_ψ	1	10	1	10	1	10	1	10
G^1	R	-	-	-	-	-	-	-	-
	F	0.22	0.11	0.22	0.11	0.22	0.22	0.22	0.11
	P	0	0	0	0	0	0	0	0
	F_1	-	-	-	-	-	-	-	-
G^2	R	1	1	1	0.5	1	0.5	1	1
	F	0	0.14	0	0.14	0	0.14	0	0.14
	P	1	0.67	1	0.5	1	0.5	1	0.67
	F_1	1	0.8	1	0.5	1	0.5	1	0.8
G^3	R	1	1	1	0.5	1	1	1	1
	F	0	0	0	0.14	0	0	0	0
	P	1	1	1	0.5	1	1	1	1
	F_1	1	1	1	0.5	1	1	1	1
G^4	R	1	1	1	1	1	0.5	1	1
	F	0	0	0	0	0	0.14	0	0
	P	1	1	1	1	1	0.5	1	1
	F_1	1	1	1	1	1	0.5	1	1

Table A6. Classification results for four-node ($M = 4$) networks for $N = 25{,}000$ samples. We present the precision (P), recall (R), fallout (F), and F_1-score for the eight arbitrary topologies of coupled Lorenz systems represented by Figure 4.

Graph	p-Value	∞		0.01		0.001		0.0001	
	σ_ψ	1	10	1	10	1	10	1	10
G^5	R	-	-	-	-	-	-	-	-
	F	0.31	0.19	0.31	0.19	0.31	0.19	0.31	0.19
	P	0	0	0	0	0	0	0	0
	F_1	-	-	-	-	-	-	-	-
G^6	R	1	0.33	1	0.33	1	0.33	1	0.33
	F	0.15	0.15	0.15	0.15	0.15	0.23	0.15	0.15
	P	0.6	0.33	0.6	0.33	0.6	0.25	0.6	0.33
	F_1	0.75	0.33	0.75	0.33	0.75	0.29	0.75	0.33
G^7	R	1	0.5	1	0.75	1	0.75	1	0.5
	F	0	0.17	0	0	0	0	0	0.17
	P	1	0.5	1	1	1	1	1	0.5
	F_1	1	0.5	1	0.86	1	0.86	1	0.5
G^8	R	1	0.75	1	0.75	1	0.75	1	0.75
	F	0	0	0	0	0	0	0	0
	P	1	1	1	1	1	1	1	1
	F_1	1	0.86	1	0.86	1	0.86	1	0.86

Table A7. Classification results for three-node ($M = 3$) networks with $N = 50{,}000$ samples. We present the precision (P), recall (R), fallout (F), and F_1-score for the eight arbitrary topologies of coupled Lorenz systems represented by Figure 4.

Graph	p-Value	∞		0.01		0.001		0.0001	
	σ_ψ	1	10	1	10	1	10	1	10
G^1	R	-	-	-	-	-	-	-	-
	F	0	0.11	0	0	0	0.11	0	0.22
	P	-	0	-	-	-	0	-	0
	F_1	-	-	-	-	-	-	-	-
G^2	R	1	0.5	1	0.5	1	0.5	1	0.5
	F	0	0.14	0	0.14	0	0.14	0	0.14
	P	1	0.5	1	0.5	1	0.5	1	0.5
	F_1	1	0.5	1	0.5	1	0.5	1	0.5
G^3	R	1	1	1	0.5	1	1	1	1
	F	0	0.14	0	0.14	0	0.14	0	0
	P	1	0.67	1	0.5	1	0.67	1	1
	F_1	1	0.8	1	0.5	1	0.8	1	1
G^4	R	1	0.5	1	1	1	0.5	1	1
	F	0	0.14	0	0	0	0.14	0	0
	P	1	0.5	1	1	1	0.5	1	1
	F_1	1	0.5	1	1	1	0.5	1	1

Table A8. Classification results for four-node ($M = 4$) networks with $N = 50{,}000$ samples. We present the precision (P), recall (R), fallout (F), and F_1-score for the eight arbitrary topologies of coupled Lorenz systems represented by Figure 4.

Graph	p-Value	∞		0.01		0.001		0.0001	
	σ_ψ	1	10	1	10	1	10	1	10
G^5	R	-	-	-	-	-	-	-	-
	F	0.19	0.062	0.19	0.19	0.19	0.12	0.19	0.12
	P	0	0	0	0	0	0	0	0
	F_1	-	-	-	-	-	-	-	-
G^6	R	1	0.33	1	0	1	0.33	1	0.33
	F	0	0.15	0	0	0	0.23	0.15	0.15
	P	1	0.33	1	-	1	0.25	0.6	0.33
	F_1	1	0.33	1	-	1	0.29	0.75	0.33
G^7	R	1	0.75	1	0.5	1	0.5	1	0.75
	F	0	0	0	0.17	0	0.083	0	0
	P	1	1	1	0.5	1	0.67	1	1
	F_1	1	0.86	1	0.5	1	0.57	1	0.86
G^8	R	1	0.75	1	0.75	1	0.75	1	0.75
	F	0	0	0	0	0	0	0	0
	P	1	1	1	1	1	1	1	1
	F_1	1	0.86	1	0.86	1	0.86	1	0.86

Table A9. Classification results for three-node ($M = 3$) networks with $N = 100{,}000$ samples. We present the precision (P), recall (R), fallout (F), and F_1-score for the eight arbitrary topologies of coupled Lorenz systems represented by Figure 4.

Graph	p-Value	∞		0.01		0.001		0.0001	
	σ_ψ	1	10	1	10	1	10	1	10
G^1	R	-	-	-	-	-	-	-	-
	F	0	0.22	0	0.11	0	0.22	0	0.11
	P	-	0	-	0	-	0	-	0
	F_1	-	-	-	-	-	-	-	-
G^2	R	1	0.5	1	1	1	1	1	1
	F	0	0.14	0	0	0	0	0	0.14
	P	1	0.5	1	1	1	1	1	0.67
	F_1	1	0.5	1	1	1	1	1	0.8
G^3	R	1	1	1	1	1	1	1	1
	F	0	0	0	0	0	0	0	0
	P	1	1	1	1	1	1	1	1
	F_1	1	1	1	1	1	1	1	1
G^4	R	1	1	1	1	1	1	1	1
	F	0	0	0	0	0	0	0	0
	P	1	1	1	1	1	1	1	1
	F_1	1	1	1	1	1	1	1	1

Table A10. Classification results for four-node ($M = 4$) networks with $N = 100{,}000$ samples. We present the precision (P), recall (R), fallout (F), and F_1-score for the eight arbitrary topologies of coupled Lorenz systems represented by Figure 4.

Graph	p-Value	∞		0.01		0.001		0.0001	
	σ_ψ	1	10	1	10	1	10	1	10
G^5	R	-	-	-	-	-	-	-	-
	F	0.19	0.062	0.19	0.062	0.19	0.19	0.19	0.12
	P	0	0	0	0	0	0	0	0
	F_1	-	-	-	-	-	-	-	-
G^6	R	1	0.33	1	0.67	1	0.33	1	0.33
	F	0	0.15	0	0.15	0	0.077	0	0.15
	P	1	0.33	1	0.5	1	0.5	1	0.33
	F_1	1	0.33	1	0.57	1	0.4	1	0.33
G^7	R	1	-	1	-	1	-	1	-
	F	0	-	0	-	0	-	0	-
	P	1	-	1	-	1	-	1	-
	F_1	1	-	1	-	1	-	1	-
G^8	R	1	0.75	1	0.75	1	0.5	1	0.75
	F	0	0	0	0	0	0.083	0	0
	P	1	1	1	1	1	0.67	1	1
	F_1	1	0.86	1	0.86	1	0.57	1	0.86

References

1. Akaike, H. Information theory and an extension of the maximum likelihood principle. In Proceedings of the Second International Symposium on Information Theory, Tsahkadsor, Armenia, USSR, 2–8 September 1971; pp. 267–281.
2. Lam, W.; Bacchus, F. Learning Bayesian belief networks: An approach based on the MDL principle. *Comput. Intell.* **1994**, *10*, 269–293.
3. de Campos, L.M. A Scoring Function for Learning Bayesian Networks Based on Mutual Information and Conditional Independence Tests. *J. Mach. Learn. Res.* **2006**, *7*, 2149–2187.
4. Sugihara, G.; May, R.; Ye, H.; Hsieh, C.H.; Deyle, E.; Fogarty, M.; Munch, S. Detecting causality in complex ecosystems. *Science* **2012**, *338*, 496–500.
5. Vicente, R.; Wibral, M.; Lindner, M.; Pipa, G. Transfer entropy—A model-free measure of effective connectivity for the neurosciences. *J. Comput. Neurosci.* **2011**, *30*, 45–67.
6. Schumacher, J.; Wunderle, T.; Fries, P.; Jäkel, F.; Pipa, G. A statistical framework to infer delay and direction of information flow from measurements of complex systems. *Neural Comput.* **2015**, *27*, 1555–1608.
7. Best, G.; Cliff, O.M.; Patten, T.; Mettu, R.R.; Fitch, R. Decentralised Monte Carlo Tree Search for Active Perception. In Proceedings of the International Workshop on the Algorithmic Foundations of Robotics (WAFR), San Francisco, CA, USA, 18–20 December 2016.
8. Cliff, O.M.; Lizier, J.T.; Wang, X.R.; Wang, P.; Obst, O.; Prokopenko, M. Delayed Spatio-Temporal Interactions and Coherent Structure in Multi-Agent Team Dynamics. *Art. Life* **2017**, *23*, 34–57.
9. Best, G.; Forrai, M.; Mettu, R.R.; Fitch, R. Planning-aware communication for decentralised multi-robot coordination. In Proceedings of the International Conference on Robotics and Automation, Brisbane, Australia, 21 May 2018.
10. Boccaletti, S.; Latora, V.; Moreno, Y.; Chavez, M.; Hwang, D.U. Complex networks: Structure and dynamics. *Phys. Rep.* **2006**, *424*, 175–308.
11. Mortveit, H.; Reidys, C. *An Introduction to Sequential Dynamical Systems*; Springer Science & Business Media: Berlin/Heidelberg, Germany, 2007.
12. Cliff, O.M.; Prokopenko, M.; Fitch, R. An Information Criterion for Inferring Coupling in Distributed Dynamical Systems. *Front. Robot. AI* **2016**, *3*, doi:10.3389/frobt.2016.00071.
13. Daly, R.; Shen, Q.; Aitken, J.S. Learning Bayesian networks: Approaches and issues. *Knowl. Eng. Rev.* **2011**, *26*, 99–157.
14. Chickering, D.M. Learning equivalence classes of Bayesian-network structures. *J. Mach. Learn. Res.* **2002**, *2*, 445–498.
15. Schwarz, G. Estimating the dimension of a model. *Ann. Stat.* **1978**, *6*, 461–464.
16. Rissanen, J. Modeling by shortest data description. *Automatica* **1978**, *14*, 465–471.
17. Ay, N.; Wennekers, T. Temporal infomax leads to almost deterministic dynamical systems. *Neurocomputing* **2003**, *52*, 461–466.
18. Ay, N. Information geometry on complexity and stochastic interaction. *Entropy* **2015**, *17*, 2432–2458.
19. Lizier, J.T.; Prokopenko, M.; Zomaya, A.Y. Information modification and particle collisions in distributed computation. *Chaos* **2010**, *20*, 037109, doi:10.1063/1.3486801.
20. Murphy, K. Dynamic Bayesian Networks: Representation, Inference and Learning. Ph.D. Thesis, UC Berkeley, Berkeley, CA, USA, 2002.
21. Kocarev, L.; Parlitz, U. Generalized synchronization, predictability, and equivalence of unidirectionally coupled dynamical systems. *Phys. Rev. Lett.* **1996**, *76*, 1816–1819.
22. Kantz, H.; Schreiber, T. *Nonlinear Time Series Analysis*; Cambridge University Press: Cambridge, UK, 2004.
23. Pearl, J. *Probabilistic Reasoning in Intelligent Systems: Networks of Plausible Inference*; Morgan Kaufmann: Burlington, MA, USA, 2014.
24. Granger, C.W.J. Investigating causal relations by econometric models and cross-spectral methods. *Econometrica* **1969**, *37*, 424–438.
25. Schreiber, T. Measuring information transfer. *Phys. Rev. Lett.* **2000**, *85*, 461–464.
26. Barnett, L.; Barrett, A.B.; Seth, A.K. Granger Causality and Transfer Entropy Are Equivalent for Gaussian Variables. *Phys. Rev. Lett.* **2009**, *103*, e238701.

27. Lizier, J.T.; Prokopenko, M. Differentiating information transfer and causal effect. *Eur. Phys. J. B* **2010**, *73*, 605–615.
28. Smirnov, D.A. Spurious causalities with transfer entropy. *Phys. Rev. E* **2013**, *87*, 042917.
29. James, R.G.; Barnett, N.; Crutchfield, J.P. Information flows? A critique of transfer entropies. *Phys. Rev. Lett.* **2016**, *116*, 238701.
30. Liang, X.S. Information flow and causality as rigorous notions *ab initio*. *Phys. Rev. E* **2016**, *94*, 052201.
31. Takens, F. Detecting strange attractors in turbulence. In *Dynamical Systems and Turbulence*; Lecture Notes in Math; Springer: Berlin/Heidelberg, Germany, 1981; Volume 898, pp. 366–381.
32. Stark, J. Delay embeddings for forced systems. I. Deterministic forcing. *J. Nonlinear Sci.* **1999**, *9*, 255–332.
33. Stark, J.; Broomhead, D.S.; Davies, M.E.; Huke, J. Delay embeddings for forced systems. II. Stochastic forcing. *J. Nonlinear Sci.* **2003**, *13*, 519–577.
34. Valdes-Sosa, P.A.; Roebroeck, A.; Daunizeau, J.; Friston, K. Effective connectivity: influence, causality and biophysical modeling. *Neuroimage* **2011**, *58*, 339–361.
35. Sporns, O.; Chialvo, D.R.; Kaiser, M.; Hilgetag, C.C. Organization, development and function of complex brain networks. *Trends Cogn. Sci.* **2004**, *8*, 418–425.
36. Park, H.J.; Friston, K. Structural and functional brain networks: From connections to cognition. *Science* **2013**, *342*, 1238411.
37. Friston, K.; Moran, R.; Seth, A.K. Analysing connectivity with Granger causality and dynamic causal modelling. *Curr. Opin. Neurobiol.* **2013**, *23*, 172–178.
38. Lizier, J.T.; Rubinov, M. *Multivariate Construction of Effective Computational Networks from Observational Data*; Preprint 25/2012; Max Planck Institute for Mathematics in the Sciences: Leipzig, Germany, 2012.
39. Sandoval, L. Structure of a global network of financial companies based on transfer entropy. *Entropy* **2014**, *16*, 4443–4482.
40. Rodewald, J.; Colombi, J.; Oyama, K.; Johnson, A. Using Information-theoretic Principles to Analyze and Evaluate Complex Adaptive Supply Network Architectures. *Procedia Comput. Sci.* **2015**, *61*, 147–152.
41. Crosato, E.; Jiang, L.; Lecheval, V.; Lizier, J.T.; Wang, X.R.; Tichit, P.; Theraulaz, G.; Prokopenko, M. Informative and misinformative interactions in a school of fish. *arXiv* **2017**, arXiv:1705.01213.
42. Kozachenko, L.; Friston, L.F.; Leonenko, N.N. Sample estimate of the entropy of a random vector. *Probl. Peredachi Inf.* **1987**, *23*, 9–16.
43. Kraskov, A.; Stögbauer, H.; Grassberger, P. Estimating mutual information. *Phys. Rev. E* **2004**, *69*, 066138.
44. Stark, J.; Broomhead, D.S.; Davies, M.E.; Huke, J. Takens embedding theorems for forced and stochastic systems. *Nonlinear Anal. Theory Methods Appl.* **1997**, *30*, 5303–5314.
45. Friedman, N.; Murphy, K.; Russell, S. Learning the structure of dynamic probabilistic networks. In Proceedings of the Fourteenth Conference on Uncertainty in Artificial Intelligence, Madison, WI, USA, 24–26 July 1998; pp. 139–147.
46. Koller, D.; Friedman, N. *Probabilistic Graphical Models: Principles and Techniques*; MIT Press: Cambridge, MA, USA, 2009.
47. Wilks, S.S. The large-sample distribution of the likelihood ratio for testing composite hypotheses. *Ann. Math. Stat.* **1938**, *9*, 60–62.
48. Barnett, L.; Bossomaier, T. Transfer entropy as a log-likelihood ratio. *Phys. Rev. Lett.* **2012**, *109*, 138105.
49. Vinh, N.X.; Chetty, M.; Coppel, R.; Wangikar, P.P. GlobalMIT: Learning globally optimal dynamic Bayesian network with the mutual information test criterion. *Bioinformatics* **2011**, *27*, 2765–2766.
50. Deyle, E.R.; Sugihara, G. Generalized theorems for nonlinear state space reconstruction. *PLoS ONE* **2011**, *6*, e18295.
51. Lloyd, A.L. The coupled logistic map: a simple model for the effects of spatial heterogeneity on population dynamics. *J. Theor. Biol.* **1995**, *173*, 217–230.
52. Lizier, J.T. JIDT: An information-theoretic toolkit for studying the dynamics of complex systems. *Front. Robot. AI* **2014**, *1*, doi:10.3389/frobt.2014.00011.
53. Silander, T.; Myllymaki, P. A simple approach for finding the globally optimal Bayesian network structure. In Proceedings of the Twenty-Second Conference on Uncertainty in Artificial Intelligence, Cambridge, MA, USA, 13–16 July 2006; pp. 445–452.
54. Ragwitz, M.; Kantz, H. Markov models from data by simple nonlinear time series predictors in delay embedding spaces. *Phys. Rev. E* **2002**, *65*, 056201.

55. Small, M.; Tse, C.K. Optimal embedding parameters: A modelling paradigm. *Physica* **2004**, *194*, 283–296.
56. Lorenz, E.N. Deterministic nonperiodic flow. *J. Atmos. Sci.* **1963**, *20*, 130–141.
57. Rössler, O.E. An equation for continuous chaos. *Phys. Lett. A* **1976**, *57*, 397–398.
58. Haken, H. Analogy between higher instabilities in fluids and lasers. *Phys. Lett. A* **1975**, *53*, 77–78.
59. Cuomo, K.M.; Oppenheim, A.V. Circuit implementation of synchronized chaos with applications to communications. *Phys. Rev. Lett.* **1993**, *71*, 65–68.
60. He, R.; Vaidya, P.G. Analysis and synthesis of synchronous periodic and chaotic systems. *Phys. Rev. A* **1992**, *46*, 7387–7392.
61. Fujisaka, H.; Yamada, T. Stability theory of synchronized motion in coupled-oscillator systems. *Prog. Theor. Phys.* **1983**, *69*, 32–47.
62. Rulkov, N.F.; Sushchik, M.M.; Tsimring, L.S.; Abarbanel, H.D. Generalized synchronization of chaos in directionally coupled chaotic systems. *Phys. Rev. E* **1995**, *51*, 980–994.
63. Acid, S.; de Campos, L.M. Searching for Bayesian network structures in the space of restricted acyclic partially directed graphs. *J. Artif. Intell. Res.* **2003**, *18*, 445–490.
64. Friston, K.; Kilner, J.; Harrison, L. A free energy principle for the brain. *J. Physiol. Paris* **2006**, *100*, 70–87.
65. Williams, P.L.; Beer, R.D. Generalized measures of information transfer. *arXiv* **2011**, arXiv:1102.1507.
66. Vakorin, V.A.; Krakovska, O.A.; McIntosh, A.R. Confounding effects of indirect connections on causality estimation. *J. Neurosci. Methods* **2009**, *184*, 152–160.
67. Spinney, R.E.; Prokopenko, M.; Lizier, J.T. Transfer entropy in continuous time, with applications to jump and neural spiking processes. *Phys. Rev. E* **2017**, *95*, 032319.
68. Hefferan, B.; Cliff, O.M.; Fitch, R. Adversarial Patrolling with Reactive Point Processes. In Proceedings of the Australasian Conference on Robotics and Automation (ACRA), Brisbane, Australia, 5–7 December 2016.
69. Prokopenko, M.; Einav, I. Information thermodynamics of near-equilibrium computation. *Phys. Rev. E* **2015**, *91*, 062143.
70. Spinney, R.E.; Lizier, J.T.; Prokopenko, M. Transfer entropy in physical systems and the arrow of time. *Phys. Rev. E* **2016**, *94*, 022135.
71. Takens, F. The reconstruction theorem for endomorphisms. *Bull. Braz. Math. Soc.* **2002**, *33*, 231–262.
72. Ay, N.; Wennekers, T. Dynamical properties of strongly interacting Markov chains. *Neural Netw.* **2003**, *16*, 1483–1497.
73. Edlund, J.A.; Chaumont, N.; Hintze, A.; Koch, C.; Tononi, G.; Adami, C. Integrated information increases with fitness in the evolution of animats. *PLoS Comput. Biol.* **2011**, *7*, e1002236.

entropy

MDPI

Article

Conformal Flattening for Deformed Information Geometries on the Probability Simplex †

Atsumi Ohara

Department of Electrical and Electronics, University of Fukui, Bunkyo, Fukui 910-8507, Japan; ohara@fuee.u-fukui.ac.jp

† This paper is an extended version of our paper published in SigmaPhi 2014, 2017 and Geometric Science of Information (GSI 2017).

Received: 20 February 2018; Accepted: 8 March 2018; Published: 10 March 2018

Abstract: Recent progress of theories and applications regarding statistical models with generalized exponential functions in statistical science is giving an impact on the movement to deform the standard structure of information geometry. For this purpose, various representing functions are playing central roles. In this paper, we consider two important notions in information geometry, i.e., invariance and dual flatness, from a viewpoint of representing functions. We first characterize a pair of representing functions that realizes the invariant geometry by solving a system of ordinary differential equations. Next, by proposing a new transformation technique, i.e., conformal flattening, we construct dually flat geometries from a certain class of non-flat geometries. Finally, we apply the results to demonstrate several properties of gradient flows on the probability simplex.

Keywords: representing functions; affine immersion; nonextensive statistical physics; invariance; dually flat structure; Legendre conjugate; gradient flow

1. Introduction

The theory of information geometry has elucidated abundant geometric properties equipped with a Riemannian metric and mutually dual affine connections. When it is applied to the study of statistical models described by the exponential family, the logarithmic function plays a significant role in giving the standard information geometric structure to the models [1,2].

Inspired by the recent progress of several areas in statistical physics and mathematical statistics [3–10] which have exploited theoretical interests and possible applications for generalized exponential families, one research direction in information geometry is pointing to constructions of deformed geometries based on the standard one, keeping its basic properties. A typical and classical example of such a deformation would be the alpha-geometry [1,2], a statistical definition of which can be regarded as a replacement of the logarithmic function by suitable power functions. Hence, for the purpose of the generalization and flexible applicability, much attention is paid to various uses of such replacements by *representing functions* as important tools [3,4,11,12].

Two major characteristics of the standard structure are dual flatness and invariance [2]. Dual flatness (or Hessian structure [13]) produces fruitful properties such as the existence of canonical coordinate systems, a pair of conjugate potential functions and the canonical divergence (relative entropy). In addition, they are connected with the Legendre duality relation, which is also fundamental in the generalization of statistical physics. On the other hand, the invariance of geometric structure is crucially valuable in developing mathematical statistics. It has been proved [14] that invariance holds for only the structure with a special triple of a Riemannian metric and a pair of mutually dual affine connections, which are respectively called the Fisher information and the alpha-connections (see Section 3 for their definitions). The study of these two characteristics from a viewpoint of representing functions would contribute to our geometrical understanding.

In this paper, we first characterize a pair of representing functions that realizes the invariant information geometric structure. Next, we propose a new transformation to obtain dually flat geometries from a certain class of non-flat information geometries, using concepts from affine differential geometry [15,16]. We call the transformation *conformal flattening*, which is a generalization of the way to realize the corresponding dually flat geometry from the alpha-geometry developed in [17,18]. As applications and easy consequences of the results, we finally show several properties of gradient flows associated with realized dually flat geometries. Focusing on geometric characteristics conserved by the transformation, we discuss the properties such as a relation between geodesics and flows, the first integral of the flows and so on. These properties are new and general. Hence, they refine the arguments of the flows in [18], where only the alpha-geometry is treated.

The paper is organized as follows. In Section 2, we introduce preliminary results, explaining several existing methods to construct the information geometric structure that includes a dually flat structure and the alpha-structure and so on. We also give a short summary of concepts from affine differential geometry, which will be used in this paper. Section 3 provides a characterization of representing functions that realize invariant geometry, i.e., the one equipped with the Fisher information and a pair of the alpha-connections. The characterization is obtained by solving a simple system of ordinary equations. In Section 4, we first obtain a certain class of information geometric structure by regarding representing functions as immersions into an ambient affine space. Then, we demonstrate the conformal flattening to realize the corresponding dually flat structure, and discuss their properties and relations with generalized entropies or escort probabilities [19]. Section 5 exhibits the geometric properties of gradient flows with respect to a conformally realized Riemannian metric. These flows are reduced to the well-known *replicator flow* [20] (Chapter 16) when we consider the standard information geometry. Suitably choosing its pay-off functions, we see that the flow follows a geodesic curve or conserves a divergence from an equilibrium. In the final section, some concluding remarks are made.

Throughout the paper, we use a probability simplex as a statistical model for the sake of simplicity.

2. Preliminaries

2.1. Information Geometry of $\mathcal{S}^n$ and $\mathbf{R}^{n+1}_+$

Let us represent an element $p \in \mathbf{R}^{n+1}$ with its components $p_i,\ i = 1, \cdots, n+1$ as $p = (p_i) \in \mathbf{R}^{n+1}$. Denote, respectively, the positive orthant by

$$\mathbf{R}^{n+1}_+ := \{p = (p_i) \in \mathbf{R}^{n+1} | p_i > 0,\ i = 1, \cdots, n+1\},$$

and the relative interior of the probability simplex by

$$\mathcal{S}^n := \left\{ p \in \mathbf{R}^{n+1}_+ \left| \sum_{i=1}^{n+1} p_i = 1 \right. \right\}.$$

Let $p(X)$ be a probability distribution of a random variable X taking a value in the finite sample space $\Omega = \{1, 2, \cdots, n, n+1\}$. We consider a set of distributions $p(X)$ with positive probabilities, i.e., $p(i) = p_i > 0,\ i = 1, \cdots, n+1$, defined by

$$p(X) = \sum_{i=1}^{n+1} p_i \delta_i(X), \quad \delta_i(j) = \delta_i^j \text{ (the Kronecker's delta)},$$

which is identified with $\mathcal{S}^n$. A statistical model in $\mathcal{S}^n$ is represented with parameters $\zeta = (\zeta^j),\ j = 1, \cdots, d \leq n$ by

$$p_\zeta(X) = \sum_{i=1}^{n+1} p_i(\zeta)\delta_i(X),$$

where each p_i is smoothly parametrized by ζ. For such a statistical model, ζ^j can also be regarded as coordinates of the corresponding submanifold in $\mathcal{S}^n$. For simplicity, we shall consider the full model, i.e., $d = n$ and the parameter set is bijective with $\mathcal{S}^n$ via $p_i(\zeta)$'s.

The information geometric structure [2] on $\mathcal{S}^n$ denoted by (g, ∇, ∇^*) is composed of the pair of *mutually dual* torsion-free affine connections ∇ and ∇^* with respect to a Riemannian metric g. If we write $\partial_i := \partial/\partial\zeta^i$, $i = 1, \cdots, n$, the mutual duality requires components of (g, ∇, ∇^*) to satisfy

$$\partial_i g_{jk} = \Gamma_{ij,k} + \Gamma^*_{ik,j}. \tag{1}$$

Let L and M be a pair of strictly monotone (i.e., one-to-one) smooth functions on the interval $(0,\ 1)$. One way of constructing such a structure (g, ∇, ∇^*) is to define the components as follows [2,11]:

$$g_{ij}(p) = \sum_{X\in\Omega} \partial_i L(p_\zeta(X))\partial_j M(p_\zeta(X)), \quad i,j = 1, \cdots, n, \tag{2}$$

$$\Gamma_{ij,k}(p) = \sum_{X\in\Omega} \partial_i\partial_j L(p_\zeta(X))\partial_k M(p_\zeta(X)), \quad i,j,k = 1, \cdots, n, \tag{3}$$

$$\Gamma^*_{ij,k}(p) = \sum_{X\in\Omega} \partial_k L(p_\zeta(X))\partial_i\partial_j M(p_\zeta(X)), \quad i,j,k = 1, \cdots, n. \tag{4}$$

In this paper, we call L and M representing functions. It is easy to verify the mutual duality (1). (Positive definiteness of g needs additional conditions.)

When the curvature tensors of both ∇ and ∇^* vanish, (g, ∇, ∇^*) is called *dually flat* [2]. It is known that (g, ∇, ∇^*) is dually flat if and only if there exist two special coordinate systems denoted by $\theta^i(p)$ and $\eta_i(p)$, $i = 1, \cdots, n$, respectively, where (θ^i) is ∇-affine, (η_i) is ∇^*-affine and they are biorthogonal, i.e.,

$$g\left(\frac{\partial}{\partial\theta^i}, \frac{\partial}{\partial\eta_j}\right) = \delta^j_i.$$

We give examples. For a real number α, define $L^{(\alpha)}(u) := 2u^{(1-\alpha)/2}/(1-\alpha)$ and $L^{(1)} := \ln u$. If we set $L(u) = L^{(\alpha)}(u)$ and $M(u) = L^{(-\alpha)}(u)$, then they derive the *alpha-structure* [2] $(g^{\mathrm{F}}, \nabla^{(\alpha)}, \nabla^{(-\alpha)})$, where g^{F} is the Fisher information and $\nabla^{(\pm\alpha)}$ are the alpha-connections (see Section 3). In particular, if we choose $\alpha = 1$, it defines the standard dually flat structure $(g^{\mathrm{F}}, \nabla^{(e)} := \nabla^{(1)}, \nabla^{(m)} := \nabla^{(-1)})$, where $\nabla^{(e)}$ and $\nabla^{(m)}$ are called the e- and m-connection, respectively [2]. Similarly, the ϕ-log geometry [3] can also be introduced in the same way by taking $L(u) = \log_\phi(u)$ and $M(u) = u$.

One traditional way to construct a general information geometric structure (g, ∇, ∇^*), without using representing functions, is by means of contrast functions (or divergences) [2,21]. In our case, let ρ be a function on $\mathcal{S}^n \times \mathcal{S}^n$ satisfying $\rho(p,r) \geq 0$, $\forall p, r \in \mathcal{S}^n$ with equality if and only if $p = r$. For a vector field ∂_i, let $(\partial_i)_p$ denote its tangent vector at p. When we define

$$g_{ij}(p) = -(\partial_i)_p(\partial_j)_r\, \rho(p,r)\big|_{p=r}, \quad i,j = 1, \cdots, n, \tag{5}$$

$$\Gamma_{ij,k}(p) = -(\partial_i)_p(\partial_j)_p(\partial_k)_r\, \rho(p,r)\big|_{p=r}, \quad i,j,k = 1, \cdots, n, \tag{6}$$

$$\Gamma^*_{ij,k}(p) = -(\partial_i)_p(\partial_j)_r(\partial_k)_r\, \rho(p,r)\big|_{p=r}, \quad i,j,k = 1, \cdots, n, \tag{7}$$

we can confirm that (1) holds. If g is positive definite, we say that ρ is a *contrast function* or a *divergence* that induces the structure (g, ∇, ∇^*).

A contrast function ρ of the form:

$$\rho(p,r) = \psi(\theta(p)) + \varphi(\eta(r)) - \sum_{i=1}^{n} \theta^i(p)\eta_i(r) \tag{8}$$

always induces the corresponding dually flat structure. Conversely, it is known [2] that if (g, ∇, ∇^*) is dually flat, then there exists the unique contrast function of the form (8) that induces the structure. Hence, it is called the *canonical divergence* of (g, ∇, ∇^*) and we say that the functions ψ and φ are *potentials*. By setting $p = r$, we see that a dually flat structure naturally gives the *Legendre duality* relations at each p, i.e., the function φ, is the Legendre conjugate of ψ satisfying

$$\eta_i = \frac{\partial \psi}{\partial \theta^i}, \quad \theta^i = \frac{\partial \varphi}{\partial \eta_i}.$$

Applying the idea of affine hypersurface theory [15] is also one of the other ways to construct the information geometric structure. Let D be the canonical flat affine connection on $\mathbf{R}^{n+1}$. Consider an immersion f from $\mathcal{S}^n$ into $\mathbf{R}^{n+1}$ and a vector field ξ on $\mathcal{S}^n$ that is transversal to the hypersurface $f(\mathcal{S}^n)$ in $\mathbf{R}^{n+1}$. Such a pair (f, ξ), called an *affine immersion*, defines a torsion-free connection ∇ and the affine fundamental form g on $\mathcal{S}^n$ via the Gauss formula as

$$D_X f_*(Y) = f_*(\nabla_X Y) + g(X,Y)\xi, \quad X, Y \in \mathcal{X}(\mathcal{S}^n), \tag{9}$$

where $\mathcal{X}(\mathcal{S}^n)$ is the set of tangent vector fields on $\mathcal{S}^n$ and f_* denotes the differential of f. By regarding g as a (pseudo-) Riemannian metric, one can discuss the realized structure (g, ∇) on $\mathcal{S}^n$.

We say that (f, ξ) is *non-degenerate* and *equiaffine* if g is non-degenerate and $D_X\xi$ is tangent to $\mathcal{S}^n$ for any $X \in \mathcal{X}(\mathcal{S}^n)$, respectively. The latter ensures that the volume element θ on $\mathcal{S}^n$ defined by

$$\theta(X_1, \cdots, X_n) = \det(f_*(X_1), \cdots, f_*(X_n), \xi), \quad X_i \in \mathcal{X}(\mathcal{S}^n)$$

is parallel to ∇ [15] (p.31). It is known [15,16] that there exists a torsion-free dual affine connection ∇^* satisfying (1) if and only if (f, ξ) is non-degenerate and equiaffine. In this case, the obtained structure (g, ∇, ∇^*) on $\mathcal{S}^n$ is not dually flat in general. However, there always exists a positive function σ and a dually flat structure $(\tilde{g}, \tilde{\nabla}, \tilde{\nabla}^*)$ on $\mathcal{S}^n$ that hold the following relations [16]:

$$\tilde{g} = \sigma g, \tag{10}$$
$$g(\tilde{\nabla}_X Y, Z) = g(\nabla_X Y, Z) - d(\ln\sigma)(Z) g(X,Y), \tag{11}$$
$$g(\tilde{\nabla}^*_X Y, Z) = g(\nabla^*_X Y, Z) + d(\ln\sigma)(X) g(Y,Z) + d(\ln\sigma)(Y) g(X,Z). \tag{12}$$

Furthermore, there exists a specific contrast function $\rho(p,r)$ for (g, ∇, ∇^*) called the *geometric divergence*. Then, a contrast function $\tilde{\rho}(p,r)$ that induces $(\tilde{g}, \tilde{\nabla}, \tilde{\nabla}^*)$ is given by the *conformal divergence* $\tilde{\rho}(p,r) = \sigma(r)\rho(p,r)$. These properties of the structure (g, ∇) realized by the non-degenerate and equiaffine immersion are called *1-conformal flatness* [16].

3. Characterization of Invariant Geometry by Representing Functions

Suppose that a pair of representing functions (L, M) defines an information geometric structure (g, ∇, ∇^*) by (2), (3) and (4). In this section, we consider the condition of (L, M) such that (g, ∇, ∇^*) is invariant. This is equivalent [2,14] to g which is the Fisher information g^{F} defined by

$$g^{\mathrm{F}}_{ij}(p) = \sum_{X \in \Omega} p_\zeta (\partial_i \ln p_\zeta)(\partial_j \ln p_\zeta) \tag{13}$$

and a pair of dual connections satisfies $\nabla = \nabla^{(\alpha)}$ and $\nabla^* = \nabla^{(-\alpha)}$ for a certain $\alpha \in \mathbf{R}$, where $\nabla^{(\alpha)}$ is the α-connection defined by

$$\Gamma^{(\alpha)}_{ij,k} = \sum_{X\in\Omega} p_\zeta \left(\partial_i\partial_j \ln p_\zeta + \frac{1-\alpha}{2}(\partial_i \ln p_\zeta)(\partial_j \ln p_\zeta) \right) (\partial_k \ln p_\zeta). \tag{14}$$

Hence, g_{ij} expressed in (2) by functions $L(u)$ and $M(u)$ coincides with the Fisher information if and only if the following equation holds:

$$\frac{dL}{du}\frac{dM}{du} = 1/u. \tag{15}$$

Similarly, we derive a condition for $\Gamma_{ij,k}$ expressed in (3) to be the α-connection. First, note that the following relations hold:

$$\partial_i \ln p_\zeta = (\partial_i p_\zeta)\frac{1}{p_\zeta}, \quad \partial_i\partial_j \ln p_\zeta = (\partial_i\partial_j p_\zeta)\frac{1}{p_\zeta} - (\partial_i p_\zeta)(\partial_j p_\zeta)\frac{1}{p_\zeta^2}. \tag{16}$$

On the other hand, we have

$$\partial_i\partial_j L(p_\zeta) = (\partial_i\partial_j p_\zeta)\frac{dL}{du}(p_\zeta) + (\partial_i p_\zeta)(\partial_j p_\zeta)\frac{d^2L}{du^2}(p_\zeta), \tag{17}$$

$$\partial_k M(p_\zeta) = (\partial_k p_\zeta)\frac{dM}{du}(p_\zeta). \tag{18}$$

Substituting (16), (17) and (18) into (3) and (14), and comparing them, we obtain (15) again and

$$\frac{d^2L}{du^2}\frac{dM}{du} = -\frac{1+\alpha}{2u^2}. \tag{19}$$

Expressing $L' := dL/du$ and $L'' := d^2L/du^2$, we have the following ODE from (15) and (19):

$$\frac{L''}{L'} = -\frac{1+\alpha}{2u}. \tag{20}$$

By integrations, we get

$$\ln L' = -\frac{(1+\alpha)}{2}\ln u + c \tag{21}$$

and

$$L(u) = c_1 u^{(1-\alpha)/2} + c_2, \quad M(u) = c_3 u^{(1+\alpha)/2} + c_4, \tag{22}$$

where c and c_i, $i = 1, \cdots, 4$ are constants with a constraint $c_1c_3 = 4/(1-\alpha^2)$. Thus, (L, M) is essentially a pair of representing functions that derives the alpha-geometry and there is only freedom of adjusting the constants for the invariance of geometry. If we require solely (15), which implies that only a Riemannian metric g is the Fisher information g^{F}, there still remains much freedom for (L, M).

4. Affine Immersion of the Probability Simplex

Now we consider the affine immersion with the following assumptions.

Assumptions:

1. The affine immersion (f, ξ) is nondegenerate and equiaffine,
2. The immersion f is given by the component-by-component and common representing function L, i.e.,
$$f : \mathcal{S}^n \ni p = (p_i) \mapsto x = (x^i) \in \mathbf{R}^{n+1}, \quad x^i = L(p_i),\ i = 1, \cdots, n+1,$$
3. The representing function $L : (0,\ 1) \to \mathbf{R}$ is sign-definite, concave with $L'' < 0$ and strictly increasing, i.e., $L' > 0$. Hence, the inverse of L denoted by E exists, i.e., $E \circ L = \mathrm{id}$.
4. Each component of ξ satisfies $\xi^i < 0$, $i = 1, \cdots, n+1$ on $\mathcal{S}^n$.

Remark 1. *From the assumption 3, it follows that $L'E' = 1$, $E' > 0$ and $E'' > 0$. Regarding sign-definiteness of L, note that we can adjust $L(u)$ to $L(u) + c$ by a suitable constant c without loss of generality since the resultant geometric structure is unchanged (See Theorem 1) by the adjustment. For a fixed L satisfying the assumption 3, we can choose ξ that meets the assumptions 1 and 4. For example, if we take $\xi^i = -|L(p_i)|$, then (f, ξ) is called* centro-affine, *which is known to be equiaffine [15] (p.37). The assumptions 3 and 4 also assure positive definiteness of g (the details are described in the proof of Theorem 1). Hence, (f, ξ) is non-degenerate and we can regard g as a Riemannian metric on $\mathcal{S}^n$.*

4.1. Conormal Vector and the Geometric Divergence

Define a function Ψ on $\mathbf{R}^{n+1}$ by

$$\Psi(x) := \sum_{i=1}^{n+1} E(x^i),$$

then $f(\mathcal{S}^n)$ immersed in $\mathbf{R}^{n+1}$ is expressed as a level surface of $\Psi(x) = 1$. Denote by $\mathbf{R}_{n+1}$ the dual space of $\mathbf{R}^{n+1}$ and by $\langle \nu, x \rangle$ the pairing of $x \in \mathbf{R}^{n+1}$ and $\nu \in \mathbf{R}_{n+1}$. The conormal vector [15] (p.57) $\nu : \mathcal{S}^n \to \mathbf{R}_{n+1}$ for the affine immersion (f, ξ) is defined by

$$\langle \nu(p), f_*(X) \rangle = 0, \ \forall X \in T_p\mathcal{S}^n, \qquad \langle \nu(p), \xi(p) \rangle = 1, \tag{23}$$

for $p \in \mathcal{S}^n$. Using the assumptions and noting the relations:

$$\frac{\partial \Psi}{\partial x^i} = E'(x^i) = \frac{1}{L'(p_i)} > 0, \quad i = 1, \cdots, n+1,$$

we have

$$\nu_i(p) := \frac{1}{\Lambda} \frac{\partial \Psi}{\partial x^i} = \frac{1}{\Lambda(p)} E'(x^i) = \frac{1}{\Lambda(p)} \frac{1}{L'(p_i)}, \quad i = 1, \cdots, n+1, \tag{24}$$

where Λ is a normalizing factor defined by

$$\Lambda(p) := \sum_{i=1}^{n+1} \frac{\partial \Psi}{\partial x^i} \xi^i = \sum_{i=1}^{n+1} \frac{1}{L'(p_i)} \xi^i(p). \tag{25}$$

Then, we can confirm (23) using the relation $\sum_{i=1}^{n+1} X^i = 0$ for $X = (X^i) \in \mathcal{X}(\mathcal{S}^n)$. Note that $v : \mathcal{S}^n \to \mathbf{R}_{n+1}$ defined by

$$v_i(p) = \Lambda(p) \nu_i(p) = \frac{1}{L'(p_i)}, \quad i = 1, \cdots, n+1,$$

also satisfies

$$\langle v(p), f_*(X) \rangle = 0, \ \forall X \in T_p\mathcal{S}^n. \tag{26}$$

Furthermore, it follows, from (24), (25) and the assumption 4, that

$$\Lambda(p) < 0, \quad \nu_i(p) < 0, \quad i = 1, \cdots, n+1,$$

for all $p \in \mathcal{S}^n$.

It is known [15] (p.57) that the affine fundamental form g can be represented by

$$g(X, Y) = -\langle \nu_*(X), f_*(Y) \rangle, \quad X, Y \in T_p\mathcal{S}^n.$$

In our case, it is calculated via (26) as

$$\begin{aligned} g(X,Y) &= -\Lambda^{-1}\langle v_*(X), f_*(Y)\rangle - X(\Lambda^{-1})\langle v, f_*(Y)\rangle \\ &= -\frac{1}{\Lambda}\sum_{i=1}^{n+1}\left(\frac{1}{L'(p_i)}\right)' L'(p_i)X^iY^i = \frac{1}{\Lambda}\sum_{i=1}^{n+1}\frac{L''(p_i)}{L'(p_i)}X^iY^i, \quad X,Y \in T_p\mathcal{S}^n. \end{aligned} \tag{27}$$

Hence, g is positive definite from the assumptions 3 and 4, and we can regard it as a Riemannian metric.

Utilizing these notions from affine differential geometry, we can introduce a geometric divergence [16] as follows:

$$\begin{aligned} \rho(p,r) &= \langle \nu(r), f(p) - f(r)\rangle = \sum_{i=1}^{n+1}\nu_i(r)(L(p_i) - L(r_i)) \\ &= \frac{1}{\Lambda(r)}\sum_{i=1}^{n+1}\frac{L(p_i) - L(r_i)}{L'(r_i)}, \qquad p,r \in \mathcal{S}^n. \end{aligned} \tag{28}$$

It is easily checked that ρ is actually a contrast function of the 1-conformally flat structure (g, ∇, ∇^*) using (5), (6) and (7).

4.2. Conformal Flattening Transformation

As is described in the preliminary section, by 1-conformally flatness there exists a positive function, i.e., conformal factor σ that relates (g, ∇, ∇^*) with a dually flat structure $(\tilde{g}, \tilde{\nabla}, \tilde{\nabla}^*)$ via the conformal transformation (10), (11) and (12). A contrast function $\tilde{\rho}$ that induces $(\tilde{g}, \tilde{\nabla}, \tilde{\nabla}^*)$ is given as the conformal divergence:

$$\tilde{\rho}(p,r) = \sigma(r)\rho(p,r), \qquad p,r \in \mathcal{S}^n. \tag{29}$$

from the geometric divergence ρ in (28).

For an arbitrary function L within our setting given by the four assumptions, we prove that we can construct a dually flat structure $(\tilde{g}, \tilde{\nabla}, \tilde{\nabla}^*)$ by choosing the conformal factor σ carefully. Hereafter, we call this transformation *conformal flattening*.

Define

$$Z(p) := \sum_{i=1}^{n+1}\nu_i(p) = \frac{1}{\Lambda(p)}\sum_{i=1}^{n+1}\frac{1}{L'(p_i)},$$

then it is negative because each $\nu_i(p)$ is negative. The conformal divergence to ρ with respect to the conformal factor $\sigma(r) := -1/Z(r)$ is

$$\tilde{\rho}(p,r) = -\frac{1}{Z(r)}\rho(p,r).$$

Theorem 1. *If the conformal factor is $\sigma = -1/Z$, then the information geometric structure $(\tilde{g}, \tilde{\nabla}, \tilde{\nabla}^*)$ on $\mathcal{S}^n$ that is transformed from the 1-conformally flat structure (g, ∇, ∇^*) via (10), (11) and (12) is dully flat.*

Furthermore, the conformal divergence $\tilde{\rho}$ that induces $(\tilde{g}, \tilde{\nabla}, \tilde{\nabla}^*)$ on $\mathcal{S}^n$ is canonical where Legendre conjugate potential functions and coordinate systems are explicitly given by

$$\theta^i(p) = x^i(p) - x^{n+1}(p) = L(p_i) - L(p_{n+1}), \quad i = 1, \cdots, n, \tag{30}$$

$$\eta_i(p) = P_i(p) := \frac{\nu_i(p)}{Z(p)} = \frac{1/L'(p_i)}{\sum_{k=1}^{n+1} 1/L'(p_k)}, \quad i = 1, \cdots, n, \tag{31}$$

$$\psi(p) = -x_{n+1}(p) = -L(p_{n+1}), \tag{32}$$

$$\varphi(p) = \frac{1}{Z(p)} \sum_{i=1}^{n+1} \nu_i(p) x^i(p) = \sum_{i=1}^{n+1} P_i(p) L(p_i). \tag{33}$$

Proof. Using given relations, we first show that the conformal divergence $\tilde{\rho}$ is the canonical divergence for $(\tilde{g}, \tilde{\nabla}, \tilde{\nabla}^*)$:

$$\begin{aligned}
\tilde{\rho}(p,r) &= -\frac{1}{Z(r)} \langle \nu(r), f(p) - f(r) \rangle = \langle P(r), f(r) - f(p) \rangle \\
&= \sum_{i=1}^{n+1} P_i(r)(x^i(r) - x^i(p)) \\
&= \sum_{i=1}^{n+1} P_i(r) x^i(r) - \sum_{i=1}^{n} P_i(r)(x^i(p) - x^{n+1}(p)) - \left(\sum_{i=1}^{n+1} P_i(r) \right) x^{n+1}(p) \\
&= \varphi(r) - \sum_{i=1}^{n} \eta_i(r) \theta^i(p) + \psi(p).
\end{aligned} \tag{34}$$

Next, let us confirm that $\partial \psi / \partial \theta^i = \eta_i$.

Since $\theta^i(p) = L(p_i) + \psi(p)$, $i = 1, \cdots, n$, we have

$$p_i = E(\theta^i - \psi), \quad i = 1, \cdots, n+1,$$

by setting $\theta^{n+1} := 0$. Hence, we have

$$1 = \sum_{i=1}^{n+1} E(\theta^i - \psi).$$

Differentiating by θ^j, we obtain

$$\begin{aligned}
0 &= \frac{\partial}{\partial \theta^j} \sum_{i=1}^{n+1} E(\theta^i - \psi) = \sum_{i=1}^{n+1} E'(\theta^i - \psi) \left(\delta^i_j - \frac{\partial \psi}{\partial \theta^j} \right) \\
&= E'(x^j) - \left(\sum_{i=1}^{n+1} E'(x^i) \right) \frac{\partial \psi}{\partial \theta^j}.
\end{aligned}$$

This implies that

$$\frac{\partial \psi}{\partial \theta^j} = \frac{E'(x^j)}{\sum_{i=1}^{n+1} E'(x^i)} = \eta_j.$$

Together with (34) and this relation, φ is confirmed to be the Legendre conjugate of ψ.

The dual relation $\partial \varphi / \partial \eta_i = \theta^i$ follows automatically from the property of the Legendre transform. □

The following corollary is straightforward because all the quantities in the theorem depend on only L:

Corollary 1. *Under the assumptions, the dually flat structure* $(\tilde{g}, \tilde{\nabla}, \tilde{\nabla}^*)$ *on* $\mathcal{S}^n$*, obtained by following the above conformal flattening, does not depend on the choice of the transversal vector* ξ.

Remark 2. *Note that the conformal metric is given by* $\tilde{g} = -g/Z$ *and is positive definite. Furthermore, the relation (12) means that the dual affine connections* ∇^* *and* $\tilde{\nabla}^*$ *are* projectively *(or* -1-conformally*)* equivalent *[15,16]. Hence,* ∇^* *is projectively flat. Furthermore, the above corollary implies that the realized affine connection* ∇ *is also projectively equivalent to the flat connection* $\tilde{\nabla}$ *if we use the centro-affine immersion, i.e.,* $\xi^i = -L(p_i)$ *[15,16]. See Proposition 3 for an application of projective equivalence of affine connections.*

Remark 3. *In our setting, conformal flattening is geometrically regarded as normalization of the conormal vector* ν*. Hence, the dual coordinates* $\eta_i(p) = P_i(p)$ *can be interpreted as a generalization of the* escort probability *[10,19] (see the following example). Similarly,* ψ *and* $-\varphi$ *might be seen as the associated Massieu function and entropy, respectively.*

Remark 4. *While the immersion* f *is composed of a representing function* L *under the assumption 2, the corresponding* M *of a single variable does not generally exist for* (g, ∇, ∇^*) *nor* $(\tilde{g}, \tilde{\nabla}, \tilde{\nabla}^*)$*. From the expressions of the Riemann metrics* g *in (27) and* $\tilde{g} = -g/Z$*, we see that the counterparts of the representing functions* $M(p_i)$ *would be, respectively,* $-\nu_i(p)$ *and* $P_i(p)$*, but note that they are multi-variable functions of* $p = (p_i)$.

4.3. Examples

If we take L to be the logarithmic function $L(t) = \ln(t)$, then the conformally flattened geometry immediately defines the standard dually flat structure $(g^{\mathrm{F}}, \nabla^{(1)}, \nabla^{(-1)})$ on the simplex $\mathcal{S}^n$. We see that $-\varphi(p)$ is the entropy, i.e., $\varphi(p) = \sum_{i=1}^{n+1} p_i \ln p_i$ and the conformal divergence is the KL divergence (relative entropy), i.e., $\tilde{\rho}(p, r) = D^{(\mathrm{KL})}(r||p) = \sum_{i=1}^{n+1} r_i(\ln r_i - \ln p_i)$.

Next, let the affine immersion (f, ξ) be defined by the following L and ξ:

$$L(t) := \frac{1}{1-q} t^{1-q}, \quad x^i(p) = \frac{1}{1-q}(p_i)^{1-q},$$

and

$$\xi^i(p) = -q(1-q)x^i(p),$$

with $0 < q$ and $q \neq 1$. We see that the immersion is centro-affine scaled by the constant factor $q(1-q)$. Then, we see that the immersion realizes the alpha-structure $(g^{\mathrm{F}}, \nabla^{(\alpha)}, \nabla^{(-\alpha)})$ on $\mathcal{S}^n$ with $q = (1+\alpha)/2$. The geometric divergence is the alpha-divergence, i.e.,

$$\rho(p, r) = \frac{4}{1-\alpha^2}\left(1 - \sum_{i=1}^{n+1} (p_i)^{(1-\alpha)/2}(r_i)^{(1+\alpha)/2}\right).$$

Following the procedure of conformal flattening described in the above, we have [17]

$$\Psi(x) = \sum_{i=1}^{n+1} ((1-q)x^i)^{1/1-q}, \quad \Lambda(p) = -q, \ (constant)$$

$$\nu_i(p) = -\frac{1}{q}(p_i)^q, \quad \sigma(p) = -\frac{1}{Z(p)} = \frac{q}{\sum_{k=1}^{n+1}(p_i)^q},$$

and obtain a dually flat structure $(\tilde{g}^{\mathrm{F}}, \tilde{\nabla}, \tilde{\nabla}^*)$ via the formulas in Theorem 1:

$$\eta_i = P_i = \frac{(p_i)^q}{\sum_{k=1}^{n+1}(p_k)^q}, \quad \theta^i = \frac{1}{1-q}(p_i)^{1-q} - \frac{1}{1-q}(p_{n+1})^{1-q} = \ln_q(p_i) - \psi(p),$$

$$\psi(p) = -\ln_q(p_{n+1}), \quad \varphi(p) = \ln_q\left(\frac{1}{\exp_q(S_q(p))}\right), \quad \tilde{g}^{\mathrm{F}} = -\frac{1}{Z(p)}g^{\mathrm{F}}.$$

Here, $\ln_q$ and $S_q(p)$ are the *q-logarithmic function* and the *Tsallis entropy* [10], respectively, defined by

$$\ln_q(t) = \frac{t^{1-q}-1}{1-q}, \quad S_q(p) = \frac{\sum_{i=1}^{n+1}(p_i)^q - 1}{1-q}.$$

Note that the escort probability appears as the dual coordinate η_i.

5. An Application to Gradient Flows on $\mathcal{S}^n$

Recall the *replicator flow* on the simplex $\mathcal{S}^n$ for given functions $f_i(p)$ defined by

$$\dot{p}_i = p_i(f_i(p) - \bar{f}(p)), \; i = 1, \cdots, n+1, \quad \bar{f}(p) := \sum_{i=1}^{n+1} p_i f_i(p), \tag{35}$$

which is extensively studied in evolutionary game theory. It is known [20] (Chapter 16) that

(i) the solution to (35) is the gradient flow that maximizes a function $V(p)$ satisfying

$$f_i = \frac{\partial V}{\partial p_i}, \; i = 1, \cdots, n+1, \tag{36}$$

with respect to the *Shahshahani metric* g^{S} (See below),

(ii) the KL divergence is a local Lyapunov function for an equilibrium called the *evolutionary stable state (ESS)* for the case of $f_i(p) = \sum_{j=1}^{n+1} a_{ij}p_j$ with $(a_{ij}) \in \mathbf{R}^{(n+1)\times(n+1)}$.

The Shahshahani metric g^{S} is defined on the positive orthant $\mathbf{R}_+^{n+1}$ by

$$g_{ij}^{\mathrm{S}}(p) = \frac{\sum_{k=1}^{n+1} p_k}{p_i}\delta_{ij}, \quad i, j = 1, \cdots, n+1.$$

Note that the Shahshahani metric induces the Fisher metric g^{F} on $\mathcal{S}^n$. Further, the KL divergence is the canonical divergence [2] of $(g^{\mathrm{F}}, \nabla^{(1)}, \nabla^{(-1)})$. Thus, the replicator dynamics (35) are closely related with the standard dually flat structure $(g^{\mathrm{F}}, \nabla^{(1)}, \nabla^{(-1)})$, which associates with exponential and mixture families of probability distributions. In addition, investigation of the flow is also important from a viewpoint of statistical physics governed by the Boltzmann–Gibbs distributions when we choose $V(p)$ as various physical quantities, e.g., free energy or entropy.

Similarly, when we consider various Legendre relations deformed by L, it would be of interest to investigate gradient flows on $\mathcal{S}^n$ for a dually flat structure $(\tilde{g}, \tilde{\nabla}, \tilde{\nabla}^*)$ or a 1-conformally flat structure (g, ∇, ∇^*). Since g and $\tilde{g}$ can be naturally extended to $\mathbf{R}_+^{n+1}$ as a diagonal form (we use the same notation for brevity):

$$g_{ij}(p) = \frac{1}{\Lambda(p)}\frac{L''(p_i)}{L'(p_i)}\delta_{ij}, \quad \tilde{g}_{ij}(p) = -\frac{1}{Z(p)}g_{ij}(p), \quad i, j = 1, \cdots, n+1 \tag{37}$$

from (27), we can define two gradient flows for $V(p)$ on $\mathcal{S}^n$. One is the gradient flow for g, which is

$$\dot{p}_i = g_{ii}^{-1}(f_i - \bar{f}^H), \quad \bar{f}^H(p) := \sum_{k=1}^{n+1} H_k(p) f_k(p), \quad H_i(p) := \frac{g_{ii}^{-1}(p)}{\sum_{k=1}^{n+1} g_{kk}^{-1}(p)}, \tag{38}$$

for $i = 1, \cdots, n+1$. It is verified that $\dot{p}$ is tangent to $\mathcal{S}^n$, i.e., $\dot{p} \in T_p\mathcal{S}^n$ and gradient of V, i.e.,

$$g(X, \dot{p}) = \sum_{i=1}^{n+1} f_i X^i - \bar{f}^H \sum_{i=1}^{n+1} X^i = \sum_{i=1}^{n+1} \frac{\partial V}{\partial p_i} X^i, \quad \forall X = (X^i) \in \mathcal{X}(\mathcal{S}^n).$$

In the same way, the other one for $\tilde{g}$ is defined by

$$\dot{p}_i = \tilde{g}_{ii}^{-1}(f_i - \tilde{f}^H), \quad \tilde{f}^H(p) := \sum_{k=1}^{n+1} H_k(p) f_k(p), \quad i = 1, \cdots, n+1. \tag{39}$$

Note that both the flows reduce to (35) when $L = \ln$.

From (37), the following consequence is immediate:

Proposition 1. *The trajectories of the gradient flow (38) and (39) starting from the same initial point coincide while velocities of time-evolutions are different by the factor-$Z(p)$.*

Taking account of the example with respect to the alpha-geometry and the conformally flattened one given in subsection 4.3, the following result shown in [18] can be regarded as a corollary of the above proposition:

Corollary 2. *The trajectories of the gradient flow (39) with respect to the conformal metric $\tilde{g}$ for $L(t) = t^{1-q}/(1-q)$ coincide with those of the replicator flow (35) while velocities of time-evolutions are different by the factor-$Z(p)$.*

Next, we particularly consider the case when $V(p)$ is a potential function or divergences. As for a gradient flow on a manifold equipped with a dually flat structure $(\tilde{g}, \tilde{\nabla}, \tilde{\nabla}^*)$, the following result is known:

Proposition 2. *[22] Consider the potential function $\psi(p)$ and the canonical divergence $\tilde{\rho}(p,r)$ of $(\tilde{g}, \tilde{\nabla}, \tilde{\nabla}^*)$ for an arbitrary prefixed point r. The gradient flows for $V(p) = \pm\psi(p)$ and $V(p) = \pm\rho(p,r)$ follow $\tilde{\nabla}^*$-geodesic curves.*

As is described in Remark 2, ∇^* and $\tilde{\nabla}^*$ are *projectively equivalent*. One geometrically interesting property of the projective equivalence is that ∇^*- and $\tilde{\nabla}^*$- geodesic curves coincide up to their parametrizations (i.e., a curve is ∇^*-pregeodesic if and only if it is $\tilde{\nabla}^*$-pregeodesic) [15] (p.17). Combining this fact with Propositions 1 and 2, we see that the following result holds:

Proposition 3. *Let $r \in \mathcal{S}^n$ be an arbitrary prefixed point. The gradient flows (38) for $V(p) = \pm\rho(p,r) = \pm\tilde{\rho}(p,r)/\sigma(r)$, $V(p) = \pm\tilde{\rho}(p,r)$ and $V(p) = \pm\psi(p)$ follow $\tilde{\nabla}^*$-geodesic curves.*

Finally, we demonstrate here another aspect of the flow (39). Let us particularly consider the following functions f_i:

$$f_i(p) := \frac{L''(p_i)}{(L'(p_i))^2} \sum_{j=1}^{n+1} a_{ij} P_j(p), \quad a_{ij} = -a_{ji} \in \mathbf{R}, \quad i,j = 1, \cdots, n+1. \tag{40}$$

Note that f_is are not integrable, i.e., non-trivial V satisfying (36) does not exist because of the anti-symmetry of a_{ij}. Hence, for this case, (39) is no longer a gradient flow. However, we can prove the following result:

Theorem 2. *Consider the flow (39) with the functions f_is defined in (40) and assume that there exists an equilibrium $r \in \mathcal{S}^n$ for the flow. Then, $\rho(p,r)$ and $\tilde{\rho}(p,r)$ are the first integral (conserved quantity) of the flow.*

Proof. By substituting (40) into $\tilde{f}^H(p)$ in (39) and using the expression of $\tilde{g}_{ii}$ in (37), we have

$$\tilde{f}^H(p) = \frac{1}{\sum_{k=1}^{n+1} L'(p_k)/L''(p_k)} \sum_{i=1}^{n+1} \frac{1}{L'(p_i)} \sum_{j=1}^{n+1} a_{ij} P_i(p).$$

By the relation $E'(x_i) = 1/L'(p_i)$ and (31), it holds that

$$\sum_{i=1}^{n+1} \frac{1}{L'(p_i)} \sum_{j=1}^{n+1} a_{ij} P_i(p) = \frac{1}{\sum_{k=1}^{n+1} E'(x_k)} \sum_{i=1}^{n+1} \sum_{j=1}^{n+1} a_{ij} E'(x_l) E'(x_j) = 0.$$

Hence, we see that $\tilde{f}^H = 0$ and the flow (39) reduces to

$$\dot{p}_i = \tilde{g}_{ii}^{-1}(p) f_i(p), \quad i = 1, \cdots, n+1. \tag{41}$$

Since r is an equilibrium point, we see from (40) that

$$\sum_{j=1}^{n+1} a_{ij} P_j(r) = 0, \quad i = 1, \cdots, n+1. \tag{42}$$

Then, using (34), (41) and (42), we have

$$\begin{aligned}
\frac{d\tilde{\rho}(p,r)}{dt} &= -\sum_{i=1}^{n+1} P_i(r) L'(p_i) \dot{p}_i \\
&= Z(p)\Lambda(p) \sum_{i=1}^{n+1} P_i(r) \frac{(L'(p_i))^2}{L''(p_i)} f_i(p) = Z(p)\Lambda(p) \sum_{i=1}^{n+1} P_i(r) \left(\sum_{j=1}^{n+1} a_{ij} P_j(p) \right) \\
&= Z(p)\Lambda(p) \sum_{i=1}^{n+1} \sum_{j=1}^{n+1} (P_i(p) - P_i(r)) a_{ij} P_j(p) \\
&= Z(p)\Lambda(p) \sum_{i=1}^{n+1} \sum_{j=1}^{n+1} (P_i(p) - P_i(r)) a_{ij} (P_j(p) - P_j(r)) = 0.
\end{aligned}$$

Thus, $\tilde{\rho}(p,r)$ is the first integral of the flow. It follows that $\rho(p,r)$ is also the first integral of the flow from the definition of conformal divergence (29). □

Remark 5. *From proposition 1, the same statement holds for the flow (38). The proposition implies the fact [20] that the KL divergence is the first integral for the replicator flow (35) with the function $f_i(p)$ in (40) defined by $L(t) = \ln t$ and $P_j(p) = p_j$.*

6. Conclusions

We have considered two important aspects of information geometric structure, i.e., invariance and dual flatness, from a viewpoint of representing functions. As for the invariance of geometry, we have proved that a pair of representing functions that derives the alpha-structure is essentially unique. On the other hand, we have shown the explicit formula of conformal flattening that transforms 1-conformally flat structures on the simplex $\mathcal{S}^n$ realized by affine immersions to the corresponding dually flat structures. Finally, we have discussed several geometric properties of gradient flows associated to two structures.

Presently, our analysis is restricted to the probability simplex, i.e., the space of discrete probability distributions. For the continuous case, the similar or related results are obtained in [23,24] without using affine immersions. Extensions of the results obtained in this paper to continuous probability space and the exploitation of relations to the literature are left for future work.

The conformal flattening can also be applied to the computationally efficient construction of a Voronoi diagram with respect to the geometric divergences [18]. Exploring the possibilities of other applications would be of interest.

Acknowledgments: Part of the results is adapted and reprinted with permission from Springer Customer Service Centre GmbH (licence No: 4294160782766): Springer Nature, *Geometric Science of Information* LNCS 10589 Nielsen, F., Barbaresco, F., Eds., (Article Name:) On affine immersions of the probability simplex and their conformal flattening, (Author:) A. Ohara, (Copyright:) Springer International Publishing AG 2017 [25]. The author is partially supported by JSPS Grant-in-Aid (C) 15K04997.

Conflicts of Interest: The author declares no conflict of interest.

References

1. Amari, S.I. *Differential-Geometrical Methods in Statistics*; Lecture Notes in Statistics Series 28; Springer: New York, NY, USA, 1985.
2. Amari, S.I.; Nagaoka, H. *Methods of Information Geometry*; Translations of Mathematical Monographs Series 191; AMS & Oxford University Press: Oxford, UK, 2000.
3. Naudts, J. Continuity of a class of entropies and relative entropies. *Rev. Math. Phys.* **2004**, *16*, 809.
4. Eguchi, S. Information geometry and statistical pattern recognition. *Sugaku Expos.* **2006**, *19*, 197.
5. Grünwald, P.D.; Dawid, A.P. Game theory, maximum entropy, minimum discrepancy, and robust Bayesian decision theory. *Ann. Statist.* **2004**, *32*, 1367.
6. Fujisawa, H.; Eguchi, S. Robust parameter estimation with a small bias against heavy contamination. *J. Multivar. Anal.* **2008**, *99*, 2053.
7. Naudts, J. The q-exponential family in statistical Physics. *Cent. Eur. J. Phys.* **2009**, *7*, 405.
8. Naudts, J. *Generalized thermostatics*; Springer: Berlin, Germany, 2010.
9. Ollila, E.; Tyler, D.; Koivunen, V.; Poor, V. Complex elliptically symmetric distributions : Survey, new results and applications. *IEEE Trans. Sig. Proc.* **2012**, *60*, 5597.
10. Tsallis, C. *Introduction to Nonextensive Statistical Mechanics: Approaching a Complex World*; Springer: Berlin, Germany, 2009.
11. Zhang, J. Divergence Function, Duality, and Convex Analysis. *Neural Comput.* **2004**, *16*, 159.
12. Wada, T.; Matsuzoe, H. Conjugate representations and characterizing escort expectations in information geometry. *Entropy* **2017**, *19*, 309.
13. Shima, H. *The Geometry of Hessian Structures*; World Scientific: Singapore, 2007.
14. Chentsov, N.N. *Statistical Decision Rules and Optimal Inference*; AMS: Providence, RI, USA, 1982.
15. Nomizu, K.; Sasaki, T. *Affine Differential Geometry*; Cambridge University Press: Cambridge, UK, 1993.
16. Kurose, T. On the divergences of 1-conformally flat statistical manifolds. *Tohoku Math. J.* **1994**, *46*, 427.
17. Ohara, A.; Matsuzoe, H.; Amari, S.I. A dually flat structure on the space of escort distributions. *J. Phys. Conf. Ser.* **2010**, *201*, 012012.
18. Ohara, A.; Matsuzoe, H.; Amari, S.I. Conformal geometry of escort probability and its applications. *Mod. Phys. Lett. B* **2012**, *26*, 1250063.
19. Tsallis, C.; Mendes, M.S.; Plastino, A.R. The role of constraints within generalized nonextensive statistics. *Physica A* **1998**, *261*, 534.
20. Hofbauer, J.; Sigmund, K. *The Theory of Evolution and Dynamical Systems: Mathematical Aspects of Selection*; Cambridge University Press: Cambridge, UK, 1988.
21. Eguchi, S. Geometry of minimum contrast. *Hiroshima Math. J.* **1992**, *22*, 631.
22. Fujiwara, A.; Amari, S.I. Gradient systems in view of information geometry. *Physica D* **1995**, *80*, 317.
23. Amari, S.I.; Ohara, A.; Matsuzoe, H. Geometry of deformed exponential families: Invariant, dually-flat and conformal geometries. *Physica A* **2012**, *391*, 4308.
24. Matsuzoe, H. Hessian structures on deformed exponential families and their conformal structures. *Diff. Geo. Appl.* **2014**, *35*, 323.
25. Ohara, A. On affine immersions of the probability simplex and their conformal flattening. In *Geometric Science of Information*; Nielsen, F., Barbaresco, F., Eds.; Springer: Berlin, Germany, 2017.

entropy

MDPI

Article

The Volume of Two-Qubit States by Information Geometry

Milajiguli Rexiti [1], Domenico Felice [2] and Stefano Mancini [3,4,*]

1 School of Advanced Studies, University of Camerino, 62032 Camerino, Italy; milajiguli.milajiguli@unicam.it
2 Max Planck Institute for Mathematics in the Sciences, Inselstrasse 22-04103 Leipzig, Germany; felice@mis.mpg.de
3 School of Science and Technology, University of Camerino, 62032 Camerino, Italy
4 INFN-Sezione di Perugia, I-06123 Perugia, Italy
* Correspondence: stefano.mancini@unicam.it; Tel.: +39-0737-402577

Received: 22 December 2017; Accepted: 22 February 2018; Published: 24 February 2018

Abstract: Using the information geometry approach, we determine the volume of the set of two-qubit states with maximally disordered subsystems. Particular attention is devoted to the behavior of the volume of sub-manifolds of separable and entangled states with fixed purity. We show that the usage of the classical Fisher metric on phase space probability representation of quantum states gives the same qualitative results with respect to different versions of the quantum Fisher metric.

Keywords: information theory; Riemannian geometry; entanglement characterization

PACS: 89.70.+c; 02.40.Ky; 03.67.Mn

1. Introduction

The volume of sets of quantum states is an issue of the utmost importance. It can help in finding separable states within all quantum states. The former are states of a composite system that can be written as convex combinations of subsystem states, in contrast to entangled states [1]. The notion of volume can also be useful when defining "typical" properties of a set of quantum states. In fact, in such a case, one uses the random generation of states according to a measure stemming from their volume [2].

Clearly, the seminal thing to do to determine the volume of a set of quantum states is to consider a metric on it. For pure quantum states, it is natural to consider the Fubini–Study metric which turns out to be proportional to the classical Fisher–Rao metric [3]. However, the situation becomes ambiguous for quantum mixed states where there is no single metric [4]. There, several measures have been investigated, each of them arising from different motivations [5]. For instance, one employed the Positive Partial Transpose criterion [1] to determine an upper bound for the volume of separable quantum states and to figure out that it is non-zero [6]. However, this criterion becomes less and less precise as the dimension increases [7]. An upper bound for the volume of separable states is also provided in [8] by combining techniques of geometry and random matrix theory. Here, separable states are approximated by an ellipsoid with respect to a Euclidean product. Following up, in Reference [9], a measure relying on the convex and Euclidean structure of states has been proposed. Such a volume measure turned out to be very effective in detecting separability for large systems. Nevertheless, for two-qubit states, the Euclidean structure is not given by the Hilbert–Schmidt scalar product. Instead, the latter is considered as a very natural measure upon the finite dimensional Hilbert spaces. For this reason, it has been employed as a natural measure in the space of density operators [6]. Such a measure revealed that the set of separable states has a non-zero volume and, in some cases, analytical lower and upper bounds have been found on it. Still based on the Hilbert–Schmidt product, a measure has been recently introduced to evaluate the volume of Gaussian states in infinite dimensional quantum systems [10]. However, it does not have a classical counterpart, as an approach based on the probabilistic structure of quantum states could have. This approach has been taken for infinite dimensional quantum systems by using information geometry [11]. This is the application of differential geometric techniques to the study of families of probabilities [12]. Thus, it can be employed whenever quantum states are represented as probability distribution functions, instead of density operators. This happens in phase space where Wigner functions representing Gaussian states were used in such a way that each class of states is associated with a statistical model which turns out to be a Riemannian manifold endowed with the well-known Fisher–Rao metric [11].

This approach could be extended to the entire set of quantum states considering the Husimi Q-function [13] instead of the Wigner function, being the former a true probability distribution function. Additionally, it could be used in finite dimensional systems by using coherent states of compact groups $SU(d)$. Here, we take this avenue for two-qubit systems. Above all, we address the question of whether such an approach gives results similar to other approaches based on the quantum version of the Fisher metric, such as the Helstrom quantum Fisher metric [14] and the Wigner-Yanase-like quantum Fisher metric [15]. We focus on states with maximally disordered subsystems and analyze the behavior of the volume of sub-manifolds of separable and entangled states with fixed purity. We show that all of the above mentioned approaches give the same qualitative results.

The layout of the paper is as follows. In Section 2, we recall the structure of a set of two-qubit states. Then, in Section 3, we present different metrics on such a space inspired by the classical Fisher–Rao metric. Section 4 is devoted to the evaluation of the volume of states with maximally disordered subsystems. Finally, we draw our conclusions in Section 5.

2. Structure of a Set of Two-Qubit States

Consider two-qubit states (labeled by 1 and 2) with associated Hilbert space $\mathcal{H} = \mathbb{C}^2 \otimes \mathbb{C}^2$. The space $\mathcal{L}(\mathcal{H})$ of linear operators acting on $\mathcal{H}$ can be supplied with a scalar product $\langle A, B \rangle = \mathrm{Tr}\left(A^{\dagger} B\right)$ to have a Hilbert–Schmidt space. Then, in such a space, an arbitrary density operator (positive and trace class operator) can be represented as follows

$$\rho = \frac{1}{4}\left(I \otimes I + \boldsymbol{r} \cdot \boldsymbol{\sigma} \otimes I + I \otimes \boldsymbol{s} \cdot \boldsymbol{\sigma} + \sum_{m,n=1}^{3} t_{mm} \sigma_m \otimes \sigma_n \right), \tag{1}$$

where I is the identity operator on $\mathbb{C}^2$, $\boldsymbol{r}, \boldsymbol{s} \in \mathbb{R}^3$, and $\boldsymbol{\sigma} := (\sigma_1, \sigma_2, \sigma_3)$ is the vector of Pauli matrices. Furthermore, the coefficients $t_{mn} := \mathrm{Tr}\left(\rho\, \sigma_m \otimes \sigma_n\right)$ form a real matrix denoted by T.

Note that $\boldsymbol{r}, \boldsymbol{s}$ are local parameters as they determine the reduced states

$$\rho_1 \;:=\; \mathrm{Tr}_2 \rho = \frac{1}{2}\left(I + \boldsymbol{r} \cdot \boldsymbol{\sigma}\right), \tag{2}$$

$$\rho_2 \;:=\; \mathrm{Tr}_1 \rho = \frac{1}{2}\left(I + \boldsymbol{s} \cdot \boldsymbol{\sigma}\right), \tag{3}$$

while the T matrix is responsible for correlations.

The number of (real) parameters characterizing ρ is 15, but they can be reduced with the following argument. Entanglement (in fact, any quantifier of it) is invariant under local unitary transformations, i.e., $U_1 \otimes U_2$ [1]. Then, without loss of generality, we can restrict our attention to states with a diagonal matrix T. To show that this class of states is representative, we can use the following fact: if a state is subjected to $U_1 \otimes U_2$ transformation, the parameters $\boldsymbol{r}, \boldsymbol{s}$ and T transform themselves as

$$\boldsymbol{r}' = O_1 \boldsymbol{r}, \tag{4}$$

$$\boldsymbol{s}' = O_2 \boldsymbol{s}, \tag{5}$$

$$T' = O_1 T O_2^{\dagger}, \tag{6}$$

where O_is correspond to U_is via

$$U_i \hat{\boldsymbol{n}} \cdot \boldsymbol{\sigma} U_i^{\dagger} = (O_i \hat{\boldsymbol{n}}) \cdot \boldsymbol{\sigma}, \tag{7}$$

being $\hat{\boldsymbol{n}}$ a versor of $\mathbb{R}^3$. Thus, given an arbitrary state, we can always choose unitaries U_1, U_2 such that the corresponding rotations will diagonalize its matrix T.

As we consider the states with diagonal T, we can identify T with the vector $\boldsymbol{t} := (t_{11}, t_{22}, t_{33}) \in \mathbb{R}^3$. Then, the following sufficient conditions are known [16]:

(i) *For any ρ, the matrix T belongs to the tetrahedron $\mathcal{T}$ with vertices* $(-1,-1,-1)$, $(-1,1,1)$, $(1,-1,1)$, $(1,1,-1)$.

(ii) *For any separable state ρ, the matrix T belongs to the octahedron $\mathcal{O}$ with vertices* $(0,0,\pm 1)$, $(0,\pm 1,0)$, $(0,0,\pm 1)$.

Still, the number of (real) parameters (9) characterizing ρ is quite large. Thus, to simplify the treatment, from now on, we focus on the states with maximally disordered subsystems, namely the states with $\boldsymbol{r}, \boldsymbol{s} = 0$. They are solely characterized by the T matrix. For these states, the following necessary and sufficient conditions are known [16]:

(i) *Any operator* (1) *with $\boldsymbol{r}, \boldsymbol{s} = 0$ and diagonal T is a state (density operator) iff T belongs to the tetrahedron $\mathcal{T}$.*

(ii) *Any state ρ with maximally disordered subsystems and diagonal T is separable iff T belongs to the octahedron $\mathcal{O}$.*

3. Fisher Metrics

The generic state with maximally disordered subsystems parametrized by $\boldsymbol{t}$ reads

$$\rho_{\boldsymbol{t}} = \frac{1}{4} \sum_{n=1}^{3} t_{nn} \sigma_n \otimes \sigma_n, \tag{8}$$

and has a matrix representation (in the canonical basis):

$$\rho_{\boldsymbol{t}} = \begin{pmatrix} \frac{t_{33}+1}{4} & 0 & 0 & \frac{t_{11}-t_{22}}{4} \\ 0 & \frac{1-t_{33}}{4} & \frac{t_{11}+t_{22}}{4} & 0 \\ 0 & \frac{t_{11}+t_{22}}{4} & \frac{1-t_{33}}{4} & 0 \\ \frac{t_{11}-t_{22}}{4} & 0 & 0 & \frac{t_{33}+1}{4} \end{pmatrix}. \tag{9}$$

We shall derive different Fisher metrics for the set of such states.

3.1. Classical Fisher Metric in Phase Space

The state (8) can also be represented in the phase space by means of the Husimi Q-function [13],

$$Q_{\boldsymbol{t}}(\boldsymbol{x}) := \frac{1}{4\pi^2} \langle \Omega_1 | \langle \Omega_2 | \rho_{\boldsymbol{t}} | \Omega_2 \rangle | \Omega_1 \rangle, \tag{10}$$

where

$$|\Omega_1\rangle = \cos\frac{\theta_1}{2}|0\rangle + e^{-i\phi_1} \sin\frac{\theta_1}{2}|1\rangle, \tag{11}$$

$$|\Omega_2\rangle = \cos\frac{\theta_2}{2}|0\rangle + e^{-i\phi_2} \sin\frac{\theta_2}{2}|1\rangle, \tag{12}$$

are $SU(2)$ coherent states [17]. Here, $\boldsymbol{x} := (\theta_1, \theta_2, \phi_1, \phi_2)$ is the vector of random variables $\theta_{1,2} \in [0, \pi)$, $\phi_{1,2} \in [0, 2\pi]$, with probability measure $d\boldsymbol{x} = \frac{1}{16\pi^2} \sin\theta_1 \sin\theta_2 d\theta_1 d\theta_2 d\phi_1 d\phi_2$.

Recall that the classical Fisher–Rao information metric for probability distribution functions Q_t is given by:

$$g_{ij}^{FR} := \int Q_t(x)\,\partial_i \log Q_t(x)\,\partial_j \log Q_t(x)\,dx, \tag{13}$$

where $\partial_i := \frac{\partial}{\partial t_{ii}}$.

Considering (8) and (10), we explicitly get

$$\begin{aligned} Q_t(\theta_1,\theta_2,\phi_1,\phi_2) &= \frac{2\sin\theta_1\sin\theta_2}{64\pi^2}\Big[t_{11}\left(\cos(\phi_1+\phi_2)+\cos(\phi_1-\phi_2)\right) \\ &\qquad -t_{22}\left(\cos(\phi_1+\phi_2)-\cos(\phi_1-\phi_2)\right)\Big] \\ &+ \frac{4}{64\pi^2}\left[t_{33}\cos\theta_1\cos\theta_2+1\right]. \end{aligned} \tag{14}$$

Unfortunately, it is not possible to arrive at an analytical expression for g_{ij}. In fact, the integral (13) can only be evaluated numerically.

Let us, however, analyze an important property of this metric.

Proposition 1. *The metric (13) is invariant under local rotations.*

Proof. From Section 2, we know that the equivalence relation $\rho \sim (U_1 \otimes U_2)\,\rho\,(U_1^\dagger \otimes U_2^\dagger)$ leads to the transformation (6) for the parameters matrix, where the correspondence between O_is and U_is, given by Equation (7), can be read as follows

$$(O_i)_{\mu\nu} = \frac{1}{2}\mathrm{Tr}\left[\sigma_\mu U_i \sigma_\nu U_i^\dagger\right]. \tag{15}$$

Therefore, the effect of the equivalence relation acts on $Q_t(x)$ of Equation (10) by changing the parameter vector $t = (t_{11}, t_{22}, t_{33})$ via an orthogonal transformation given by (15). That is

$$Q_t(x) \mapsto Q_{t'}(x), \tag{16}$$

where $t' = Ot$, or explicitly

$$\begin{aligned} t'_{11} &= O_{11}t_{11} + O_{12}t_{22} + O_{13}t_{33}, \\ t'_{22} &= O_{21}t_{11} + O_{22}t_{22} + O_{23}t_{33}, \\ t'_{33} &= O_{31}t_{11} + O_{32}t_{22} + O_{33}t_{33}. \end{aligned} \tag{17}$$

Let us now see how the entries of the Fisher–Rao metric (13) change under an orthogonal transformation of T given in terms of (15). To this end, consider (13) written for t' as follows

$$\begin{aligned} \tilde{g}_{ij}^{FR} &= \int Q_{t'}(x)\,\partial'_i \log Q_{t'}(x)\,\partial'_j \log Q_{t'}(x)\,dx \\ &= \int \frac{1}{Q_{t'}(x)}\partial'_i Q_{t'}(x)\,\partial'_j Q_{t'}(x)\,dx, \end{aligned} \tag{18}$$

where $\partial'_i := \frac{\partial}{\partial t'_{ii}}$. From (17), we recognize the functional relation $t'_{ii} = t'_{ii}(t)$; therefore, we have

$$\partial_i Q_{t'}(x) = \partial'_1 Q_{t'}(x)\,\partial_i t'_{11} + \partial'_2 Q_{t'}(x)\,\partial_i t'_{22} + \partial'_3 Q_{t'}(x)\,\partial_i t'_{33} \tag{19}$$

Finally, from (18) and (19), we arrive at

$$g_{ij}^{FR} = \sum_{m,n=1}^{3} \tilde{g}_{mn}^{FR} O_{in} O_{jm}, \tag{20}$$

and as a consequence we immediately obtain $g^{FR} = O\, \tilde{g}^{FR}\, O^{\top}$. $\square$

3.2. Quantum Fisher Metrics

It is remarkable that (13) is the unique (up to a constant factor) monotone Riemannian metric (that is, the metric contracting under any stochastic map) in the class of probability distribution functions (classical or commutative case) [18]. This is not the case in an operator setting (quantum case), where the notion of Fisher information has many natural generalizations due to non-commutativity [4]. Among the various generalizations, two are distinguished.

The first natural generalization of the classical Fisher information arises when one formally generalizes the expression (13). This was, in fact, first done in a quantum estimation setting [14]. To see how this happens, note that in a symmetric form, the derivative of Q_t reads

$$\partial_i Q_t = \frac{1}{2}\left(\partial_i \log Q_t \cdot Q_t + Q_t \cdot \partial_i \log Q_t\right). \tag{21}$$

In Equation (13), replacing the integration by trace, Q_t by ρ_t, and the logarithmic derivative $\partial_i \log Q_t$ by the symmetric logarithmic derivative L_i determined by

$$\partial_i \rho_t = \frac{1}{2}\left(L_i \rho_t + \rho_t L_i\right), \tag{22}$$

we come to the quantum Fisher information (derived via the symmetric logarithmic derivative)

$$g^{H}_{ij}(\rho_\theta) := \frac{1}{2}\mathrm{Tr}\left(L_i L_j \rho_t + L_j L_i \rho_t\right). \tag{23}$$

Using (9) to solve (22) yields

$$L_1 = \begin{pmatrix} \frac{t_{11}-t_{22}}{(t_{11}-t_{22})^2-(t_{33}+1)^2} & 0 & 0 & \frac{t_{33}+1}{-(t_{11}-t_{22})^2+(t_{33}+1)^2} \\ 0 & \frac{t_{11}+t_{22}}{(t_{11}+t_{22})^2-(t_{33}-1)^2} & \frac{t_{33}-1}{(t_{11}+t_{22})^2-(t_{33}-1)^2} & 0 \\ 0 & \frac{t_{33}-1}{(t_{11}+t_{22})^2-(t_{33}-1)^2} & \frac{t_{11}+t_{22}}{(t_{11}+t_{22})^2-(t_{33}-1)^2} & 0 \\ \frac{t_{33}+1}{-(t_{11}-t_{22})^2+(t_{33}+1)^2} & 0 & 0 & \frac{t_{11}-t_{22}}{(t_{11}-t_{22})^2-(t_{33}+1)^2} \end{pmatrix}, \tag{24}$$

$$L_2 = \begin{pmatrix} \frac{t_{22}-t_{11}}{(t_{11}-t_{22})^2-(t_{33}+1)^2} & 0 & 0 & \frac{t_{33}+1}{(t_{11}-t_{22})^2-(t_{33}+1)^2} \\ 0 & \frac{t_{11}+t_{22}}{(t_{11}+t_{22})^2-(t_{33}-1)^2} & \frac{t_{33}-1}{(t_{11}+t_{22})^2-(t_{33}-1)^2} & 0 \\ 0 & \frac{t_{33}-1}{(t_{11}+t_{22})^2-(t_{33}-1)^2} & \frac{t_{11}+t_{22}}{(t_{11}+t_{22})^2-(t_{33}-1)^2} & 0 \\ \frac{t_{33}+1}{(t_{11}-t_{22})^2-(t_{33}+1)^2} & 0 & 0 & \frac{t_{22}-t_{11}}{(t_{11}-t_{22})^2-(t_{33}+1)^2} \end{pmatrix}, \tag{25}$$

$$L_3 = \begin{pmatrix} \frac{t_{33}+1}{-(t_{11}-t_{22})^2+(t_{33}+1)^2} & 0 & 0 & \frac{t_{11}-t_{22}}{(t_{11}-t_{22})^2-(t_{33}+1)^2} \\ 0 & \frac{1-t_{33}}{(t_{11}+t_{22})^2-(t_{33}-1)^2} & -\frac{t_{11}+t_{22}}{(t_{11}+t_{22})^2-(t_{33}-1)^2} & 0 \\ 0 & -\frac{t_{11}+t_{22}}{(t_{11}+t_{22})^2-(t_{33}-1)^2} & \frac{1-t_{33}}{(t_{11}+t_{22})^2-(t_{33}-1)^2} & 0 \\ \frac{t_{11}-t_{22}}{(t_{11}-t_{22})^2-(t_{33}+1)^2} & 0 & 0 & \frac{t_{33}+1}{-(t_{11}-t_{22})^2+(t_{33}+1)^2} \end{pmatrix}. \tag{26}$$

Then, from Equation (23), we obtain

$$g^H = \frac{1}{\Delta}\begin{pmatrix} 1-\|\boldsymbol{t}\|^2-2t_{11}t_{22}t_{33} & (1+\|\boldsymbol{t}\|^2-2t_{33}^2)t_{33}+2t_{11}t_{22} & (1+\|\boldsymbol{t}\|^2-2t_{22}^2)\,t_{22}+2t_{11}t_{33} \\ (1+\|\boldsymbol{t}\|^2-2t_{33}^2)t_{33}+2t_{11}t_{22} & 1-\|\boldsymbol{t}\|^2-2t_{11}t_{22}t_{33} & (1+\|\boldsymbol{t}\|^2-2t_{11}^2)\,t_{11}+2t_{22}t_{33} \\ (1+\|\boldsymbol{t}\|^2-2t_{22}^2)\,t_{22}+2t_{11}t_{33} & (1+\|\boldsymbol{t}\|^2-2t_{11}^2)\,t_{11}+2t_{22}t_{33} & 1-\|\boldsymbol{t}\|^2-2t_{11}t_{22}t_{33} \end{pmatrix}, \tag{27}$$

where

$$\Delta := \left((t_{11}+t_{22})^2-(1-t_{33})^2\right)\left((t_{11}-t_{22})^2-(1+t_{33})^2\right). \tag{28}$$

It follows

$$\det g^H = \frac{1}{\Delta}. \tag{29}$$

The second generalization of the classical Fisher information arises by noticing that (13) can be equivalently expressed as

$$g_{ij}^{FR} = 4\int \partial_i\sqrt{Q_t\,(x)}\partial_j\sqrt{Q_t\,(x)}\,dx. \tag{30}$$

Replacing the integration by trace, and the parameterized probabilities Q by a parameterized density operator ρ, we can arrive at [15]

$$g_{ij}^{WY} := 4\mathrm{Tr}\left[\left(\partial_i\sqrt{\rho_t}\right)\left(\partial_j\sqrt{\rho_t}\right)\right]. \tag{31}$$

The superscript indicates the names of Wigner and Yanase since Equation (31) is motivated by their work on skew information [19].

The following proposition relates g^{WY} and g^H.

Proposition 2. *If* $[\rho_t, L_i]=0, \forall i$, *then* $g^{WY}=g^H$.

Proof. If $[\rho_t, L_i]=0$, $i=1,2,3$ from (22) it follows $\partial_i\rho_t=\rho_t L_i$ and $g_{ij}^H=\mathrm{Tr}\left[\rho_t L_i L_j\right]$. It is also

$$\begin{aligned} g_{ij}^{WY} &= \mathrm{Tr}\left[\rho_t^{-1/2}\left(\partial_i\rho_t\right)\rho_t^{-1/2}\left(\partial_j\rho_t\right)\right] && (32)\\ &= \mathrm{Tr}\left[\rho_t^{-1/2}\rho_t L_i\rho_t^{-1/2}\rho_t L_j\right] && (33)\\ &= \mathrm{Tr}\left[\rho_t^{1/2}L_i\rho_t^{1/2}L_j\right] && (34)\\ &= \mathrm{Tr}\left[\rho_t L_i L_j\right]=g_{ij}^H, && (35)\end{aligned}$$

where in (32) we have used the definition (31); in (33) we have used the result $\partial_i\rho_t=\rho_t L_i$; and in (35) the commutativity of ρ_t and L_is. □

By referring to (9) and (24)–(26), we can see that the conditions of Proposition 2 are satisfied, hence hereafter we shall consider $g^{WY}=g^H$.

4. Volume of States with Maximally Disordered Subsystems

The volume of two-qubit states with maximally disordered subsystems will be given by

$$V=\int_{\mathcal{T}}\sqrt{\det g}\,dt, \tag{36}$$

where the tetrahedron $\mathcal{T}$ is characterized by equations:

$$\begin{aligned} 1-t_{11}-t_{22}-t_{33} &\geq 0,\\ 1-t_{11}+t_{22}+t_{33} &\geq 0,\\ 1+t_{11}-t_{22}+t_{33} &\geq 0,\\ 1+t_{11}+t_{22}-t_{33} &\geq 0. \end{aligned} \tag{37}$$

(If the Fisher metric is degenerate, then the volume is meaningless, given that $\det g=0$. This reflects on the parameters describing the states. Indeed, at least one of them should be considered as depending on the others. Thus, one should restrict the attention to a proper submanifold.)

Analogously, the volume of two-qubit separable states with maximally disordered subsystems will be given by

$$V_s = \int_{\mathcal{O}} \sqrt{\det g}\, dt, \tag{38}$$

where the octahedron $\mathcal{O}$ is characterized by equations:

$$\begin{aligned}
1 - t_{11} - t_{22} - t_{33} &\geq 0, \\
1 + t_{11} - t_{22} - t_{33} &\geq 0, \\
1 + t_{11} + t_{22} - t_{33} &\geq 0, \\
1 - t_{11} + t_{22} - t_{33} &\geq 0, \\
1 - t_{11} - t_{22} + t_{33} &\geq 0, \\
1 + t_{11} - t_{22} + t_{33} &\geq 0, \\
1 + t_{11} + t_{22} + t_{33} &\geq 0, \\
1 - t_{11} + t_{22} + t_{33} &\geq 0.
\end{aligned} \tag{39}$$

Concerning the classical Fisher metric, as a consequence of the result $g^{FR} = O\, \tilde{g}^{FR} O^\top$ (Proposition 1), we have that the volume computed as $\int \sqrt{\det g^{FR}}\, dt$ is invariant under orthogonal transformations of the parameters matrix T.

However, the integral (13) can only be performed numerically. To this end, we have generated 10^3 points randomly distributed inside the tetrahedron $\mathcal{T}$. On each of these points, the integral (13) has been numerically evaluated, hence the value of the function $\sqrt{\det g^{FR}}$ determined on a set of discrete points. After that, data has been interpolated and a smooth function $\sqrt{\det g^{FR}}(t_{11}, t_{22}, t_{33})$ obtained. This has been used to compute the quantities (36), (38) and their ratio, resulting as $V = 0.168$, $V_s = 0.055$, $V_s/V = 0.327$.

The same volumes computed by the quantum Fisher metric result in $V = \pi^2$, $V_s = (4 - \pi)\pi$ and their ratio is $V_s/V = (4 - \pi)/\pi \approx 0.27$.

It would also be instructive to see how the ratio of the volume of separable states and the volume of total states varies versus the purity. This latter quantity turns out to be

$$P = \mathrm{Tr}\left(\rho^2\right) = \frac{1}{4}\left(1 + \|t\|^2\right) \tag{40}$$

Thus, fixing a value of $P \in [\frac{1}{4}, 1]$ amounts to fixing a sphere $\mathcal{S}$ centered in the origin of $\mathbb{R}^3$ and with a radius $\sqrt{4P - 1}$. Hence, the volume of states with fixed purity will be given by

$$V(P) = \int_{\mathcal{T} \cap \mathcal{S}} \sqrt{\det g}\, dt, \tag{41}$$

while the volume of separable states with fixed purity will be given by

$$V_s(P) = \int_{\mathcal{O} \cap \mathcal{S}} \sqrt{\det g}\, dt. \tag{42}$$

As a consequence, we can obtain the ratio

$$R(P) = \frac{V_s(P)}{V(P)}, \tag{43}$$

as a function of purity P. Such a ratio is plotted in Figure 1 as a solid (resp. dashed) line for the classical (resp. quantum) Fisher metric. There, it is shown that by increasing the purity, the volume of separable states diminishes until it becomes a null measure. In fact, this happens already at purity $P = 1/2$. However, when the purity is low enough (below $1/3$), all states are separable ($R = 1$). Above all, the two curves show the same qualitative behavior. This means that the usage of classical Fisher information on the phase space representation of quantum states (as a true probability distribution function) is able to capture the main feature of the geometry of quantum states.

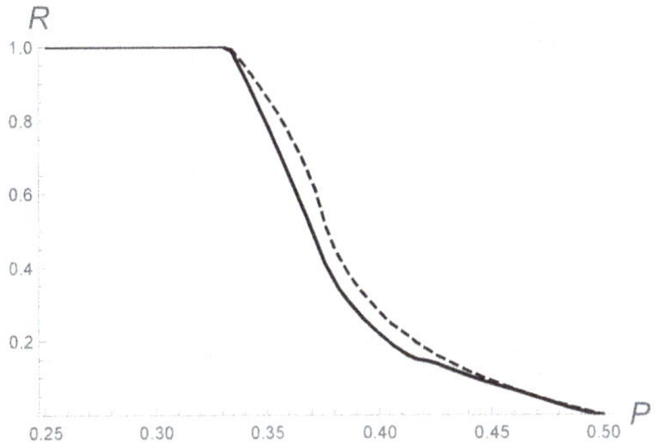

Figure 1. Ratio of volumes (43) vs purity *P*. Solid line refers to the classical Fisher metric; dashed line refers to the quantum Fisher metric.

5. Conclusions

In conclusion, we have investigated the volume of the set of two-qubit states with maximally disordered subsystems by considering their phase space representation in terms of probability distribution functions and by applying the classical Fisher information metric to them. The results have been contrasted with those obtained by using quantum versions of the Fisher metric. Although the absolute values of volumes of separable and entangled states turn out to be different in the two approaches, their ratios are comparable. Above all, the behavior of the volume of sub-manifolds of separable and entangled states with fixed purity is shown to be almost the same in the two approaches.

Thus, we can conclude that classical Fisher information in phase space is able to capture the features of the volume of quantum states. The question then arise as to which aspects such an approach will be unable to single out with respect to the purely quantum one. Besides that, our work points out other interesting issues within the quantum metrics. In fact, in the considered class of two-qubit states, the Helstrom and the Wigner–Yanase-like quantum Fisher metrics coincide. This leads us to ask the following questions: For which more general class of states in finite dimensional systems are the two equal? When they are not equal, what is their order relation? These investigations are left for future works.

Additionally, it is worth noting that our approach to the volume of states by information geometry offers the possibility to characterize quantum logical gates. In fact, given a logical gate (unitary) G on two-qubit states, the standard entangling power is defined as [20]:

$$\mathfrak{E}(G) := \int E\left(G|\psi\rangle\right) d\mu\left(|\psi\rangle\right),$$

where E is an entanglement quantifier [1] and the overall average is the product states $|\psi\rangle = |\psi_1\rangle \otimes |\psi_2\rangle$ according to a suitable measure $\mu\left(|\psi\rangle\right)$. It is then quite natural to take the measure induced by the Haar measure on $SU(2) \otimes SU(2)$ [21]. However, $\mathfrak{E}(G)$ is not usable on the subset of states with maximally disordered subsystems as it is restricted to pure states. Here, our measure comes into play which leads to the following:

$$\mathfrak{E}(G) := \int_{\mathcal{O}} E\left(G\rho_t G^{\dagger}\right) \sqrt{\det g}\, dt.$$

In turn, this paves the way to a general formulation that involves the average overall separable states by also including parameters $\boldsymbol{r}, \boldsymbol{s}$. Clearly, this would provide the most accurate characterization of the entangling potentialities of G.

Finally, our approach can be scaled up to three or more qubit, but since analytical calculations will soon become involved, one should consider families of states with a low number of parameters, e.g., those proposed in [22]. Nevertheless, such families can provide geometrical insights for more general cases.

Acknowledgments: The work of M.R. is supported by China Scholarship Council.

Author Contributions: The authors have equally contributed to the manuscript. They all have read and approved its final version.

Conflicts of Interest: The authors declare no conflict of interest.

References

1. Horodecki, R.; Horodecki, P.; Horodecki, M.; Horodecki, K. Quantum entanglement. *Rev. Mod. Phys.* **2009**, *81*, 865.
2. Dahlsten, O.C.O.; Lupo, C.; Mancini, S.; Serafini, A. Entanglement typicality. *J. Phys. A Math. Theor.* **2014**, *47*, 363001.
3. Facchi, P.; Kulkarni, R.; Man'ko, V.I.; Marmo, G.; Sudarshan, E.C.G.; Ventriglia, F. Classical and Quantum Fisher Information in the Geometrical Formulation of Quantum Mechanics. *Phys. Lett. A* **2010**, *374*, 4801–4803.
4. Petz, D. Monotone metrics on matrix spaces. *Linear Algebra Appl.* **1996**, *244*, 81–96.
5. Bengtsson, I.; Życzkowski, K. *Geometry of Quantum States*; Cambridge University Press: Cambridge, UK, 2006.
6. Życzkowski, K.; Horodecki, P.; Sanpera, A.; Lewenstein, M. Volume of the set of separable states. *Phys. Rev. A* **1998**, *58*, 883.
7. Ye, D. On the Bures volume of separable quantum states. *J. Math. Phys.* **2009**, *50*, 083502.
8. Szarek, S.J. Volume of separable states is super-doubly-exponentially small in the number of qubits. *Phys. Rev. A* **2005**, *72*, 032304.
9. Aubrun, G.; Szarek, S.J. Tensor products of convex sets and the volume of separable states on N qudits. *Phys. Rev. A* **2006**, *73*, 022109.
10. Link, V.; Strunz, W.T. Geometry of Gaussian quantum states. *J. Phys. A Math. Theor.* **2015**, *48*, 275301.
11. Felice, D.; Hà Quang, M.; Mancini, S. The volume of Gaussian states by information geometry. *J. Math. Phys.* **2017**, *58*, 012201.
12. Amari, S.; Nagaoka, H. *Methods of Information Geometry*; Oxford University Press: Oxford, UK, 2000.
13. Husimi, K. Some formal properties of the density matrix. *Proc. Phys. Math. Soc. Jpn.* **1940**, *22*, 264–314.
14. Helstrom, C.W. Quantum detection and estimation theory. *J. Stat. Phys.* **1969**, *1*, 231–252.
15. Luo, S.-L. Fisher information of wavefunctions: Classical and quantum. *Chin. Phys. Lett.* **2006**, *23*, 3127.
16. Horodecki, R.; Horodecki, M. Information-theoretic aspects of quantum inseparability of mixed states. *Phys. Rev. A* **1996**, *54*, 1838.
17. Klauder, J.R.; Skagerstam, B.S. *Coherent states*; World Scientific: Singapore, 1985.
18. Chentsov, N.N. *Statistical decision rules and optimal inferences*; American Mathematical Society: Providence, RI, USA, 1982.
19. Wigner, E.P.; Yanase, M.M. Information contents of distributions. *Proc. Nat. Acad. Sci. USA* **1963**, *49*, 910–918.
20. Zanardi, P.; Zalka, C.; Faoro, L. Entangling power of quantum evolutions. *Phys. Rev. A* **2000**, *62*, 030301(R).
21. Nourmandipour, A.; Tavassoly, M.K.; Mancini, S. The entangling power of a "glocal" dissipative map. *Quantum Inf. Comput.* **2016**, *16*, 969–981.
22. Altafini, C. Tensor of coherences parameterization of multiqubit density operators for entanglement characterization. *Phys. Rev. A* **2004**, *69*, 012311.

entropy

MDPI

Article

Collective Motion of Repulsive Brownian Particles in Single-File Diffusion with and without Overtaking

Takeshi Ooshida [1,*], Susumu Goto [2] and Michio Otsuki [2]

[1] Department of Mechanical and Physical Engineering, Tottori University, Tottori 680-8552, Japan
[2] Graduate School of Engineering Science, Osaka University, Toyonaka, Osaka 560-8531, Japan; goto@me.es.osaka-u.ac.jp (S.G.); otsuki@me.es.osaka-u.ac.jp (M.O.)
* Correspondence: ooshida@damp.tottori-u.ac.jp

Received: 7 June 2018; Accepted: 16 July 2018; Published: 2 August 2018

Abstract: Subdiffusion is commonly observed in liquids with high density or in restricted geometries, as the particles are constantly pushed back by their neighbors. Since this "cage effect" emerges from many-body dynamics involving spatiotemporally correlated motions, the slow diffusion should be understood not simply as a one-body problem but as a part of collective dynamics, described in terms of space–time correlations. Such collective dynamics are illustrated here by calculations of the two-particle displacement correlation in a system of repulsive Brownian particles confined in a (quasi-)one-dimensional channel, whose subdiffusive behavior is known as the single-file diffusion (SFD). The analytical calculation is formulated in terms of the Lagrangian correlation of density fluctuations. In addition, numerical solutions to the Langevin equation with large but finite interaction potential are studied to clarify the effect of overtaking. In the limiting case of the ideal SFD without overtaking, correlated motion with a diffusively growing length scale is observed. By allowing the particles to overtake each other, the short-range correlation is destroyed, but the long-range weak correlation remains almost intact. These results describe nested space–time structure of cages, whereby smaller cages are enclosed in larger cages with longer lifetimes.

Keywords: caged dynamics; stochastic processes; collective motion; single-file diffusion; normal and anomalous diffusion; displacement correlation; overtaking; hopping rate; label variable; Dean–Kawasaki equation

1. Introduction

Particles in dense liquids are hindered from free motion, being constantly pushed back by their neighbors. This is often described as a "cage" that confines each particle. The cage effect makes the motion subdiffusive and, in certain cases, leads to the glass transition [1,2].

To be specific, let us consider a system consisting of Brownian particles with a nearly hardcore interaction. The position vector of the i-th particle, $\mathbf{r}_i = \mathbf{r}_i(t)$, is governed by the Langevin equation

$$m\ddot{\mathbf{r}}_i = -\mu\dot{\mathbf{r}}_i - \frac{\partial U}{\partial \mathbf{r}_i} + \mu\mathbf{f}_i(t), \tag{1a}$$

with m and μ denoting the mass and the drag coefficient of the particle, $\mu\mathbf{f}_i(t)$ representing the thermal fluctuating force, and the interaction being prescribed as

$$U = U(\mathbf{r}_1, \mathbf{r}_2, \ldots) = \sum_{(j,k)} V_{jk} \tag{1b}$$

in terms of the pair potential V_{jk}. Among the fundamental statistical quantities characterizing this system is the mean square displacement (MSD), i.e., the second moment of the displacement

$\mathbf{R}_i = \mathbf{r}_i(t) - \mathbf{r}_i(0)$. If the interaction through U is negligible, each particle diffuses freely so that $\langle \mathbf{R}_i^2 \rangle$ grows in proportion to t (for timescales longer than m/μ). This occurs when the colloidal fluid modeled by Equation (1) is dilute enough. In contrast, particles in a denser colloidal fluid are hindered from free motion by the cage effect, so that the growth of $\langle \mathbf{R}_i^2 \rangle$ is much slower. As is illustrated in Figure 1a, every particle in such a system is almost arrested in a "cage" consisting of its neighbors. In extreme cases, the system ceases to be fluid and becomes a kind of amorphous solid, referred to as colloidal glass [1].

Although it is true that the cage effect suppresses the growth of MSD on the whole, the details are rather complicated [2]. The behavior of the MSD in dense liquids reflects at least three aspects of caged dynamics: nearly free motion within the cage for a short time, possible drift of the cage enclosing the particle at a longer timescale, and hopping of the particle out of the cage as a rare event. Proper characterization of these processes requires space–time description, typically in terms of some four-point space–time correlation [2–4], as the cage effect actually emerges from many-body dynamics involving collective motions of numerous particles correlated both spatially and temporally.

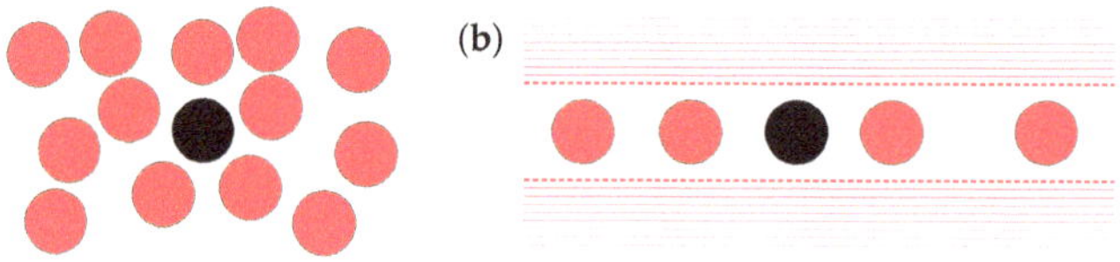

Figure 1. (**a**) Schematic description of a cage in a dense liquid, consisting of the surrounding particles that hinder free motion of the enclosed particle. (**b**) A (quasi-)one-dimensional model of the cage effect, with the particles confined in a narrow channel.

In search of insight into the theoretical treatment of such collective motions, here, we take note of the one-dimensional (1D) system illustrated in Figure 1b, following several authors who studied it as a simplified model of the cage effect [4–10]. The slow dynamics of such a 1D system are known by the name of single-file diffusion (SFD). In what we call the ideal SFD, every particle is eternally trapped within the "cage" formed by their neighbors. The MSD in the ideal SFD is known to grow subdiffusively as $\langle R_i^2 \rangle \propto \sqrt{t}$ [11–14] (for 1D systems, we write R_i instead of $\mathbf{R}_i$). The subdiffusion in SFD emerges from collective motion of particles [5,10,15] and is also detected as a negative longtime tail in the velocity autocorrelation [16,17], indicating that the particle is pushed back by its neighbors. The importance of the collective motion is understood by considering the origin of the effective stochastic equation for a single particle in SFD: the one-body equation (yielding the negative velocity autocorrelation) is actually based on the collective dynamics described in terms of the fluctuating density field [18].

Focusing on the collective motions in SFD, the group of the present authors has noticed the usefulness of the *displacement correlation* $\langle R_i R_j \rangle$ [4,10,19]. It is a kind of four-point space–time correlation that probes both the time scale and length scale; the definition of the displacement includes t, while the spatial scale is included as the mean distance between the two particles (i, j). In the ideal SFD in which the particles are forbidden to overtake each other, the displacement correlation has been calculated both analytically and numerically [4,10,20]. The calculated displacement correlation revealed collective motions behind the slow diffusion in SFD, in contrast to free diffusion in which $\langle R_i R_j \rangle$ vanishes (unless $i = j$). The formalism for analytical calculation of the displacement correlation can be extended to the case of two-dimensional (2D) colloidal liquids [10,21], which reproduces some numerical findings in 2D systems, such as vortical cooperative motion, with negative velocity autocorrelation being a manifestation of the cage effect. One of the delicate points in this extension is that the cage effect in 2D liquids cannot be infinitely strong, in the sense that eventually the particles can escape from the 2D cage. The escape from the cage is an important process, which requires further investigation.

Methodologically, it should be noted that one of the most powerful approaches to SFD, employing the relation between the position of the tagged particle and the density fluctuation [10,12,17,22], has been formulated in reliance on the assumption that overtaking is completely forbidden. To extend this formalism to the case of "non-ideal SFD"—in which overtaking is allowed—is a challenging problem, which is the main objective of the present work. Most of the existing works on SFD with overtaking have reported numerical simulations [23–27], while analytical results are quite rare. In the exceptional case of lattice SFD, the method of vacancy dynamics [15,16] has been applied to quasi-1D geometries allowing some kind of overtaking [28,29]. The analysis of lattice SFD with overtaking, however, is not readily extensible to the cases with a continuous space coordinate.

In the present work, we discuss how the analytical results on the displacement correlation in SFD [4,10,20] are modified, if the particles are allowed to overtake each other and thereby escape from the quasi-1D cage as a rare event. The probability of the escape is regulated by the height of the potential barrier, denoted by $V_{\max}$, so that $V_{\max} \to +\infty$ and $V_{\max} \to 0$ correspond to the ideal SFD and the free diffusion, respectively. Some numerical solutions for finite $V_{\max}$ were included in our previous work [19], but analytical calculations were limited to the ideal case without overtaking, for the very reason that overtaking was difficult to take into account in the theoretical framework based on the density fluctuation. Now, the effect of a non-zero overtaking rate on the displacement correlation will be shown analytically as a main result of the present work.

Logical presentation of the main results in Section 4 requires a considerable amount of review in preparatory sections. For this reason, the paper is organized as follows: In Section 2, after the governing equation of the 1D system is specified and the collective motion is illustrated in a space–time diagram, we define some basic concepts and variables, such as the displacement correlation, overtaking, the fluctuating density field $\rho(x,t)$, and the label variable ξ, clarifying their background. In particular, the kinematics of overtaking are discussed in Subsection 2.5. The usage of the label variable is a keystone for the analytical calculation of the displacement correlation [4,10,30], as is reviewed in Section 3 in the case of SFD without overtaking. The displacement correlation in this case is expressed in terms of a similarity variable, implying a nested space–time structure of cages. Subsequently, we proceed to the main topic in Section 4, in which we incorporate the effect of overtaking into the calculation of the displacement correlation by considering the dynamics of overtaking in terms of the $\Xi_i(t)$ prepared in Subsection 2.5. It is shown analytically and confirmed numerically that the infrequent overtaking events destroy the short-range correlation, while the long-range weak correlation remains almost intact. The final section is allotted for discussion and concluding remarks.

2. Formulation and Background

2.1. Specification of the System

We consider a 1D system of Brownian particles with short-range repulsive interaction, confined in a narrow channel, as is depicted in Figure 1b. With the position of the i-th particle denoted by $X_i = X_i(t)$, the system is governed by the 1D Langevin equation:

$$m\ddot{X}_i = -\mu\dot{X}_i - \frac{\partial}{\partial X_i}\sum_{j<k} V(X_k - X_j) + \mu f_i(t), \tag{2}$$

where m and μ represent the mass and the drag coefficient of the particle, respectively. The system contains N particles within the length L. Posing the periodic boundary condition, $X_{i+N} = X_i + L$, we consider the limit of $N \to \infty$ with the density $\rho_0 \stackrel{\text{def}}{=} N/L$ kept constant.

The thermal fluctuating force, $\mu f_i(t)$, is characterized by the variance,

$$\langle f_i(t) f_j(t') \rangle = \frac{2k_{\mathrm{B}}T}{\mu}\,\delta_{ij}\delta(t-t'), \tag{3}$$

with T denoting the temperature of the medium. The whole system is assumed to be at thermal equilibrium, which implies spacial homogeneity and temporal steadiness.

The interaction between the particles is expressed by the pairwise potential, $V(r)$. We could choose any family of $V(r)$ that interpolates between the limiting case of $V = 0$ and the opposite limit of the hardcore potential,

$$V_{\mathrm{HC}}(r) = \begin{cases} \infty & (|r| \le \sigma) \\ 0 & (|r| > \sigma) \end{cases} \tag{4}$$

with the diameter σ. Here, we choose

$$V(r) = \begin{cases} V_{\max}\left(1 - \dfrac{|r|}{\sigma}\right)^2 & (|r| \le \sigma) \\ 0 & (|r| > \sigma) \end{cases} \tag{5}$$

which is parametrized by the barrier height $V_{\max}$. We also tested some other potentials [19], only to find that the basic behavior of the 1D system is qualitatively unaffected by different choices of $V(r)$. Preference was given to Equation (5) merely because its hard sphere limit ($V_{\max} \gg k_{\mathrm{B}}T$) has been studied systematically [31] in the context of 3D glassy dynamics.

In regard to the system governed by Equation (2), we refer to the case of $V_{\max} \to +\infty$ as the *ideal SFD*, in which every particle is eternally caged by its neighbors. Large but finite values of $V_{\max}$ allow the particle to exchange positions with one of its neighbors as a rare event which we call *overtaking* (borrowing the word from traffic flow). The ideal SFD means SFD without overtaking, and we may say "non-ideal SFD" referring to the case of finite $V_{\max}/k_{\mathrm{B}}T$. Note that the description of non-ideal SFD with the 1D equation (2) can be interpreted as modeling a quasi-1D system [19,23–27] in which, typically, the position vector $\mathbf{r}_i = (X_i, Y_i)$ is governed by Equation (1a) with the potential term

$$U = U(\mathbf{r}_1, \mathbf{r}_2, \ldots) = \sum_{(j,k)} V_{jk} + \sum_j V_{\mathrm{ex}}(Y_j), \tag{6}$$

where $V_{\mathrm{ex}} = V_{\mathrm{ex}}(y)$ denotes the external confinement potential such that $V_{\mathrm{ex}}(\pm\infty) \to +\infty$. In this description, $V(X_k - X_j)$ in Equation (2) represents the free energy of the subsystem consisting of the neighboring particles j and k.

The specification of the system by Equations (2), (3) and (5), supplemented with the periodic boundary condition, involves some dimensional constants. As the basic scales of the length and the time, we take the particle diameter (σ) and the corresponding diffusive time (σ^2/D), where $D = k_{\mathrm{B}}T/\mu$ is the diffusion constant of a free Brownian particle. A finite value of mass, such that $m/\mu : \sigma^2/D = 1:1$, is specified for computational ease, unless specified otherwise. The system size, L, must be infinitely large; though, in numerical computations, we must specify some finite values for it. For later convenience, we introduce $\ell_0 \stackrel{\mathrm{def}}{=} L/N = 1/\rho_0$ which has the dimension of length. The nondimensional barrier height, $V_{\max}/k_{\mathrm{B}}T$, has an effect on the dynamics through the overtaking frequency, as will be discussed later.

In numerical simulations, the system is equilibrated by a preparatory run started at $t = -\mathcal{T}_{\mathrm{w}}$. Subsequently, for a reason clarified in the next subsection, the particles are renumbered consecutively in the sense that

$$X_0 < X_1 < \cdots < X_i < X_{i+1} < \cdots < X_N \; (= X_0 + L) \tag{7}$$

at $t = 0$. It should be noted that $\mathcal{T}_{\mathrm{w}}$ must be longer than $\max t$ for sufficient equilibration [4].

2.2. Spatiotemporally Correlated Motion in SFD

As a graphic depiction of collective motions in SFD, let us examine Figure 2, in which a numerical solution of Equation (2) in the case of the ideal SFD is represented as worldlines in the (x, t)-plane. To visualize the correlation of the worldlines, we measured the displacement for each particle i,

$$R_i = R_i(t) \stackrel{\text{def}}{=} X_i(t) - X_i(0), \tag{8}$$

for the time interval from 0 to $t = 2^n \times 10\sigma^2/D$ (with $n = 1, 2, \ldots$), in accordance with Ref. [4]. If $R_i(t) > 5\sigma$, the position of the particle is marked with a filled circle (•); if $R_i(t) < -5\sigma$, it is marked with an open square (□). As the time difference (t) increases, a string of the same kind of symbol is formed, expressing a cluster of particles moving together in the same direction.

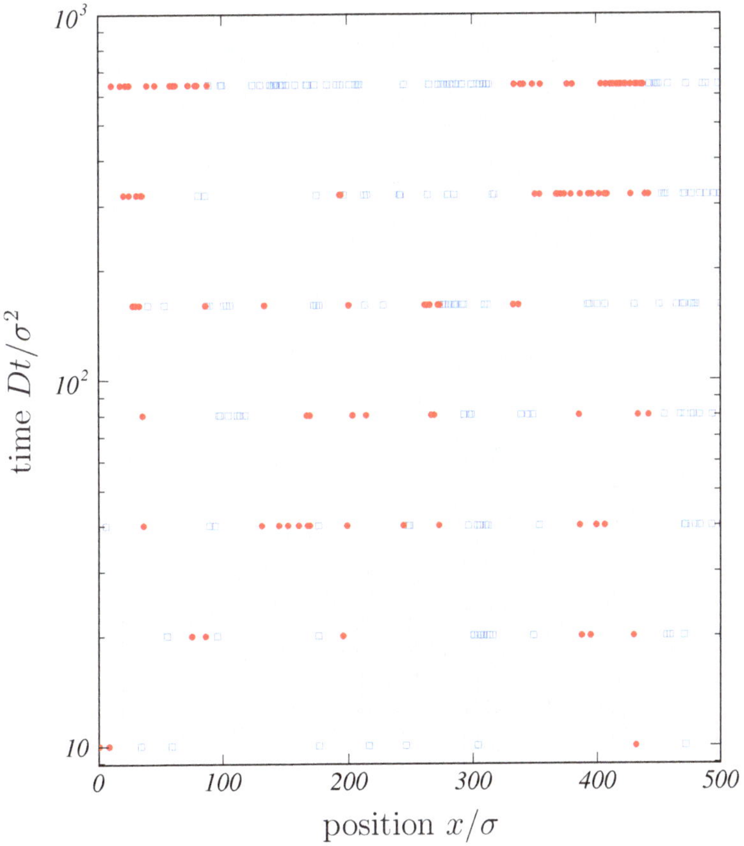

Figure 2. A space–time diagram representing cooperative motion in SFD. A numerical solution to Equation (2), calculated for $\rho_0 = N/L = 0.20\,\sigma^{-1}$, is plotted as worldlines in the (x, t)-plane (note that the t-axis is on a logarithmic scale). The symbols • and □ mark particles displaced (by more than 5σ) rightward and leftward, respectively.

The formation of clusters, visually shown in Figure 2, is quantified by calculating the displacement correlation $\langle R_i R_j \rangle$. Note that the consecutive numbering in Equation (7) is needed to make $\langle R_i R_j \rangle$ meaningful as a function of $j - i$ ($= \Delta$) and t. Since R_i and R_j have the same sign within the same cluster, their product must be positive if the distance in the numbering, $\Delta = j - i$, is small, while $\langle R_i R_j \rangle$ for distant particles (with $\ell_0 \Delta$ greater than the correlation length) is expected to vanish. The average, denoted by $\langle\ \rangle$, is taken over the initial condition and the Langevin noise. Computationally, the displacement correlation is calculated as

$$\langle R_i R_j \rangle = \langle R_i R_{i+\Delta} \rangle = \frac{1}{N} \sum_i \langle R_i R_{i+\Delta} \rangle . \tag{9}$$

2.3. Continuum Description

Our theoretical approach to the displacement correlation, $\langle R_i R_j \rangle$, is based on a continuum description of the dynamics of Brownian particles. As fluctuating hydrodynamic fields describing the temporally coarse-grained dynamics of $\{X_i\}_{i=1,2,...}$ for timescales longer than m/μ, one may take the density fields

$$\rho(x,t) = \sum_i \rho_i(x,t), \quad \rho_i(x,t) = \delta(x - X_i(t)), \tag{10}$$

and their fluxes

$$Q(x,t) = \sum_i Q_i(x,t), \quad Q_i(x,t) = \rho_i(x,t)\dot{X}_i(t). \tag{11}$$

Note that the delta function in Equation (10) should be regarded as a blunted one, as a result of the coarse-graining (see §II-B in Ref. [21] and references therein).

The density field, $\rho(x,t)$, is governed by the Dean–Kawasaki equation [32–37], which can be presented as a set of equations of the following form:

$$\partial_t \rho + \partial_x Q = 0, \tag{12a}$$

$$Q = -D\left(\partial_x \rho + \frac{\rho}{k_B T}\partial_x U\right) + \sum_i \rho_i(x,t) f_i(t), \tag{12b}$$

$$U = U[\rho](x) = \int V_{\text{eff}}(x - x')\rho(x')\mathrm{d}x'. \tag{12c}$$

The effective potential, V_{eff} in Equation (12c), is determined by the condition that the density fluctuation, described by Equations (12), should be consistent with the static structure factor,

$$S(k) \overset{\text{def}}{=} \frac{1}{N} \sum_i \sum_j \langle \exp\left[\mathrm{i}k\left(X_j - X_i\right)\right] \rangle \quad (k \neq 0), \tag{13}$$

determined directly from Equation (2). More specifically, with the Fourier mode of the density and its correlation defined as

$$\hat{\rho}(k,t) \overset{\text{def}}{=} \int e^{\mathrm{i}kx}\rho(x,t)\mathrm{d}x = \sum_j \exp\left[\mathrm{i}kX_j(t)\right], \quad F(k,t) \overset{\text{def}}{=} \frac{1}{N}\langle \hat{\rho}(k,t)\hat{\rho}(-k,0)\rangle \quad (k \neq 0),$$

evidently, $F(k, t = 0)$ must be equal to $S(k)$, and the initial decay of $F(k,t)$ is known to be exponential, as is shown in Ref. [38] as Equation (4.144). In our notation, it reads

$$F(k,t) = S(k)e^{-D^{\mathrm{c}}k^2 t} \qquad (\text{for } t \ll \sigma^2/D), \tag{14}$$

with $D^{\mathrm{c}} = D^{\mathrm{c}}(k) = D/S(k)$ referred to as the (short-time) collective diffusion coefficient [38,39]. Once V_{eff} is determined so as to make the linearized dynamics of Equations (12) consistent with Equation (14), we may redefine $S(k)$ by the ratio of D to $D^{\mathrm{c}}(k)$.

2.4. Label Variable

Now let us outline the main idea that makes it possible to calculate $\langle R_i R_j \rangle$ analytically on the ground of Equations (12) [4,10,19,30]. The point is to introduce a new variable, ξ, referred to as the *label variable*, to incorporate the notion of particle tracking into the mathematical formalism that links $\langle R_i R_j \rangle$ with Equations (12).

The necessity of the new variable for particle tracking is understood by noticing how a more straightforward approach, based on $\rho_i(x,t)$ in Equation (10), is confronted by a difficulty. In principle, $\langle R_i R_j \rangle$ can be obtained from the Fourier mode of $\rho_i(x,t)$, because the definition

$$\hat{\rho}_i(k,t) \stackrel{\text{def}}{=} \int e^{ikx} \rho_i(x,t) \mathrm{d}x = \exp\left[ikX_i(t)\right] \tag{15}$$

implies $\hat{\rho}_i(k,t)\hat{\rho}_i(-k,0) = \exp(ikR_i)$, and therefore

$$\langle \hat{\rho}_i(k,t)\hat{\rho}_i(-k,0)\hat{\rho}_j(\pm k,t)\hat{\rho}_j(\mp k,0) \rangle = \left\langle e^{ik(R_i \pm R_j)} \right\rangle = 1 - \frac{k^2}{2}\left\langle (R_i \pm R_j)^2 \right\rangle + \cdots. \tag{16}$$

If the correlation on the left side is successfully evaluated by nonlinear analysis of the equation governing ρ_i and ρ_j, analogous to Equation (12b) and later shown as Equation (48), then $\langle R_i R_j \rangle$ can be obtained from the power series on the right side. This is insurmountable, unfortunately, as is evident from the complication encountered in the apparently easier problem of evaluating $\langle \hat{\rho}_i(k,t)\hat{\rho}_i(-k,0) \rangle$ in SFD [8,9].

The difficulty originates from the choice of the standard field representation with the independent variables (x,t), referred to as the Eulerian description (according to the terminology of fluid mechanics [40,41]). Since $\langle R_i R_j \rangle$ is treated as a kind of *four-point* space–time correlation [42,43], it implies a *four-body* correlation in the Eulerian description, as is seen in Equation (16). However, this difficulty can be avoided by switching to another way referred to as the Lagrangian description [30,40,41], which means to include particle tracking mechanism in the definition of the fields and their correlations. In this way, the displacement correlation can be treated simply as a *two-body* Lagrangian correlation of the field [4,10,19].

Deferring a concrete calculation of $\langle R_i R_j \rangle$ until the next section, here we only define the label variable ξ to lay the foundation for it. Instead of (x,t) for the standard space–time coordinate system, we introduce a stretchable coordinate system (ξ,t), requiring $\xi = \xi(x,t)$ to satisfy the convective equation,

$$(\rho\partial_t + Q\partial_x)\xi = 0, \tag{17}$$

which states that ξ should be convected with the velocity $u = Q/\rho$. To satisfy Equation (17), we *define* ξ as a solution to

$$(\rho, Q) = (\partial_x \xi, \ -\partial_t \xi). \tag{18}$$

This is solvable because of the continuity equation (12a), with the solution determined uniquely by some initial condition, such as $\xi(X_0(t_0), t_0) = 0$. Subsequently, by inverting the mapping $(x,t) \mapsto \xi$, we obtain the coordinate system with the independent variables (ξ,t) [4,10,19,21,30]. It should be emphasized that we take Equation (18), not Equation (17), as the definition of the mapping between ξ and x. In other words, we do not *solve* Equation (17) in the usual sense of the word; rather, Equation (17) is satisfied as a consequence of Equation (18). In this way, we avoid complication of the attempt to define $\xi = \xi(x,t)$ directly with Equation (17), which would require specification of the initial condition We also remind the readers that the delta function in the definition of ρ is a blunted one, as has been noted immediately after Equations (10) and (11).

Using the label variable ξ, defined in this way, we can calculate $\langle R_i R_j \rangle$ analytically. The (ξ,t) coordinate system has an advantage of making it easy to trace the worldlines, such as the ones plotted in Figure 2, because ξ is expected to keep the same value if we follow the identical particle. In order to see it, we define

$$\Xi_i(t) \stackrel{\text{def}}{=} \xi(X_i(t),t) \tag{19}$$

as a function of the particle number i and the time t, and consider its t-derivative [30,44]. Provided that the conventional chain rule is valid, the time-derivative of $\Xi_i(t)$ is

$$\frac{\mathrm{d}\Xi_i(t)}{\mathrm{d}t} = \dot{X}_i(t)\left.\frac{\partial\xi}{\partial x}\right|_{x=X_i} + \left.\frac{\partial\xi}{\partial t}\right|_{x=X_i} = \left(\rho\dot{X}_i - Q\right)\big|_{x=X_i}, \tag{20}$$

where the definition of ξ in Equation (18) is taken into account. The expression on the right side of Equation (20) vanishes unless the i-th particle overlaps with another. In the ideal SFD, in which the particles never overlap, $\Xi_i(t)$ is none other than the numbering i; this is the key ingredient for the analytical calculation of $\langle R_i R_j \rangle$ in the ideal SFD.

In the absence of overtaking, the particles can move only as a result of changes in inter-particle spaces, which is illustrated schematically in Figure 3a as a migrating "vacancy" causing correlated motion of particles. The overtaking allows another kind of motion, illustrated in Figure 3b, which does not require migration of a vacancy. Before proceeding to a concrete calculation of $\langle R_i R_j \rangle$, let us discuss how to describe an overtaking event within the framework of the (ξ, t) coordinate system.

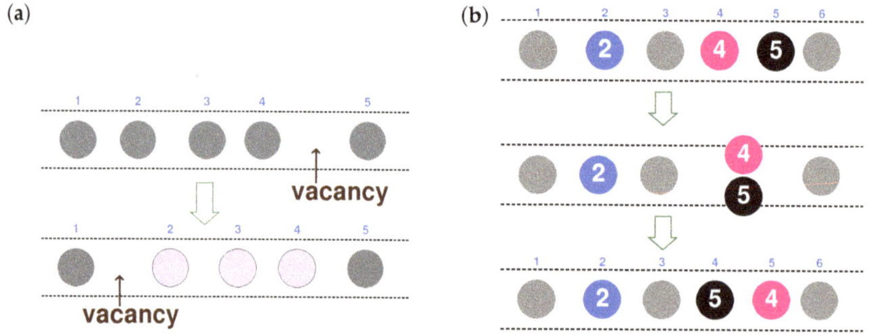

Figure 3. Schematic illustration of two kinds of processes in the 1D system under consideration. The small digits represent the values of the label variable ξ, while the larger digits are numbers to identify the particles. (**a**) Fluctuation of the inter-particle space without overtaking, interpretable as the migration of a vacancy. (**b**) An overtaking event, in which two particles exchange their labels.

2.5. *Kinematics of Overtaking Events*

In the presence of overtaking, $\Xi_i(t) = \xi(X_i(t), t)$ is not necessarily equal to i. In this case, we must distinguish between them and discuss correspondence among three variables: the numbering ($i \in \mathbb{Z}$), the label variable ($\xi \in \mathbb{R}$), and the position (x).

We still define ξ by Equation (18) without regard to overtaking. The constant of integration is chosen appropriately so that $\Xi_i(t) = \xi(X_i(t), t)$ has an integer value unless the i-th particle overlaps with another. On this premise, the mapping from $i \in \mathbb{Z}$ to $\Xi_i(t) = \xi(X_i(t), t) \in \mathbb{Z}$ is injective, which means, so to speak, a kind of exclusion principle whereby different particles must carry different values of ξ.

The overtaking process is a transition from one such mapping to another. Since the effect of overtaking on the mapping $i \mapsto \Xi_i(t)$ is local, as is readily seen from Equation (18), we may identify the overtaking event with a process in which two particles exchange the value of their label. In the case of a pair of particles (i, j), with t_1 and t_2 denoting points in time immediately before and after the exchange, this process is described as

$$\Xi_i(t_2) = \Xi_j(t_1), \quad \Xi_j(t_2) = \Xi_i(t_1). \tag{21}$$

To justify Equation (21), we consider what occurs in the time interval from t_1 to t_2. Since the particles may overlap in the course of overtaking, the value of $\Xi_i(t)$ is not limited to $\mathbb{Z}$ but rather belongs to $\mathbb{R}$. Using ρ and ρ_i given in Equation (10), we rewrite Equation (20) as

$$\frac{\mathrm{d}\Xi_i}{\mathrm{d}t} = \int (\rho \dot{X}_i - Q)\, \rho_i \mathrm{d}x = \sum_j \int (\rho_j \dot{X}_i - Q_j)\, \rho_i \mathrm{d}x \\ = \sum_j \int (\rho_j Q_i - \rho_i Q_j)\, \mathrm{d}x, \tag{22}$$

which implies that $\Xi_i(t)$ is locally conserved. Note that only overlapping particles contribute to the integral. If no particle overlaps the i-th one, then the expression on the right side of Equation (20) vanishes so that Ξ_i remains constant. If only one particle, say the j-th one, overlaps particle i, Equation (22) gives

$$\frac{\mathrm{d}\Xi_i}{\mathrm{d}t} = \int (\rho_j Q_i - \rho_i Q_j)\, \mathrm{d}x = -\frac{\mathrm{d}\Xi_j}{\mathrm{d}t}. \tag{23}$$

By integrating Equation (23) over the time interval from t_1 to t_2 and taking the conditions such as $\Xi_i(t_1) \in \mathbb{Z}$ into account, we find

$$\Xi_i(t_2) - \Xi_i(t_1) = -\Xi_j(t_2) + \Xi_j(t_1) \in \mathbb{Z},$$

so that only two cases are possible: one is the case of unsuccessful or retracted overtaking, with the same value of (Ξ_i, Ξ_j) restored after the interaction, and the other case corresponds to Equation (21) that describes (successful) overtaking. Note that the other possibilities are eliminated by the "exclusion principle" for the configurations before and after the overtaking process. As a rare case, three-body or four-body interactions might be possible, but it suffices to approximate such a case with a sequence of two-body exchange processes.

Changes of $\Xi_i(t) = \xi(X_i(t), t)$ and $X_i(t) = x(\Xi_i(t), t)$ are exemplified in Figure 3. Since $\xi = \xi(x, t)$ satisfies $\partial\xi/\partial x = \rho$ by definition in Equation (18), $\{\Xi_i\}$ is always spatially consecutive, as is shown in Figure 3 with small digits. The important point is that this spatial consecutiveness of $\{\Xi_i\}$ holds true even if the numbering is inverted by overtaking. In other words, ξ labels the order in the file and not the particles themselves. In the case of Figure 3b, the particles are initially numbered consecutively so that $\Xi_i = i$ for all i, until particles 4 and 5 start to overlap. During overtaking, the values of Ξ_4 and Ξ_5 evolve according to Equation (23) with $(i, j) = (4, 5)$, while the labels for other particles, such as Ξ_2, Ξ_3 and Ξ_6, remain unchanged. Finally, after overtaking, particle 4 carries the label $\Xi_4 = 5$, while label 4 is now carried by particle 5 so that $\Xi_5 = 4$.

In the next section, we start with the case of the ideal SFD, in which the requirement of $(\mathrm{d}/\mathrm{d}t)\Xi_i(t) = 0$ is fulfilled so that ξ simply plays the role of particle numbering. Later, we consider the temporal change of $\Xi_i(t)$ to allow for overtaking in Section 4.

3. Displacement Correlation in SFD without Overtaking

3.1. Analytical Calculation of Displacement Correlation

Here we review the analytical calculation of $\langle R_i R_j \rangle$ in the ideal SFD [4,10,19], in which Ξ_i is independent of t.

Out of the three equations composing the Dean–Kawasaki equation (12), the continuity equation (12a) is already included in $\xi = \xi(x, t)$, or its inverse mapping $x = x(\xi, t)$, through Equation (18). To relate $x = x(\xi, t)$ to the remaining two equations, we solve Equation (12b) for $u = Q/\rho$, which should be equal to $u = \partial_t x(\xi, t)$. Noticing that $u(\xi, t)$ is the (negative) flux of $1/\rho = \partial_\xi x(\xi, t)$, we introduce [30]

$$\psi = \psi(\xi, t) \stackrel{\mathrm{def}}{=} \rho_0 \frac{\partial x}{\partial \xi} - 1 \tag{24}$$

so that $1/\rho = \ell_0(1+\psi)$. The fluctuating field ψ can be interpreted as a continuum representation of migrating vacancies [16] or elongation of a chain [45]. As is illustrated schematically in Figure 3a, migration of a vacancy can give rise to correlated motion of particles.

Once the dynamics of ψ are known from Equations (12), the remaining task for the calculation of $\langle R_i R_j \rangle$ is only a problem of kinematics—given the field ψ, how can we find the displacement? This is readily solved in a Fourier representation,

$$\psi(\xi,t) = \sum_k \check{\psi}(k,t)e^{-\mathrm{i}k\xi}, \quad \check{\psi}(k,t) = \int e^{\mathrm{i}k\xi}\psi(\xi,t)\frac{\mathrm{d}\xi}{N} \qquad \left(\text{with } \frac{k}{2\pi/N} \in \mathbb{Z}\right), \tag{25}$$

allowing us to express $x = x(\xi,t)$ as an antiderivative of ψ:

$$x = x(\xi,t) = \ell_0\xi + \ell_0 \sum_k e^{-\mathrm{i}k\xi}\frac{\check{\psi}(k,t)}{-\mathrm{i}k} + X_{\mathrm{G}}(t), \tag{26}$$

where $X_{\mathrm{G}}(t)$ corresponds to the center-of-mass motion which should be negligible in the limit of $N \to +\infty$ [10]. The positions of the i-th or j-th particle are then obtained by substituting Ξ_i or Ξ_j into ξ in Equation (26), which readily yields

$$R_i = x(\Xi_i,t) - x(\Xi_i,0) = \ell_0 \sum_k e^{-\mathrm{i}k\Xi_i}\frac{\check{\psi}(k,t)-\check{\psi}(k,0)}{-\mathrm{i}k} \tag{27}$$

in the absence of overtaking (i.e., $\mathrm{d}\Xi_i/\mathrm{d}t = 0$). To calculate $\langle R_i R_j \rangle$, we multiply Equation (27) by its duplicate with (i,k) replaced by $(j,-k)$. Subsequently, taking it into account that $\langle \check{\psi}(k,t)\check{\psi}(k',t') \rangle$ generally vanishes unless $k+k'=0$ (due to the space-translation symmetry of the system), we find

$$\begin{aligned}\langle R_i R_j \rangle &= \ell_0^2 \sum_k \frac{e^{-\mathrm{i}k(\Xi_j-\Xi_i)}\left\langle [\check{\psi}(k,t)-\check{\psi}(k,0)]\,[\check{\psi}(-k,t)-\check{\psi}(-k,0)] \right\rangle}{k^2} \\ &\to \frac{\ell_0^2}{\pi}\int_{-\infty}^{\infty} e^{-\mathrm{i}k\Delta}\frac{C_\psi(k,0)-C_\psi(k,t)}{k^2}\mathrm{d}k \qquad (N\to\infty),\end{aligned} \tag{28}$$

where we have defined

$$C_\psi(k,t) \stackrel{\text{def}}{=} N\left\langle \check{\psi}(k,t)\check{\psi}(-k,0) \right\rangle \tag{29}$$

and used $\Xi_j - \Xi_i = j - i = \Delta$. We refer to Equation (28) as the Alexander–Pincus formula [10,12], which relates the displacement correlation to C_ψ. Since C_ψ is a *two-body* correlation, it is much more tractable than the four-body correlation in Equation (16).

To allow concrete calculation of C_ψ, Equations (12b) and (12c) are rewritten as an equation for $\check{\psi}(k,t)$ in the following form:

$$\partial_t\check{\psi}(k,t) = -D_*^{\mathrm{c}}k^2\check{\psi}(k,t) + \sum_{p+q+k=0} \mathcal{V}_k^{pq}\check{\psi}(-p,t)\check{\psi}(-q,t) + O(\check{\psi}^3) + \rho_0\check{f}_{\mathrm{L}}(k,t). \tag{30}$$

The statistics of the random force term are specified as

$$\rho_0^2\left\langle \check{f}_{\mathrm{L}}(k,t)\check{f}_{\mathrm{L}}(-k',t') \right\rangle = \frac{2D_*}{N}k^2\delta_{kk'}\delta(t-t'), \tag{31}$$

where $D_* = D/\ell_0^2$. The coefficient of the linear term is

$$D_*^{\mathrm{c}} = \frac{D^{\mathrm{c}}}{\ell_0^2} = D_*\left(1 + \frac{2\sin\rho_0\sigma k}{k}\right), \tag{32}$$

which also gives $S = D/D^{c} = D_{*}/D_{*}^{c}$, and the coefficients $\mathcal{V}_{k}^{pq}$ in the nonlinear term are found in Refs. [4,30].

Within the linear approximation to Equation (30), C_{ψ} is readily calculated as

$$C_{\psi}(k,t) = Se^{-D_{*}^{c}k^{2}t}. \tag{33}$$

Although a nonlinear theory producing a correction term to be added to Equation (33) is also possible [4], here we ignore this correction, giving priority to the simpler expression in Equation (33).

The displacement correlation is obtained by substituting Equation (33) into the Alexander–Pincus Formula (28) and evaluating the integral. As the contribution from the longwave modes is dominant, $S = S(k)$ and $D_{*}^{c} = D_{*}/S(k)$ can be replaced by their limiting values for $k \to +0$. Thus, $\langle R_i R_j \rangle$ is obtained explicitly as a function of Δ and t, expressible in terms of a similarity variable

$$\theta \stackrel{\text{def}}{=} \frac{\ell_0 \Delta}{\lambda(t)} = \frac{\Delta}{2\sqrt{D_{*}^{c}t}}, \qquad \lambda(t) = 2\sqrt{D^{c}t}\,, \tag{34}$$

as [4,10,19]

$$\frac{\langle R_i R_j \rangle}{2S\ell_0^2\sqrt{(D_{*}^{c}/\pi)t}} = e^{-\theta^2} - \sqrt{\pi}\,|\theta|\,\mathrm{erfc}\,|\theta| \stackrel{\text{def}}{=} \varphi(\theta), \tag{35}$$

so that $\langle R_i R_j \rangle = K\sqrt{t}\,\varphi(\theta)$ where $K = 2S\ell_0^2\sqrt{D_{*}^{c}/\pi}$. The dynamical correlation length $\lambda(t)$ grows diffusively, which seems to be consistent with the observation of growing clusters in Figure 2.

Note that infrared and ultraviolet cutoffs are *not* necessary in Equation (28), as the integrand is regular around $k = 0$ and decays for $k \to \pm\infty$ (algebraically, but fast enough). This should not be confused with the infrared divergence of $x(\xi, t)$ in Equation (26) in the limit of $L \to \infty$.

3.2. Particle Simulation in the Absence of Overtaking

The theoretical prediction in Equation (35) is tested in Figure 4a, by plotting the values of the displacement correlation, $\langle R_i R_j \rangle$, obtained from numerical simulation of the ideal SFD. The computed values are plotted in terms of rescaled variables; according to Equation (35), a plot of $\langle R_i R_j \rangle/(K\sqrt{t})$ against the similarity variable θ should give a single master curve for all values of the time interval t. The ideal SFD was simulated by solving Equation (2) numerically with a very high barrier (we chose $V_{\max} = 50k_{\mathrm{B}}T$). The system size and the density were specified as $N = 10^4$ and $\rho_0 = N/L = 0.2\,\sigma^{-1}$ so that $L = 5N\sigma$.

The three kinds of symbols in Figure 4a correspond to three different values of t. The plots for all these values of t are seen to be reducible to a single master curve given by $\varphi(\,\cdot\,)$ in Equation (35), supporting the prediction of the analytical calculation.

To be precise, a small but finite discrepancy is found at $\theta = 0$ for $t = 10\sigma^2/D$. One may be tempted to explain this short-time discrepancy simply as indicating a lack of time to establish correlated motion, because a particle, on average, takes time on the order of $1/D_{*}^{c}$ to encounter its neighbors. Unfortunately, this argument appears too simple to explain the numerical result in Figure 4a in which a positive correlation for $\theta \neq 0$ (i.e., $|\Delta| \geq 1$) has already been established at $t = 10\sigma^2/D$. We note, on the other hand, that the discrepancy can be ascribed to the nonlinear term in Equation (30) which was ignored in the previous subsection. It is shown that inclusion of the nonlinear term gives a correction to Equation (35), which is significant only for $|\theta| \ll 1$ and $D_{*}^{c}t < 1$ [4], and the sign of the correction for the MSD is negative [4,10]. Thus, the short-distance correlations are affected by nonlinear coupling of $\check{\psi}$, while correlations over a long distance seem to be tractable with a linear theory.

It should be also noted that the theoretical predictions discussed here are based on the Dean–Kawasaki equation in which inertia is completely ignored, while the particle-based simulation is performed with finite m/μ. To check for consistency between the numerical simulation and the theoretical predictions, three cases with different values of m/μ are compared in Figure 4b. It is seen that the change in m/μ does not make any remarkable difference. This is in agreement with the general expectation that the Langevin dynamics on time scales longer than m/μ are basically independent of the inertia, because the momentum can be eliminated by temporal coarse-graining [46–48].

In regard to correlations over long distances, one might be tempted to suppose that the static structure factor, $S(k)$ in Equation (13), could be helpful in the detection of such long-ranged correlations. This point was discussed in Ref. [19], leading to the conclusion that the structure of the collective motion is not properly captured by the static structure factor. It is for this reason that we focused on $\langle R_i R_j \rangle$ rather than $S(k)$.

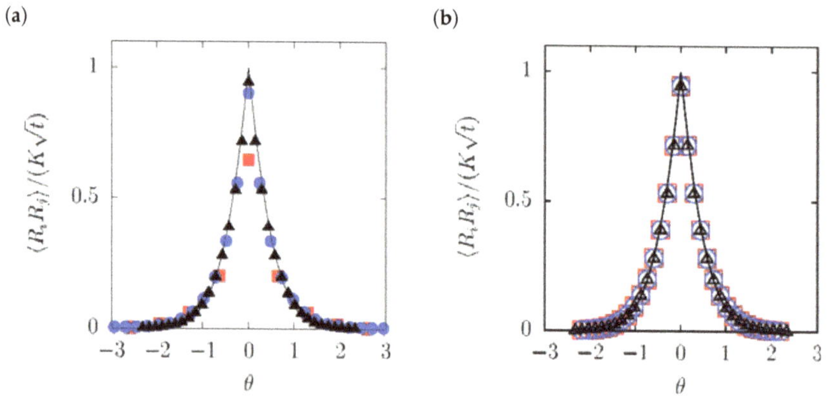

Figure 4. Comparison of the theoretical prediction (35) for the ideal single-file diffusion (SFD) with a particle-based simulation. (**a**) Displacement correlation for different values of the time interval t. The red squares (■), the blue circles (•) and the black triangles (▲) represent the numerical values of $\langle R_i R_j \rangle / (K\sqrt{t})$ at the time intervals $t/(\sigma^2/D) = 10$, 70 and 200, respectively. The ratio $m/\mu{:}\sigma^2/D$ was chosen to be 1:1. The values are plotted against $\theta = \ell_0 \Delta / \lambda(t)$, i.e., the distance rescaled with $\lambda(t)$. The thin, solid line shows the theoretical master curve given by $\varphi(\,\cdot\,)$ in Equation (35). (**b**) Displacement correlation for different values of m/μ. The open squares (□) represent the numerical result for $m/\mu{:}\sigma^2/D$ = 1:1, the open circles (○) for 1:2, and the open triangles (△) for 1:5. The time interval was chosen to be $t = 200\,\sigma^2/D$.

4. Effects of Overtaking on Displacement Correlation

Having reviewed the analytical calculation of the displacement correlation $\langle R_i R_j \rangle$, which leads to Equation (35) for the ideal SFD, let us now consider effects of overtaking that were ignored in the previous section. We start with numerical observation, noticing how the behavior of $\langle R_i R_j \rangle$ deviates from Equation (35) due to overtaking. This deviation is then compared with a modified theory in which overtaking is allowed for.

4.1. Particle Simulation of SFD with Overtaking

The governing equations of the system, namely Equations (2), (3) and (5), contain a parameter $V_{\max}$ representing the barrier height. This parameter, $V_{\max}$, regulates the frequency of overtaking, if the other parameters are kept unchanged.

Two extreme cases are already known theoretically: the case of $V_{\max} \to \infty$ implying the ideal SFD in which overtaking is completely forbidden, and $V_{\max} = 0$ corresponding to free diffusion in which overtaking is always allowed. In the ideal SFD, there is a positive correlation between displacements of two particles, as is shown in Equation (35), while in free diffusion, displacements of two particles are totally uncorrelated, as is easily seen by proving

$$\langle R_i R_j \rangle = \begin{cases} 2Dt + O(Dm/\mu) & (i = j) \\ 0 & (i \neq j) \end{cases} \tag{36}$$

for $V_{\text{max}} = 0$.

Between these two extreme cases, there are cases of finite values of V_{max}, allowing overtaking with some probability. Three such cases are shown in Figure 5, where the computed values of $\langle R_i R_j \rangle$ at $t = 200\sigma^2/D$ are plotted against $\Delta = j - i$. In the case of the lowest barrier, $V_{\text{max}} = k_B T$, the plot is similar to Equation (36) in that $\langle R_i R_j \rangle$ almost vanishes for $i \neq j$; instead, the MSD ($i = j$) is greater than in the other two cases, indicating that the particles are diffusing almost freely. The case of the highest barrier with $V_{\text{max}} = 5k_B T$ resembles the ideal SFD, although a close inspection reveals a slight deviation from Equation (35) as a result of overtaking that occurs at a very small rate.

The intermediate case with $V_{\text{max}} = 3k_B T$ is interesting. At large distances, the same correlation is observed in the case of $V_{\text{max}} = 3k_B T$ as in the case of $V_{\text{max}} = 5k_B T$ (and as in the ideal SFD). In contrast, at $\Delta = \pm 1$ and ± 2, the correlation in the case of $V_{\text{max}} = 3k_B T$ is remarkably smaller than that for $V_{\text{max}} = 5k_B T$. The decrease in the displacement correlation and the increase in MSD must be attributed to overtaking.

The numerical observation on the effects of a finite V_{max} may be understood intuitively, if $\langle R_i R_j \rangle$ is regarded as representing a nested structure of cages with different radii. From this point of view, the plot for $V_{\text{max}} = 3k_B T$ in Figure 5 can be interpreted as describing the breakdown of inner cages, while the outer ones persist (at least until $t = 200\sigma^2/D$). To elevate this pictorial idea to a quantitative theory on collective motion in SFD with overtaking, we raise the question: How can we modify Equation (35) allowing for overtaking?

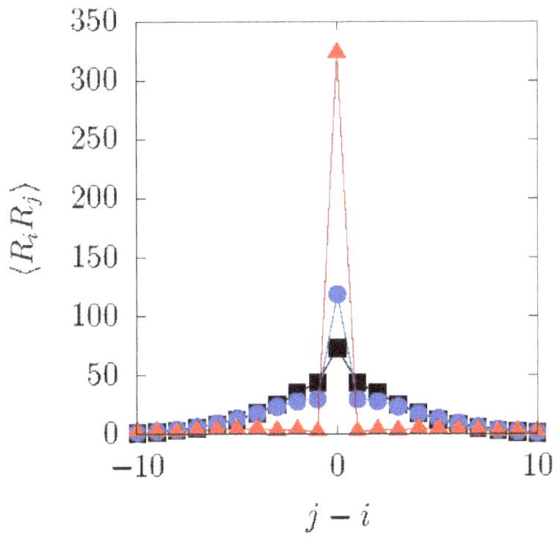

Figure 5. Two-particle displacement correlation, $\langle R_i R_j \rangle$ (nondimensionalized with σ^2), numerically obtained for three different values of V_{max}. The red triangles (▲), the blue circles (•) and the black squares (■) represent the data for $V_{\text{max}} = k_B T$, $3k_B T$ and $5k_B T$, respectively. The time interval is fixed at $t = 200\sigma^2/D$. The density and the system size are the same as in Figure 4.

4.2. Theory of Displacement Correlation in SFD with Overtaking

To find out how Equation (35) should be modified by overtaking, let us re-examine its derivation process. A crucial step is found in Equation (27) where R_i is obtained on the assumption that Ξ_i is independent of t. This is the point at which the "no overtaking" rule was enforced [30,44].

Overtaking is incorporated into the theory though temporal changes of $\Xi_i(t)$ [44]. The kinematics of overtaking were discussed in Subsection 2.5: as is illustrated in Figure 3, two particles exchange their labels according to Equation (21).

In order to recalculate the displacement correlation, now we need to consider the dynamics of overtaking. In principle, the stochastic dynamics of $\Xi_i(t)$ should be determined from "first principles" through Equation (22). As a crude approximation, however, here we adopt a phenomenological description characterized by the parameter ν_α, referred to as the *overtaking frequency* or the *hopping rate*.

The dynamics of $\Xi_i(t)$ is modeled as a simple Markovian process in which the labels of neighboring particles are exchanged at a rate of ν_α. More precisely speaking, for every pair (i,j) such that $\Xi_j(t_1) = \Xi_i(t_1) \pm 1$, their labels are exchanged according to Equation (21) with a probability of $1 - \nu_\alpha \Delta t$ in every short time interval, $\Delta t = t_2 - t_1$. This process is essentially equivalent to what is known as "Amida-kuji" (Amitabha's lottery) or the "ladder lottery" [49,50]. In regard to the dynamics of a single tagged particle, the Amida-kuji process is merely a diffusion on a 1D lattice [49,51] such that

$$[\Xi_i(t) - \Xi_i(0)]^2 = 2\nu_\alpha t. \tag{37}$$

This is readily shown by solving the master equation for the one-tag diffusion propagator [51],

$$\partial_t P(a,t) = \nu_\alpha \left[P(a+1,t) + P(a-1,t) - 2P(a,t)\right], \tag{38}$$

where $P(a,t) = P(0,0;a,t)$ denotes the probability that the tagged particle, the 0-th one such that $\Xi_0(0) = 0$, carries the label $\Xi_0(t) = a$ at the time t.

In order to calculate the contribution of overtaking to the displacement correlation $\langle R_i R_j \rangle$, we need to know the two-body diffusion propagator, $P(i,j,0;a,b,t)$, which represents the probability that the tagged particles, initially labelled with $\Xi_i(0) = i$ and $\Xi_j(0) = j$, are found to carry $\Xi_i(t) = a$ and $\Xi_j(t) = b$ at the time t. Assuming $j - i = \Delta > 0$ without loss of generality, we write the master equation for $P_{a,b} = P(i, i+\Delta, 0; a, b, t)$ with $a \neq b$ as

$$\partial_t P_{a,b} = \nu_\alpha \left(P_{a+1,b} + P_{a-1,b} + P_{a,b+1} + P_{a,b-1} - 4P_{a,b}\right) + (\delta_{a+1,b} + \delta_{a,b+1})\nu_\alpha \left(P_{b,a} + P_{a,b}\right) \tag{39}$$

and $P_{a,a} = 0$. This is more complicated than Equation (38) but still solvable in a Fourier representation (as is shown in Appendix A), so that two-body correlations of $\Xi_i(t)$ are given in terms of the modified Bessel function in the limit of $N \to \infty$.

Now, we are prepared for calculation of $\langle R_i R_j \rangle$. From Equation (26) and $R_i = x(\Xi_i, t) - x(\Xi_i, 0)$, we find

$$\begin{aligned} R_i &= \ell_0 \sum_{k \neq 0} \frac{e^{-ik\Xi_j(t)} \check{\psi}(k,t) - e^{-ik\Xi_j(0)} \check{\psi}(k,0)}{-ik} + \ell_0 \left[\Xi_i(t) - \Xi_i(0)\right] \\ &= \ell_0 \sum_{k \neq 0} \frac{e^{-ik\Xi_i^0}}{-ik} \left[e^{-ik\delta\Xi_i} \check{\psi}(k,t) - \check{\psi}(k,0)\right] + \ell_0 \delta\Xi_i, \end{aligned} \tag{40}$$

where we have defined

$$\Xi_i^0 \overset{\text{def}}{=} \Xi_i(0), \quad \delta\Xi_i = \delta\Xi_i(t) \overset{\text{def}}{=} \Xi_i(t) - \Xi_i^0$$

for the sake of brevity. Following the same line of argument as in the derivation of Equation (28), on the assumption that $\check{\psi}$ and Ξ_i are uncorrelated, we have

$$\begin{aligned} \langle R_i R_j \rangle = \ell_0^2 \sum_{k \neq 0} \frac{e^{ik(\Xi_j^0 - \Xi_i^0)}}{k^2} \left\langle \left[e^{-ik\delta\Xi_i(t)} \check{\psi}(k,t) - \check{\psi}(k,0)\right] \left[e^{ik\delta\Xi_j(t)} \check{\psi}(-k,t) - \check{\psi}(-k,0)\right] \right\rangle \\ + \ell_0^2 \left\langle \delta\Xi_i \delta\Xi_j \right\rangle. \end{aligned} \tag{41}$$

The summand can be expanded as

$$
\begin{aligned}
&\left\langle \left[e^{-\mathrm{i}k\delta\Xi_i} \check{\psi}(k,t) - \check{\psi}(k,0) \right] \left[e^{\mathrm{i}k\delta\Xi_j} \check{\psi}(-k,t) - \check{\psi}(-k,0) \right] \right\rangle \\
&= \left\langle e^{\mathrm{i}k(\delta\Xi_j - \delta\Xi_i)} \check{\psi}(k,t)\check{\psi}(-k,t) \right\rangle - \left\langle e^{-\mathrm{i}k\delta\Xi_i} \check{\psi}(k,t)\check{\psi}(-k,0) \right\rangle - \left\langle e^{\mathrm{i}k\delta\Xi_j} \check{\psi}(-k,t)\check{\psi}(k,0) \right\rangle \\
&\quad + \langle \check{\psi}(k,0)\check{\psi}(-k,0) \rangle \\
&\simeq \left(1 + \left\langle e^{\mathrm{i}k(\delta\Xi_j - \delta\Xi_i)} \right\rangle\right) \langle \check{\psi}(k,t)\check{\psi}(-k,t) \rangle - 2\mathrm{Re} \left\langle e^{-\mathrm{i}k\delta\Xi_i} \right\rangle \langle \check{\psi}(k,t)\check{\psi}(-k,0) \rangle,
\end{aligned}
\tag{42}
$$

again, with the assumption that $\check{\psi}$ and Ξ_i are uncorrelated. The terms including the exponentials of $\delta\Xi_i$ can be evaluated analytically on the basis of solutions to the master equations (38) and (39).

In this way, after some calculation, we obtain the displacement correlation. For $i = j$, we have

$$
\frac{\left\langle {R_i}^2 \right\rangle}{\ell_0^2} = 2S\sqrt{\frac{D'_* t}{\pi}} + 2\nu_\alpha t, \tag{43}
$$

where $D'_* = D^{\mathrm{c}}_* + \nu_\alpha$. Note that the last term, $2\nu_\alpha t$, originates from Equation (37). In the case of $i \neq j$, using $\varphi(\,\cdot\,)$ defined in Equation (35), we obtain

$$
\frac{\langle R_i R_j \rangle}{\ell_0^2} = S\left[2\sqrt{\frac{D'_* t}{\pi}}\, \varphi\!\left(\frac{|j-i|}{\sqrt{4D'_* t}} \right) - \sqrt{\frac{2\nu_\alpha t}{\pi}}\, \varphi\!\left(\frac{|j-i|}{\sqrt{8\nu_\alpha t}} \right) \right] + \langle \delta\Xi_i \delta\Xi_j \rangle . \tag{44}
$$

The last term ($\Delta = j - i \geq 1$ without loss of generality) needs to be evaluated with the two-body diffusion propagator in the form of Equation (A13) in Appendix A, which yields

$$
\langle \delta\Xi_i \delta\Xi_{i+\Delta} \rangle = -2\nu_\alpha t\, e^{-4\nu_\alpha t} \left[I_{\Delta-1}(4\nu_\alpha t) + I_\Delta(4\nu_\alpha t) \right] + \left(\Delta - \frac{1}{2} \right) e^{-4\nu_\alpha t} \sum_{n=\Delta}^{\infty} I_n(4\nu_\alpha t), \tag{45}
$$

with $I_n(\,\cdot\,)$ denoting the n-th modified Bessel function. It is easy to verify that, in the limit of $\nu_\alpha \to 0$, Equations (43) and (44) are reduced to the ideal case in Equation (35).

The theoretical prediction in Equations (43)–(45) is compared with a result of our particle simulation in Figure 6. With the barrier height and the density chosen as $V_{\max} = 3k_{\mathrm{B}}T$ and $\rho_0 = N/L = 0.2\,\sigma^{-1}$ ($N = 10^4$ and $L = 5N\sigma$), we calculated $\langle R_i R_j \rangle$ and delineated it for three different values of t. The same rescaled variables were used as in Figure 4a: namely, $\langle R_i R_j \rangle / (K\sqrt{t})$ is plotted against θ, with K given immediately below Equation (35). The hopping rate was evaluated numerically and estimated to be $\nu_\alpha = 0.0057\,D/\sigma^2$ (see Appendix B), which was used to plot Equations (43)–(45) as theoretical curves in Figure 6. The prediction for the ideal SFD ($\nu_\alpha = 0$) in Equation (35) is also included with a broken line.

In regard to the difference between the particle simulation and theory for the ideal SFD, Figure 6 exhibits qualitatively the same behavior as was observed in Figure 5—the difference due to overtaking is remarkable only for small θ and occurs in such a way that, except for the self part ($i = j$, i.e., the MSD), the numerical values of the displacement correlation are smaller than the prediction for the ideal SFD in Equation (35). This means that the effect of overtaking should manifest itself as a negative correction to $\langle R_i R_j \rangle$ for $i \neq j$. In this sense, the present theory modifies Equation (35) in the right direction, as the theoretical curve in Figure 6 predicts *smaller* values of $\langle R_i R_j \rangle$ for $i \neq j$ in comparison to Equation (35).

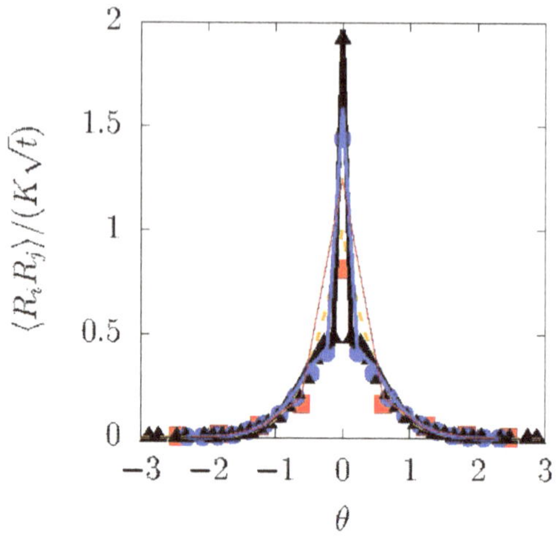

Figure 6. Comparison of the theoretical predictions of $\langle R_i R_j \rangle$ with particle-based numerical calculations of SFD with overtaking ($V_{\text{max}} = 3k_{\text{B}}T$). The three symbols have the same meaning as in Figure 4 with the axes rescaled also in the same way: $\langle R_i R_j \rangle / (K\sqrt{t})$ is plotted against θ. The thinnest red line, the thin blue line, and the solid black line represent Equation (44) that allows for overtaking ($\nu_\alpha = 0.0057\, D/\sigma^2$), evaluated at the values of t corresponding to the three kinds of symbols, namely, at $t/(\sigma^2/D) = 10, 70$, and 200, respectively. The theory without overtaking [$\nu_\alpha = 0$, i.e., Equation (35)] is shown by a broken line.

5. Discussion and Concluding Remarks

We have presented calculations of two-particle displacement correlation, $\langle R_i R_j \rangle$, in a 1D or quasi-1D system of Brownian particles with repulsive interaction, i.e., in SFD with or without overtaking, as an illustrative model of collective dynamics associated with the cage effect. In the ideal SFD (without overtaking), correlated motion with a diffusively growing length scale, $\lambda = 2\sqrt{D^c t}$, was observed. Subsequently, we studied how this result is modified by overtaking; it was shown both numerically and analytically that the overtaking processes destroy short-range correlations alone, leaving long-range correlations nearly intact. This behavior of $\langle R_i R_j \rangle$, evidenced in Figures 5 and 6, suggests a spatiotemporally nested structure of cages, such that smaller cages are enclosed in larger cages with longer lifetimes.

The main objective of the present work was to shed theoretical light on overtaking in SFD, by extending an analytical theory of SFD to the case of non-ideal SFD in which overtaking is allowed. The analytical theory is based on the method of the label variable ξ. The Lagrangian correlation of the field ψ links $\langle R_i R_j \rangle$ to the Dean–Kawasaki equation (12) that describes the fluctuating density field, while overtaking is taken into account through $\delta\Xi_i(t)$. The main analytical result is represented by Equations (43) and (44). This result is reasonably consistent with the numerical behavior of $\langle R_i R_j \rangle$ in Figure 6. A linear solution to the transformed Dean–Kawasaki equation (30) seems to suffice for the description of the outer cages (long scales). Contrastively, the inner cages are not only affected by overtaking via $\delta\Xi_i(t)$ but also subject to nonlinear coupling of $\check{\psi}$, as is suggested by the deviation from the linear theory in Figure 4a.

In spite of the reasonable agreement between the analytical and numerical results, however, there are at least two issues that need to be discussed and probably improved in the future. Firstly, the hopping rate (ν_α) could be obtained from βV_{max} and other parameters in a more first-principle-oriented manner, as opposed to the numerical fitting adopted here. Secondly, the validity of the decoupling approximation in Equation (42) is questionable.

In regard to the first issue, we note that the numerical fit for ν_α in Equation (A14) in Appendix B is not simply given by an Arrhenius-like expression, $e^{-\beta V_{\max}}$ [46], but includes a prefactor that depends on ρ_0 in a nontrivial way. Although accurate computation of overtaking is difficult and may require improvement in the numerical scheme, which is outside the scope of the present work, the numerical prefactor is interesting enough to motivate theoretical attempts to explain it. While $V_{\max}$ in Equation (2) corresponds to the Helmholtz free energy barrier in quasi-1D systems governed by Equations (1a) and (6) [10,27], the hopping rate is rather related to a barrier in the Gibbs free energy [52]. The ρ_0-dependent prefactor is also reminiscent of the escape probability in predator–prey problems on a 2D lattice [53]. Theoretical evaluation of the hopping rate (ν_α) in the present system will be quite suggestive in a wider context, such as that of 2D colloidal liquids.

In fact, correspondence between the 2D dynamics of colloids and the non-ideal SFD may deserve serious consideration. Displacement correlations in 2D colloidal liquids [54] were recently calculated analytically with the method of the label variable [21]. A linear analysis of the transformed Dean–Kawasaki equation was found to suffice to explain displacement correlations at larger scales, while a phenomenological correction was needed for behavior at smaller scales. The 2D dynamics involve dilatational and rotational modes; the former modes change the local density and correspond to $\psi(\xi, t)$ in SFD, while the latter (or, to be precise, their short-wave components) allow the particles to escape from the cage and correspond to $\delta\Xi_i(t)$ in this sense. More intuitively speaking, the overtaking event in Figure 3b can be understood as a small vortex involving two particles, namely 4 and 5.

The second issue concerns the approximation of treating $\check{\psi}$ and $\delta\Xi_i$ in Equation (42) as uncorrelated, which seems to have made the behavior of Equations (43) and (44) quantitatively incorrect for shorter distances. While a quantitative test of Equation (44) is already given in Figure 6, the MSD in Equation (43) requires more discussion. The validity of Equation (43) could be checked by way of

$$D_\alpha \stackrel{\text{def}}{=} \lim_{t\to\infty} \frac{\langle R^2 \rangle}{2t}, \tag{46}$$

which can be computed numerically and compared with analytical predictions. It seems, however, that the ν_α-dependence of D_α is in dispute. Hahn and Kärger [23] asserted $D_\alpha \propto \nu_\alpha$, while Mon and Percus [24] claimed $D_\alpha \propto \nu_\alpha^{1/2}$. If Equation (43) is taken literally, it predicts that the longtime behavior of MSD is dominated by the second term on the right side and makes essentially the same prediction as Hahn and Kärger [23]; unfortunately, it contradicts the numerical results of Mon and Percus [24], at least for a certain range of ν_α. On the other hand, the reasoning about the origin of $D_\alpha \propto \nu_\alpha^{1/2}$ by Mon and Percus [24] is unsatisfactory, as it seems to lack connection with the collective motion.

As a possible scenario for reconciliation, we may conjecture that the correlation between $\check{\psi}$ and $\delta\Xi_i$, ignored in the present theory, makes a difference to the short-range behavior of $\langle R_i R_j \rangle$ and in the MSD as its limiting case. It seems physically plausible that the collapse of smaller cages may be influenced by density fluctuations with long wavelengths which correlate $\check{\psi}$ and $\delta\Xi_i$. If this correlation modifies the term containing S in Equation (43) so as to give

$$\frac{\langle R^2 \rangle}{\ell_0^2} \simeq \left(\frac{2}{\sqrt{\pi}} S + c\sqrt{\nu_\alpha t} \right) \sqrt{D'_* t + 2\nu_\alpha t} \tag{47}$$

with some constant c, then it predicts D_α to be a sum of terms proportional to $\sqrt{\nu_\alpha}$ and ν_α. This form might be consistent at once with Hahn and Kärger [23] and with Mon and Percus [24], depending on the values of parameters.

All these issues originate from the phenomenological treatment of overtaking. Improvement upon the present analysis, going beyond the "Amida-kuji" or random exchange approximation, will need to be grounded on the integral in Equation (22) that gives $\mathrm{d}\Xi_i/\mathrm{d}t$. Its systematic treatment will allow calculation of ν_α, and it will also make it possible to take the correlation between $\check{\psi}$ and Ξ_i into account.

The integrand in Equation (22) is given, in principle, by a solution of the Dean equation [32] for the single particle density, $\rho_i = \rho_i(x,t)$, and its flux, Q_i:

$$\partial_t \rho_i(x,t) + \partial_x Q_i = 0 \tag{48a}$$

$$Q_i = -D\left[\partial_x \rho_i + \frac{\rho_i}{k_\mathrm{B}T}\partial_x \sum_j V(x - X_j(t))\right] + \rho_i f_i. \tag{48b}$$

Technically difficult though it might be, this strategy seems quite natural, as it is consistent with the treatment of ψ based on the Dean–Kawasaki equation (12). Development of this strategy for systematic treatment of overtaking effects on SFD will provide useful insights into cage-breaking events in 2D and 3D colloidal glasses.

Author Contributions: T.O. designed the basic plan of the research. All the three authors worked together for the theoretical part. M.O. performed numerical calculations and S.G. verified them. The manuscript was written by T.O. and checked by all the other authors.

Funding: This work was supported by JSPS KAKENHI Grant Numbers JP-15K05213 and JP-18K03459.

Acknowledgments: The authors thank Hajime Yoshino, Takeshi Kawasaki, Takeshi Matsumoto, Sheida Ahmadi and Richard Bowles for valuable comments and discussions. The first author (Ooshida) expresses his cordial gratitude to the organizers and the participants in the Special Session A1 ("Stochastic Processes in Complex Environment" organized by Julian Talbot and Carlos Mejía-Monasterio) in the SigmaPhi2017 conference held in Kerkira, Greece, at which he presented outline of this work and received many helpful comments.

Conflicts of Interest: The authors declare no conflict of interest.

Abbreviations

The following abbreviations are used in this manuscript:

1D	one-dimensional
2D	two-dimensional
3D	three-dimensional
MSD	mean square displacement
SFD	single-file diffusion

Appendix A. Two-Body Propagator in the Amida-Kuji Process

Here we outline solution of the master equation (39) for the two-body diffusion propagator in the Amida-kuji process. Note that most of the symbols in this Appendix (such as λ, μ and σ) are unrelated to the homonymous ones used in the main text.

To introduce some notation, we begin with a simpler problem of the one-body diffusion propagator governed by Equation (38). Its solution is found to be expressible in terms of

$$\Phi_n(s) \stackrel{\text{def}}{=} \int_{-\pi}^{\pi} \exp\left[in\lambda + s(\cos\lambda - 1)\right]\frac{d\lambda}{2\pi} = e^{-s}I_n(s) \qquad (n \in \mathbb{Z},\ s \in \mathbb{R}), \tag{A1}$$

where I_n denotes the n-th modified Bessel function. This is obtained by looking for a solution in Fourier representation,

$$P(a,t) = \sum_l \hat{P}_l(t)e^{\mathrm{i}a\lambda}, \quad \hat{P}_l(t) \propto e^{-\sigma t}, \quad \lambda = \frac{2\pi l}{N},$$

which is found to satisfy Equation (38) if $\sigma = 2\nu_\alpha(1 - \cos\lambda)$. The amplitude of $\hat{P}_l$ is determined by the Fourier representation of the initial condition,

$$P(a,0) = \delta_{a,0} = \frac{1}{N}\sum_l e^{\mathrm{i}a\lambda},$$

so that the solution is

$$P(a,t) = \frac{1}{N}\sum_{l} e^{\mathrm{i}a\lambda - \sigma t} \to \Phi_a(2\nu_\alpha t) \tag{A2}$$

in the limit of $N \to \infty$. This solution allows calculating

$$\left\langle [\delta\Xi_i(t)]^p \right\rangle = \sum_{a=-\infty}^{+\infty} a^p P(a,t);$$

in particular, for $p = 2$ we obtain Equation (37).

The two-body problem in Equation (39) is more complicated. As a counterpart of $\Delta = j - i$, we introduce $d = b - a$. Although a general solution could be sought in Fourier representation as $P_{a,b} = \sum_l \tilde{P}_{l,d}(t)e^{\mathrm{i}a\lambda}$, here we concentrate on the mode with $l = 0$, introducing

$$Q_d = Q(\Delta, d, t) \stackrel{\text{def}}{=} \sum_a P(i, i+\Delta, 0; a, a+d, t); \tag{A3}$$

this is sufficient for the present purpose, as the knowledge of Q_d will allow us to calculate

$$\left\langle \left[\delta\Xi_j(t) - \delta\Xi_i(t)\right]^p \right\rangle = \sum_d (d-\Delta)^p Q(\Delta, d, t). \tag{A4}$$

In terms of Q_d, the master equation (39) is rewritten as

$$\partial_t Q_d = 2\nu_\alpha \left(Q_{d+1} + Q_{d-1} - 2Q_d\right) + (\delta_{d,1} + \delta_{d,-1})\nu_\alpha \left(Q_1 + Q_{-1}\right) \qquad (d \neq 0), \tag{A5}$$

while Q_0 vanishes identically. The initial condition can be expressed as

$$Q_d|_{t=0} = Q(\Delta, d, 0) = \delta_{\Delta,d} = \frac{1}{N}\sum_m e^{\mathrm{i}(d-\Delta)\mu} \tag{A6}$$

where $\mu = 2\pi m/N$, the summation ranges from $-N/2$ to $N/2$ (with the endpoints included by half), and Δ is taken to be positive without loss of generality.

To solve Equation (A5), we assume

$$Q_d = \begin{cases} \frac{1}{N}\sum_m A_m e^{\mathrm{i}d\mu} e^{-\sigma_m t} & (d > 0) \\ 0 & (d = 0) \\ \frac{1}{N}\sum_m B_m e^{\mathrm{i}d\mu} e^{-\sigma_m t} & (d < 0) \end{cases} \tag{A7}$$

with $A_{-m} = A_m^*$ and $B_{-m} = B_m^*$. By considering the cases of $|d| \geq 2$ in Equation (A5), we find

$$\sigma_m = \sigma_{-m} = 2\nu_\alpha \left(2 - \varepsilon - \varepsilon^*\right) \tag{A8}$$

with $\varepsilon \stackrel{\text{def}}{=} e^{\mathrm{i}\mu}$. Subsequently, evaluation of Equation (A5) for $d = \pm 1$ gives

$$(2-\varepsilon)A_m + (2-\varepsilon^*)A_{-m} = \varepsilon^* B_m + \varepsilon B_{-m} \tag{A9a}$$
$$(2-\varepsilon^*)B_m + (2-\varepsilon)B_{-m} = \varepsilon A_m + \varepsilon^* A_{-m}, \tag{A9b}$$

which also implies $A_m + A_{-m} = B_m + B_{-m}$.

Although the initial condition (A6) seems to suggest $A_m = B_m = \varepsilon^{-\Delta}$, this naive choice does not satisfy Equations (A9). Instead, we assume

$$A_m = \varepsilon^{-\Delta} + C, \qquad B_m = \varepsilon^{-\Delta} + C^*, \tag{A10}$$

where $C = C(\varepsilon)$ is a polynomial of ε (consisting of terms with non-negative powers only) so that

$$\sum_m C(\varepsilon)\varepsilon^d = 0$$

for $d > 0$. Substituting Equation (A10) into Equations (A9) yields

$$2(1-\varepsilon)C + 2(1-\varepsilon^*)C^* = (-2+\varepsilon+\varepsilon^*)\left(\varepsilon^{\Delta}+\varepsilon^{-\Delta}\right), \tag{A11}$$

which is satisfied if we choose

$$C = \frac{1}{2}\left(\varepsilon^{-1}-1\right)\varepsilon^{\Delta}. \tag{A12}$$

Note that this is indeed a polynomial, since $\Delta \geq 1$. Thus Q_d is obtained in the form of Equation (A7), with A_m and B_m given by Equations (A10) and (A12), where $\varepsilon = e^{\mathrm{i}\mu} = e^{2\pi \mathrm{i} m/N}$.

In the limit of $N \to \infty$, the summations in Equation (A7) are evaluated as integrals, expressible in terms of $\Phi_n(s)$ in Equation (A1). The result reads

$$Q_d = \begin{cases} \Phi_{d-\Delta}(4\nu_\alpha t) - \frac{1}{2}\Phi_{d+\Delta}(4\nu_\alpha t) + \frac{1}{2}\Phi_{d+\Delta-1}(4\nu_\alpha t) & (d>0) \\ 0 & (d=0) \\ \frac{1}{2}\Phi_{d-\Delta}(4\nu_\alpha t) + \frac{1}{2}\Phi_{d-\Delta+1}(4\nu_\alpha t) & (d<0). \end{cases} \tag{A13}$$

This result allows us to calculate the moments of $\delta\Xi_j - \delta\Xi_i$ in Equation (A4). In particular, using the second moment, we can also calculate $\langle \delta\Xi_i \delta\Xi_j \rangle$ as

$$\langle \delta\Xi_i \delta\Xi_j \rangle = \frac{1}{2}\left[\left\langle \delta\Xi_i^2 \right\rangle + \left\langle \delta\Xi_j^2 \right\rangle - \left\langle (\delta\Xi_j - \delta\Xi_i)^2 \right\rangle\right],$$

where $\langle \delta\Xi_i^2 \rangle = \langle \delta\Xi_j^2 \rangle = 2\nu_\alpha t$ according to Equation (37). After some rearrangement with the recurrence formula for the modified Bessel function, we arrive at Equation (45).

Appendix B. Numerical Evaluation of the Hopping Rate

Within the approximation of the present work, the hopping rate ν_α is regarded as a constant whose value depends on the control parameters of the system. The problem of how to specify ν_α might be trivial if the lattice dynamics were adopted instead of the present system with the continuous space, because the probability of overtaking, i.e., the ratio of ν_α to the collision frequency, would be given as a part of dynamical rules governing the lattice system. More consideration is required in the present case, because overtaking is not given as an elementary process but determined as a result of the Langevin dynamics regulated by the interaction potential $V(r)$ with $r \in \mathbb{R}$.

For the present, we have determined the function $\nu_\alpha = \nu_\alpha(\rho_0, V_{\max})$ by solving Equation (2) numerically, with $V(r)$ given in Equation (5); for numerical ease, the mass is chosen to be finite so that $m/\mu : \sigma^2/D = 1:1$. For each run with some specific values of $(\rho_0, V_{\max})$, we counted overtaking events that had occurred since the time 0, and plotted the cumulative number of overtaking, divided by N, against t. Thus we obtain a plot analogous to Figure 4(a) in Ref. [25] that can be fitted with a straight line, whose slope gives ν_α. The values of ν_α, collected in this way as a function of $(\rho_0, V_{\max})$, are fitted with an Arrhenius-like expression, with a prefactor that depends both on $V_{\max}$ and ρ_0 [44]:

$$\nu_\alpha = D\left(a_0\frac{\rho_0}{\sigma} + a_1\rho_0^2\beta V_{\max}\right)e^{-\beta V_{\max}}, \quad \beta = \frac{1}{k_B T}, \tag{A14}$$

where $a_0 \approx 1/2$ and $a_1 \approx 1/6$.

The numerical values of ν_α reported here, however, should not be regarded as decisive. We must be cautious about difficulty in precise numerical calculation of rare events such as overtaking. In comparison to other quantities, computation of ν_α may be sensitive to the choice of the numerical scheme, and in the present calculation, based on a Verlet-like scheme, the computed values of ν_α seem to depend on the ratio $m/\mu{:}\sigma^2/D$. A systematic study of this issue will be expected in the future.

References

1. Pusey, P.N.; van Megen, W. Observation of a glass transition in suspensions of spherical colloidal particles. *Phys. Rev. Lett.* **1987**, *59*, 2083. [CrossRef] [PubMed]
2. Berthier, L.; Biroli, G. Theoretical perspective on the glass transition and amorphous materials. *Rev. Mod. Phys.* **2011**, *83*, 587–645. [CrossRef]
3. Glotzer, S.C.; Novikov, V.N.; Schrøder, T.B. Time-dependent, four-point density correlation function description of dynamical heterogeneity and decoupling in supercooled liquids. *J. Chem. Phys.* **2000**, *112*, 509–512. [CrossRef]
4. Ooshida, T.; Goto, S.; Matsumoto, T.; Nakahara, A.; Otsuki, M. Analytical calculation of four-point correlations for a simple model of cages involving numerous particles. *Phys. Rev. E* **2013**, *88*, 062108.
5. Rallison, J.M. Brownian diffusion in concentrated suspensions of interacting particles. *J. Fluid Mech.* **1988**, *186*, 471–500. [CrossRef]
6. Lefèvre, A.; Berthier, L.; Stinchcombe, R. Spatially heterogeneous dynamics in a model for granular compaction. *Phys. Rev. E* **2005**, *72*, 010301(R). [CrossRef] [PubMed]
7. Pal, P.; O'Hern, C.S.; Blawzdziewicz, J.; Dufresne, E.R.; Stinchcombe, R. Minimal model for kinetic arrest. *Phys. Rev. E* **2008**, *78*, 011111. [CrossRef] [PubMed]
8. Abel, S.M.; Tse, Y.L.S.; Andersen, H.C. Kinetic theories of dynamics and persistent caging in a one-dimensional lattice gas. *Proc. Natl. Acad. Sci. USA* **2009**, *106*, 15142–15147. [CrossRef] [PubMed]
9. Miyazaki, K. Garasu Ten'i to Môdo Ketsugô Riron (Glass Transition and Mode-Coupling Theory). *Bussei Kenkyû* **2007**, *88*, 621–720. (In Japanese)
10. Ooshida, T.; Goto, S.; Matsumoto, T.; Otsuki, M. Insights from Single-File Diffusion into Cooperativity in Higher Dimensions. *Biophys. Rev. Lett.* **2016**, *11*, 9–38. [CrossRef]
11. Harris, T.E. Diffusion with "collisions" between particles. *J. Appl. Probab.* **1965**, *2*, 323–338. [CrossRef]
12. Alexander, S.; Pincus, P. Diffusion of labeled particles on one-dimensional chains. *Phys. Rev. B* **1978**, *18*, 2011. [CrossRef]
13. Hahn, K.; Kärger, J. Propagator and mean-square displacement in single-file systems. *J. Phys. A Math. Gen.* **1995**, *28*, 3061–3070. [CrossRef]
14. Kollmann, M. Single-file Diffusion of Atomic and Colloidal Systems: Asymptotic Laws. *Phys. Rev. Lett.* **2003**, *90*, 180602. [CrossRef] [PubMed]
15. Illien, P.; Bénichou, O.; Mejía-Monasterio, C.; Oshanin, G.; Voituriez, R. Active Transport in Dense Diffusive Single-File Systems. *Phys. Rev. Lett.* **2013**, *111*, 038102. [CrossRef] [PubMed]
16. Van Beijeren, H.; Kehr, K.W.; Kutner, R. Diffusion in concentrated lattice gases. III. Tracer diffusion on a one-dimensional lattice. *Phys. Rev. B* **1983**, *28*, 5711–5723. [CrossRef]
17. Taloni, A.; Lomholt, M.A. Langevin formulation for single-file diffusion. *Phys. Rev. E* **2008**, *78*, 051116. [CrossRef] [PubMed]
18. Taloni, A.; Marchesoni, F. Interacting single-file system: Fractional Langevin formulation versus diffusion-noise approach. *Biophys. Rev. Lett.* **2014**, *9*, 381–396. [CrossRef]
19. Ooshida, T.; Goto, S.; Matsumoto, T.; Otsuki, M. Displacement correlation as an indicator of collective motion in one-dimensional and quasi-one-dimensional systems of repulsive Brownian particles. *Mod. Phys. Lett. B* **2015**, *29*, 1550221. [CrossRef]
20. Majumdar, S.N.; Barma, M. Two-tag correlation functions in one-dimensional lattice gases. *Physica A* **1991**, *177*, 366–372. [CrossRef]
21. Ooshida, T.; Goto, S.; Matsumoto, T.; Otsuki, M. Calculation of displacement correlation tensor indicating vortical cooperative motion in two-dimensional colloidal liquids. *Phys. Rev. E* **2016**, *94*, 022125. [CrossRef] [PubMed]
22. Krapivsky, P.L.; Mallick, K.; Sadhu, T. Large Deviations in Single-File Diffusion. *Phys. Rev. Lett.* **2014**, *113*, 078101. [CrossRef] [PubMed]
23. Hahn, K.; Kärger, J. Deviations from the normal time regime of single-file diffusion. *J. Phys. Chem. B* **1998**, *102*, 5766–5771. [CrossRef]

24. Mon, K.; Percus, J. Self-diffusion of fluids in narrow cylindrical pores. *J. Chem. Phys.* **2002**, *117*, 2289–2292. [CrossRef]
25. Lucena, D.; Tkachenko, D.; Nelissen, K.; Misko, V.R.; Ferreira, W.P.; Farias, G.A.; Peeters, F.M. Transition from single-file to two-dimensional diffusion of interacting particles in a quasi-one-dimensional channel. *Phys. Rev. E* **2012**, *85*, 031147. [CrossRef] [PubMed]
26. Siems, U.; Kreuter, C.; Erbe, A.; Schwierz, N.; Sengupta, S.; Leiderer, P.; Nielaba, P. Non-monotonic crossover from single-file to regular diffusion in micro-channels. *Sci. Rep.* **2012**, *2*, 1015. [CrossRef] [PubMed]
27. Wanasundara, S.N.; Spiteri, R.J.; Bowles, R.K. A transition state theory for calculating hopping times and diffusion in highly confined fluids. *J. Chem. Phys.* **2014**, *140*, 024505. [CrossRef] [PubMed]
28. Kutner, R.; van Beijeren, H.; Kehr, K.W. Diffusion in concentrated lattice gases. VI. Tracer diffusion on two coupled linear chains. *Phys. Rev. B* **1984**, *30*, 4382–4391. [CrossRef]
29. Bénichou, O.; Illien, P.; Oshanin, G.; Sarracino, A.; Voituriez, R. Diffusion and Subdiffusion of Interacting Particles on Comblike Structures. *Phys. Rev. Lett.* **2015**, *115*, 220601. [CrossRef] [PubMed]
30. Ooshida, T.; Goto, S.; Matsumoto, T.; Nakahara, A.; Otsuki, M. Continuum Theory of Single-File Diffusion in Terms of Label Variable. *J. Phys. Soc. Jpn.* **2011**, *80*, 074007.
31. Berthier, L.; Witten, T.A. Compressing nearly hard sphere fluids increases glass fragility. *Europhys. Lett.* **2009**, *86*, 10001. [CrossRef]
32. Dean, D.S. Langevin equation for the density of a system of interacting Langevin processes. *J. Phys. A Math. Gen.* **1996**, *29*, L613. [CrossRef]
33. Kawasaki, K. Stochastic model of slow dynamics in supercooled liquids and dense colloidal suspensions. *Phys. A* **1994**, *208*, 35–64. [CrossRef]
34. Kawasaki, K. Microscopic Analyses of the Dynamical Density Functional Equation of Dense Fluids. *J. Stat. Phys.* **1998**, *93*, 527–546. [CrossRef]
35. Das, S.P. *Statistical Physics of Liquids at Freezing and Beyond*; Cambridge University Press: New York, NY, USA, 2011.
36. Das, S.P.; Yoshimori, A. Coarse-grained forms for equations describing the microscopic motion of particles in a fluid. *Phys. Rev. E* **2013**, *88*, 043008. [CrossRef] [PubMed]
37. Kim, B.; Kawasaki, K.; Jacquin, H.; van Wijland, F. Equilibrium dynamics of the Dean–Kawasaki equation: Mode-coupling theory and its extension. *Phys. Rev. E* **2014**, *89*, 012150. [CrossRef] [PubMed]
38. Nägele, G. On the dynamics and structure of charge-stabilized suspensions. *Phys. Rep.* **1996**, *272*, 215–372. [CrossRef]
39. Dhont, J.K.G. *An Introduction to Dynamics of Colloids*; Elsevier: Amsterdam, The Netherlands, 1996.
40. Lamb, H. *Hydrodynamics*, 6th ed.; Cambridge University Press: Cambridge, UK, 1932.
41. Landau, L.D.; Lifshitz, E.M. Fluid Mechanics. In *Theoretical Physics*; Butterworth-Heinemann: Oxford, UK, 1987; Volum 6.
42. Toninelli, C.; Wyart, M.; Berthier, L.; Biroli, G.; Bouchaud, J.P. Dynamical susceptibility of glass formers: Contrasting the predictions of theoretical scenarios. *Phys. Rev. E* **2005**, *71*, 041505. [CrossRef] [PubMed]
43. Flenner, E.; Szamel, G. Long-Range Spatial Correlations of Particle Displacements and the Emergence of Elasticity. *Phys. Rev. Lett.* **2015**, *114*, 025501. [CrossRef] [PubMed]
44. Ooshida, T.; Otsuki, M. Effects of Cage-Breaking Events in Single-File Diffusion on Elongation Correlation. *J. Phys. Soc. Jpn.* **2017**, *86*, 113002. [CrossRef]
45. Spohn, H. Nonlinear Fluctuating Hydrodynamics for Anharmonic Chains. *J. Stat. Phys.* **2014**, *154*, 1191–1227. [CrossRef]
46. Kramers, H.A. Brownian motion in a field of force and the diffusion model of chemical reactions. *Physica* **1940**, *7*, 284–304. [CrossRef]
47. Sancho, J.M.; Miguel, M.S.; Dürr, D. Adiabatic Elimination for Systems of Brownian Particles with Nonconstant Damping Coefficients. *J. Stat. Phys.* **1982**, *28*, 291–305. [CrossRef]
48. Sekimoto, K. Temporal Coarse Graining for Systems of Brownian Particles with Non-Constant Temperature. *J. Phys. Soc. Jpn.* **1999**, *68*, 1448–1449. [CrossRef]
49. Inoue, Y. Statistical analysis on Amida-kuji. *Phys. A Stat. Mech. Appl.* **2006**, *369*, 867–876. [CrossRef]
50. Yamanaka, K.; Nakano, S.I. Enumeration, Counting, and Random Generation of Ladder Lotteries. *IEICE Trans. Inf. Syst.* **2017**, *100*, 444–451. [CrossRef]

51. Kitahara, K. *Hi-heikô Kei no Kagaku II: Kanwa Katei no Tôkei Rikigaku [Science of Non-Equilibrium Systems II: Statistical Mechanics of Relaxation Processes]*; Kôdansha: Tokyo, Japan, 1994. (In Japanese)
52. Ahmadi, S.; Bowles, R.K. Diffusion in quasi-one-dimensional channels: A small system n, p, T transition state theory for hopping times. *J. Chem. Phys.* **2017**, *146*, 154505. [CrossRef] [PubMed]
53. Oshanin, G.; Vasilyev, O.; Krapivsky, P.L.; Klafter, J. Survival of an evasive prey. *Proc. Nat. Acad. Sci. USA* **2009**, *106*, 13696–13701. [CrossRef] [PubMed]
54. Doliwa, B.; Heuer, A. Cooperativity and spatial correlations near the glass transition: Computer simulation results for hard spheres and disks. *Phys. Rev. E* **2000**, *61*, 6898–6908. [CrossRef]

entropy

MDPI

Article

Strong- and Weak-Universal Critical Behaviour of a Mixed-Spin Ising Model with Triplet Interactions on the Union Jack (Centered Square) Lattice

Jozef Strečka

Department of Theoretical Physics and Astrophysics, Institute of Physics, Faculty of Science, P. J. Šafárik University, Park Angelinum 9, 040 01 Košice, Slovak Republic; jozef.strecka@upjs.sk

Received: 30 December 2017; Accepted: 26 January 2018; Published: 29 January 2018

Abstract: The mixed spin-1/2 and spin-S Ising model on the Union Jack (centered square) lattice with four different three-spin (triplet) interactions and the uniaxial single-ion anisotropy is exactly solved by establishing a rigorous mapping equivalence with the corresponding zero-field (symmetric) eight-vertex model on a dual square lattice. A rigorous proof of the aforementioned exact mapping equivalence is provided by two independent approaches exploiting either a graph-theoretical or spin representation of the zero-field eight-vertex model. An influence of the interaction anisotropy as well as the uniaxial single-ion anisotropy on phase transitions and critical phenomena is examined in particular. It is shown that the considered model exhibits a strong-universal critical behaviour with constant critical exponents when considering the isotropic model with four equal triplet interactions or the anisotropic model with one triplet interaction differing from the other three. The anisotropic models with two different triplet interactions, which are pairwise equal to each other, contrarily exhibit a weak-universal critical behaviour with critical exponents continuously varying with a relative strength of the triplet interactions as well as the uniaxial single-ion anisotropy. It is evidenced that the variations of critical exponents of the mixed-spin Ising models with the integer-valued spins S differ basically from their counterparts with the half-odd-integer spins S.

Keywords: mixed-spin ising model; triplet interaction; weak-universal critical behaviour

PACS: 05.50.+q; 75.10.Hk; 75.40.Cx

---

1. Introduction

One of the most important concepts elaborated in the theory of phase transitions and critical phenomena is universality hypothesis, which states that a critical behaviour does not depend on specific details of a model but only upon its spatial dimensionality, symmetry and number of components of the relevant order parameter. The foremost consequence of the universality hypothesis is that the critical behaviour of very different models may be characterized by the same set of critical exponents and one says that the models with the identical set of critical exponents belong to the same universality class. However, there exists a few exactly solved models whose critical exponents do depend on the interaction parameters and thus contradict the usual universality hypothesis [1]. The spin-1/2 Ising model with a three-spin (triplet) interaction on planar lattices belongs to paradigmatic exactly solved models of this type. As a matter of fact, the exact solutions for the spin-1/2 Ising model with the triplet interaction gave rigorous proof for different sets of critical exponents on different planar lattices [2–7]. More specifically, the critical exponent α for the specific heat fundamentally differs when this model is defined on centered square lattice ($\alpha = 1/2$) [2], triangular lattice ($\alpha = 2/3$) [3–5], decorated triangular [6], honeycomb and diced lattices [7] ($\alpha \approx 0$, logarithmic singularity). In addition, the spin-1/2 Ising model with the triplet interaction on a kagomé lattice [8] does not display a phase transition at all.

In the present work, we will consider and exactly solve the mixed spin-1/2 and spin-S Ising model with the triplet interaction on the Union Jack (centered square) lattice by establishing a rigorous mapping correspondence with the symmetric (zero-field) eight-vertex model. The investigated model generalizes the model originally proposed and examined by Urumov [9] when accounting for the additional uniaxial single-ion anisotropy acting on the spin-S atoms. It will be demonstrated hereafter that the critical exponents of the mixed spin-1/2 and spin-S Ising model with the triplet interaction on the centered square lattice fundamentally depend on the interaction anisotropy, the uniaxial single-ion anisotropy, as well as, the spin parity.

2. Model and Exact Solution

Let us introduce the mixed spin-1/2 and spin-S Ising model with pure three-spin (triplet) interactions on a centered square lattice defined through the Hamiltonian:

$$\mathcal{H} = -J_1 \sum_{i,j}^{\triangledown} S_{i,j}\sigma_{i,j}\sigma_{i+1,j} - J_2 \sum_{i,j}^{\triangleleft} S_{i,j}\sigma_{i+1,j}\sigma_{i+1,j+1} - J_3 \sum_{i,j}^{\triangle} S_{i,j}\sigma_{i,j+1}\sigma_{i+1,j+1} - J_4 \sum_{i,j}^{\triangleright} S_{i,j}\sigma_{i,j}\sigma_{i,j+1} - D\sum_{i,j} S_{i,j}^2, \quad (1)$$

whereas the spin-1/2 atoms (light blue circles in Figure 1) represented by the Ising spin variables $\sigma_{i,j} = \pm 1/2$ are placed at corners of a square lattice, the spin-S atoms (dark blue circles in Figure 1) are situated in the middle of square plaquettes. The Hamiltonian (1) takes into account four different triplet interactions J_1, J_2, J_3 and J_4 within down-, left-, up- and right-pointing triangles, respectively, in addition to the uniaxial single-ion anisotropy D acting on the spin-S atoms.

Figure 1. A schematic illustration of the mixed spin-1/2 (light blue) and spin-S (dark blue) Ising model on a centered square lattice. Four different colors are used to distinguish triplet interactions J_1, J_2, J_3 and J_4 within down-, left-, up- and right-pointing triangles, respectively.

The Hamiltonian (1) can be alternatively rewritten as a sum of cell Hamiltonians $\mathcal{H} = \sum_{i,j}^{\square} \mathcal{H}_{i,j}$, whereas the cell Hamiltonian $\mathcal{H}_{i,j}$ involves all interactions terms depending on the central spin $S_{i,j}$:

$$\mathcal{H}_{i,j} = -J_1\, S_{i,j}\sigma_{i,j}\sigma_{i+1,j} - J_2\, S_{i,j}\sigma_{i+1,j}\sigma_{i+1,j+1} - J_3\, S_{i,j}\sigma_{i,j+1}\sigma_{i+1,j+1} - J_4\, S_{i,j}\sigma_{i,j}\sigma_{i,j+1} - DS_{i,j}^2. \quad (2)$$

The partition function of the mixed spin-1/2 and spin-S Ising model with triplet interactions on a centered square lattice can be then cast into the following form:

$$\mathcal{Z} = \sum_{\{\sigma_{i,j}\}} \prod_{i,j} \sum_{S_{i,j}=-S}^{S} \exp(-\beta\mathcal{H}_{i,j}) = \sum_{\{\sigma_{i,j}\}} \prod_{i,j} \omega(\sigma_{i,j}, \sigma_{i+1,j}, \sigma_{i+1,j+1}, \sigma_{i,j+1}), \quad (3)$$

where the summation $\sum_{\{\sigma_{i,j}\}}$ runs over all available spin configurations of the spin-1/2 atoms, $\beta = 1/(k_{\mathrm{B}}T)$, k_{B} is Boltzmann's constant, T is the absolute temperature and the expression ω denotes the Boltzmann's weight obtained after tracing out degrees of freedom of the central spin-S atom:

$$\omega(a,b,c,d) = \sum_{n=-S}^{S} \exp(\beta D n^2) \cosh\left[\beta n \left(J_1\, ab + J_2\, bc + J_3\, cd + J_4\, da\right)\right]. \tag{4}$$

An invariance of the Boltzmann's factor $\omega(a,b,c,d) = \omega(-a,-b,-c,-d)$ implies that there exist at most eight different Boltzmann's weights obtained from Equation (4) by considering all 16 spin configurations of the four corner spins involved therein. Hence, it follows that one may establish two-to-one mapping correspondence between a spin configuration and a relevant graph representation of the equivalent eight-vertex model on a dual square lattice according to the scheme shown in Figure 2. A solid line is drawn on a respective edge of a dual square lattice lying in between two unequally aligned neighbouring spins, while a broken line is drawn otherwise. It turns out, moreover, that the effective Boltzmann's weights obtained after inserting all possible spin configurations of the four corner spins into Equation (4) are pairwise equal to each other:

$$\begin{aligned}
\omega_1(+,+,+,+) = \omega_2(+,-,+,-) &= \sum_{n=-S}^{S} \exp(\beta D n^2) \cosh\left[\frac{\beta n}{4}\left(J_1 + J_2 + J_3 + J_4\right)\right], \\
\omega_3(+,-,-,+) = \omega_4(+,+,-,-) &= \sum_{n=-S}^{S} \exp(\beta D n^2) \cosh\left[\frac{\beta n}{4}\left(J_1 - J_2 + J_3 - J_4\right)\right], \\
\omega_5(-,+,+,+) = \omega_6(+,+,-,+) &= \sum_{n=-S}^{S} \exp(\beta D n^2) \cosh\left[\frac{\beta n}{4}\left(J_1 - J_2 - J_3 + J_4\right)\right], \\
\omega_7(+,+,+,-) = \omega_8(+,-,+,+) &= \sum_{n=-S}^{S} \exp(\beta D n^2) \cosh\left[\frac{\beta n}{4}\left(J_1 + J_2 - J_3 - J_4\right)\right],
\end{aligned} \tag{5}$$

which means that the mixed-spin Ising model with triplet interactions on a centered square lattice is equivalent with the symmetric (zero-field) eight-vertex model exactly solved by Baxter [10,11]. Owing to this fact, one may easily prove an exact mapping relationship between the partition functions of the mixed-spin Ising model with triplet interactions on a centered square lattice and the zero-field eight-vertex model on a dual square lattice:

$$\mathcal{Z}(\beta, J_1, J_2, J_3, J_4, D) = 2\mathcal{Z}_{8-\text{vertex}}(\omega_1, \omega_3, \omega_5, \omega_7). \tag{6}$$

It is apparent from the mapping relation in Equation (6) between the partition functions that the mixed-spin Ising model with triplet interactions on a centered square lattice becomes critical only if the corresponding zero-field eight-vertex model becomes critical as well. Bearing this in mind, the critical points of the mixed-spin Ising model with triplet interactions on a centered square lattice can be readily obtained from Baxter's critical condition [10,11] when the explicit form of the effective Boltzmann's weights in Equation (5) is taken into consideration:

$$\omega_1 + \omega_3 + \omega_5 + \omega_7 = 2\max\{\omega_1, \omega_3, \omega_5, \omega_7\}. \tag{7}$$

It should be stressed that the critical exponents for the specific heat, magnetization, susceptibility and correlation length satisfy Suzuki's weak-universal hypothesis [12] and can be calculated from:

$$\alpha = \alpha' = 2 - \pi/\mu, \qquad \beta = \pi/16\mu, \qquad \gamma = \gamma' = 7\pi/8\mu, \qquad \nu = \nu' = \pi/2\mu, \tag{8}$$

where $\tan(\mu/2) = (\omega_1\omega_3/\omega_5\omega_7)^{1/2}$ on assumption that $\omega_1 = \max\{\omega_1, \omega_3, \omega_5, \omega_7\}$.

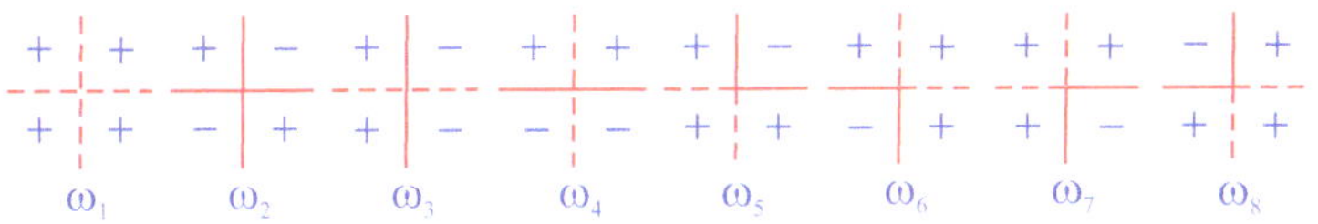

Figure 2. A schematic representation of two-to-one mapping correspondence between the Ising spin configurations and line coverings of the equivalent eight-vertex model on a dual square lattice (the sign $\pm$ marks $\sigma = \pm 1/2$).

The exact mapping equivalence with the zero-field eight-vertex model can be alternatively proven by exploiting the spin representation of the eight-vertex model. For this purpose, the effective Boltzmann factor in Equation (4) can be replaced via the generalized star-square transformation [13–15] schematically drawn in Figure 3:

$$\begin{aligned} \omega &= \sum_{n=-S}^{S} \exp(\beta D n^2) \cosh\left[\beta n (J_1 \sigma_{i,j}\sigma_{i+1,j} + J_2 \sigma_{i+1,j}\sigma_{i+1,j+1} + J_3 \sigma_{i+1,j+1}\sigma_{i,j+1} + J_4 \sigma_{i,j+1}\sigma_{i,j})\right] \\ &= R_0 \exp(\beta R_1 \sigma_{i,j}\sigma_{i+1,j+1} + \beta R_2 \sigma_{i+1,j}\sigma_{i,j+1} + \beta R_4 \sigma_{i,j}\sigma_{i+1,j}\sigma_{i+1,j+1}\sigma_{i,j+1}). \end{aligned} \tag{9}$$

The physical meaning of the generalized star-square transformation in Equation (9) lies in replacing spin degrees of freedom related to the central spin-S atom through the effective pair interactions (R_1, R_2) and the effective quartic interaction (R_4) between the four enclosing spin-1/2 atoms (see Figure 3). The star-square transformation in Equation (9) must hold for any spin state of the four enclosing spin-1/2 atoms and this "self-consistency" condition unambiguously determines so far unspecified mapping parameters:

$$R_0 = (\omega_1\omega_3\omega_5\omega_7)^{1/4}, \ \beta R_1 = \ln\left(\frac{\omega_1\omega_7}{\omega_3\omega_5}\right), \ \beta R_2 = \ln\left(\frac{\omega_1\omega_5}{\omega_3\omega_7}\right), \ \beta R_4 = 4\ln\left(\frac{\omega_1\omega_3}{\omega_5\omega_7}\right). \tag{10}$$

The star-square transformation in Equation (9) establishes a rigorous mapping correspondence between the partition function of the mixed-spin Ising model with triplet interactions on a centered square lattice and the partition function of the spin-1/2 Ising model on two inter-penetrating square lattices with the effective pair interactions (R_1, R_2) and the effective quartic interaction (R_4):

$$\mathcal{Z}(\beta, J_1, J_2, J_3, J_4, D) = R_0^{2N} \mathcal{Z}_{8-\text{vertex}}(\beta, R_1, R_2, R_4).$$

It has been proven previously that the spin-1/2 Ising model defined on two inter-penetrating square lattices coupled together by means of the quartic interaction is nothing but the Ising representation of the zero-field eight-vertex model on a square lattice [16,17]. In this way, we have afforded alternative proof for an exact mapping equivalence between the mixed-spin Ising model with triplet interactions on a centered square lattice and the zero-field eight-vertex model on a square lattice.

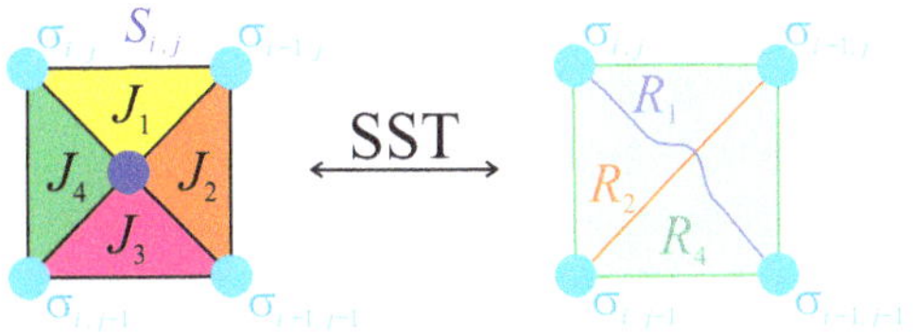

Figure 3. A schematic representation of the generalized star-square transformation, which replaces spin degrees of freedom of the central spin-S atom through two effective pair interactions (R_1, R_2) and the effective quartic interaction (R_4) between the four enclosing spin-1/2 atoms.

3. Results and Discussion

In this section, let us discuss the most interesting results for the mixed spin-1/2 and spin-S Ising model with triplet interactions on a centered square lattice depending on the interaction anisotropy, the uniaxial single-ion anisotropy and the spin magnitude S. For the sake of simplicity, our further attention will be restricted to four particular cases to be further referred to as:

- Model A: $J \equiv J_1 = J_2 = J_3 = J_4$,
- Model B: $J \equiv J_1, J' \equiv J_2 = J_3 = J_4$,
- Model C: $J \equiv J_1 = J_3, J' \equiv J_2 = J_4$,
- Model D: $J \equiv J_1 = J_2, J' \equiv J_3 = J_4$,

which will be separately treated in the following subsections. For better illustration, the four aforementioned special cases of the mixed spin-1/2 and spin-S Ising model with triplet interactions on a centered square lattice are schematically drawn in Figure 4, where different colors are used for distinguishing triplet interactions of different size. In addition, our subsequent discussion will be henceforth restricted to the mixed spin-1/2 and spin-S Ising model with both positive triplet interactions ($J > 0$, $J' > 0$), because the other meaningful particular case with both negative triplet interactions ($J < 0$, $J' < 0$) displays the identical critical behavior because $J \to -J$, $J' \to -J'$ interchange merely causes a trivial change in a relative orientation of the nearest-neighbor spins.

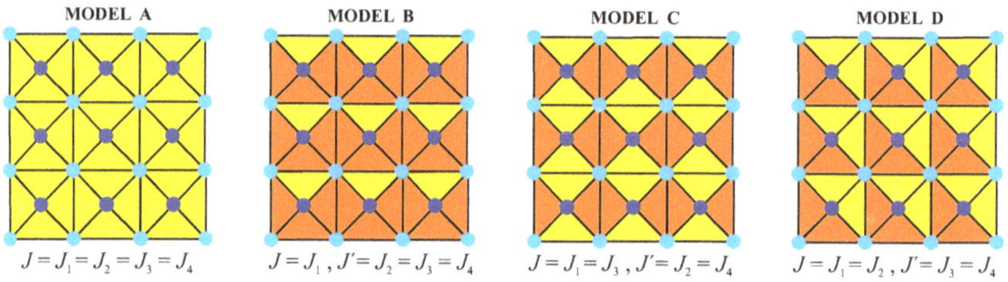

Figure 4. The four particular cases of the mixed spin-1/2 and spin-S Ising model with triplet interactions on a centered square lattice, which will be comprehensively studied in the following subsections. Different colors are used for distinguishing triplet interactions of different size.

3.1. Model A ($J \equiv J_1 = J_2 = J_3 = J_4$)

Let us begin with a detailed analysis of critical behaviour of the mixed spin-1/2 and spin-S Ising model with unique triplet interaction on a centered square lattice, which represents a very special case due to the isotropic nature of the triplet interactions $J \equiv J_1 = J_2 = J_3 = J_4$. Under this condition, one gains from Equation (5) just two different Boltzmann's weights:

$$\omega_1 = \sum_{n=-S}^{S} \exp(\beta D n^2) \cosh(\beta n J), \qquad \omega_3 = \omega_5 = \omega_7 = \sum_{n=-S}^{S} \exp(\beta D n^2), \tag{11}$$

where the Boltzmann's weights in Equation (11) evidently satisfy the inequality $\omega_1 \geq \omega_3$. The ordered state emergent at low enough temperature can be accordingly found from the lowest-energy states entering the Boltzmann's weights $\omega_1(+,+,+,+)$ and $\omega_2(+,-,+,-)$. As a result, the spontaneous order has a four-fold degeneracy, because the nodal spin-1/2 atoms display either a ferromagnetic or antiferromagnetic long-range order and the central spins are oriented in such a way that either zero or two down spins appear within each triangular plaquette. This statement will remain true also for the other three particular models treated hereafter. The critical condition in Equation (7) of the mixed spin-1/2 and spin-S Ising model with unique triplet interaction on a centered square lattice simplifies owing to a validity of the inequality $\omega_1 \geq \omega_3$ to the following form:

$$\omega_1 = 3\omega_3 \quad \Rightarrow \quad \sum_{n=-S}^{S} \exp(\beta_c D n^2) \cosh(\beta_c n J) = 3 \sum_{n=-S}^{S} \exp(\beta_c D n^2), \tag{12}$$

where $\beta_c = 1/(k_B T_c)$ and T_c marks the critical temperature. It follows from Equation (8) that the critical exponents remain constant along the whole critical line in Equation (12), irrespective of the spin magnitude S:

$$\tan(\mu/2) = \sqrt{\omega_1/\omega_3} = \sqrt{3} \quad \Rightarrow \quad \alpha = \alpha' = 1/2, \ \ \beta = 3/32, \ \ \gamma = \gamma' = 21/16, \ \ \nu = \nu' = 3/4, \tag{13}$$

where their size is identical with the ones predicted for the spin-1/2 Ising model with the unique triplet interaction on a centered square lattice, i.e. the so-called Hintermann–Merlini model [2]. The critical temperature obtained from the numerical solution of the critical condition in Equation (12) is plotted in Figure 5 against the uniaxial single-ion anisotropy for several spin magnitudes S. Although the critical temperature monotonically decreases with decreasing of the uniaxial single-ion anisotropy regardless of the spin size S, there is fundamental difference in the critical behaviour of the mixed-spin Ising models with integer and half-odd-integer spins S, respectively. Namely, the critical temperature of the former mixed-spin systems becomes zero for $D/J < -1$ in accordance with presence of the disordered ground state, which appears due to energetic favoring of the nonmagnetic spin state $S = 0$ of the integer-valued spins. On the other hand, the critical temperature of the latter mixed-spin systems tends towards the critical temperature of the Hintermann–Merlini model [2] $k_B T_c/J = 1/4 \ln(1 + \sqrt{2}) \approx 0.2836\ldots$, which is achieved in the asymptotic limit $D/J \to -\infty$ (but practically already at $D/J \approx -1$) due to energetic favoring of two lowest-valued states $S = \pm 1/2$ of the half-odd-integer spins.

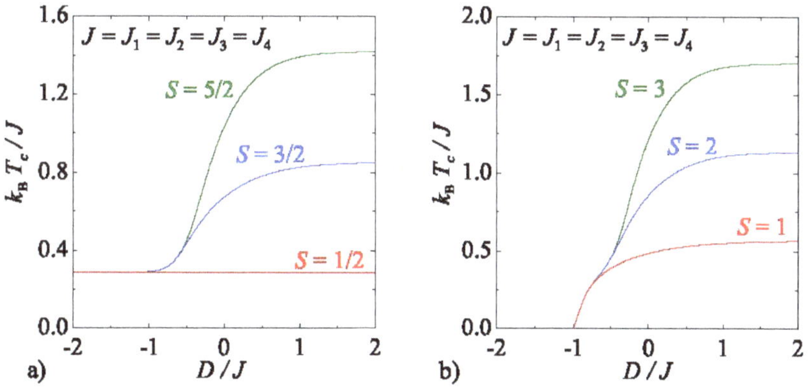

Figure 5. The critical temperature of the model A as a function of the uniaxial single-ion anisotropy by considering: (**a**) half-odd-integer spins S; and (**b**) integer spins S.

3.2. Model B ($J \equiv J_1$, $J' \equiv J_2 = J_3 = J_4$)

Next, let us relax the condition of the isotropic triplet interactions by considering the model B, where one triplet interaction (say $J \equiv J_1$) differs from the other three ($J' \equiv J_2 = J_3 = J_4$). Even under this constraint, one still gets from Equation (5) just two different Boltzmann's weights:

$$\omega_1 = \sum_{n=-S}^{S} \exp(\beta D n^2) \cosh\left[\frac{\beta n}{4}(J + 3J')\right], \quad \omega_3 = \omega_5 = \omega_7 = \sum_{n=-S}^{S} \exp(\beta D n^2) \cosh\left[\frac{\beta n}{4}(J - J')\right], \tag{14}$$

which are however slightly more complicated due to the interaction anisotropy. It is evident from Equation (14) that the Boltzmann's weights still satisfy the inequality $\omega_1 \geq \omega_3$, which affords the following critical condition:

$$\omega_1 = 3\omega_3 \quad \Rightarrow \sum_{n=-S}^{S} \exp(\beta_c D n^2) \cosh\left[\frac{\beta_c n}{4}(J + 3J')\right] = 3\sum_{n=-S}^{S} \exp(\beta_c D n^2) \cosh\left[\frac{\beta_c n}{4}(J - J')\right]. \quad (15)$$

It should be pointed out, moreover, that the critical exponents are still constants independent of the spin magnitude and the interaction anisotropy as given by Equation (13).

The critical frontiers of the model B are illustrated in Figure 6 for two different spin values, which demonstrate typical critical behaviour of the mixed-spin systems with half-odd-integer and integer spins S, respectively. It can be seen in Figure 6 that the critical temperature of the investigated mixed-spin system rises steadily with increasing of the interaction ratio J'/J both for the integer as well as half-odd-integer spins. However, it is worth remarking that the critical value of the uniaxial single-ion anisotropy needed for an onset of the disordered ground state of the mixed-spin systems with integer spins S shifts towards more negative values upon strengthening of the interaction ratio J'/J.

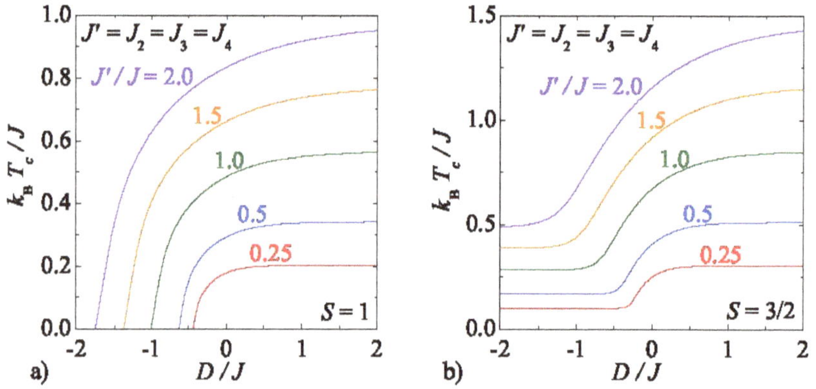

Figure 6. The critical temperature of the model B as a function of the uniaxial single-ion anisotropy by considering several values of the interaction anisotropy J'/J and two different spin magnitudes: (**a**) $S = 1$; (**b**) $S = 3/2$.

3.3. Model C ($J \equiv J_1 = J_3, J' \equiv J_2 = J_4$)

Now, let us turn our attention to a critical behaviour of the model C with the triplet interactions $J \equiv J_1 = J_3$, $J' \equiv J_2 = J_4$, which are pairwise equal in triangles lying in opposite to each other within elementary square cells (see Figure 4). In this particular case, one gets from Equation (5) three different Boltzmann's weights:

$$\omega_1 = \sum_{n=-S}^{S} \exp(\beta D n^2) \cosh\left[\frac{\beta n}{2}(J + J')\right], \qquad \omega_3 = \sum_{n=-S}^{S} \exp(\beta D n^2) \cosh\left[\frac{\beta n}{2}(J - J')\right],$$
$$\omega_5 = \omega_7 = \sum_{n=-S}^{S} \exp(\beta D n^2). \quad (16)$$

It directly follows from Equation (16) that the Boltzmann's weights obey the inequality $\omega_1 \geq \omega_3 \geq \omega_5$ and, hence, the critical condition $\omega_1 = \omega_3 + 2\omega_5$ can be explicitly written as follows:

$$\sum_{n=-S}^{S} \exp(\beta_c D n^2) \cosh\left[\frac{\beta_c n}{2}(J + J')\right] = \sum_{n=-S}^{S} \exp(\beta_c D n^2) \cosh\left[\frac{\beta_c n}{2}(J - J')\right] + 2\sum_{n=-S}^{S} \exp(\beta_c D n^2). \quad (17)$$

Besides, the Boltzmann's weights in Equation (16) imply that the critical exponents given by Equation (8) may display striking dependence along the critical line determined by Equation (17) on the spin magnitude, the uniaxial single-ion anisotropy as well as the interaction anisotropy according to the formulas:

$$\tan\left(\frac{\mu}{2}\right) = \frac{\sqrt{\omega_1 \omega_3}}{\omega_5} \quad \Rightarrow \quad \alpha = \alpha' = 2 - \frac{\pi}{\mu}, \quad \beta = \frac{\pi}{16\mu}, \quad \gamma = \gamma' = \frac{7\pi}{8\mu}, \quad \nu = \nu' = \frac{\pi}{2\mu}. \quad (18)$$

For illustrative purposes, Figures 7a and 8a display phase boundaries of the model C for the specific spin values $S = 1$ and $3/2$, respectively. As one can see, the same general trends can be observed in the relevant dependencies of the critical temperature on the uniaxial single-ion anisotropy and the interaction anisotropy as previously discussed for the model B. However, the model C exhibits along the displayed critical lines continuously varying critical exponents unlike the model B with the strong-universal (constant) critical exponents. For instance, it can be found in Figure 7b that the critical exponent α for the specific heat monotonically increases with increasing of the uniaxial single-ion anisotropy for the integer-valued spins $S = 1$, whereas the observed increase is the greater the higher the interaction anisotropy is. In addition, it is quite surprising that the critical exponents for the model C with the half-odd-integer spins $S = 3/2$ exhibit completely different weak-universal critical behaviour. Although the critical exponent α for the specific heat is shifted towards higher values upon increasing of the interaction anisotropy too, but this time the critical exponent α displays a more peculiar nonmonotonous dependence on the uniaxial single-ion anisotropy with a round minimum whose depth basically depends on the interaction anisotropy.

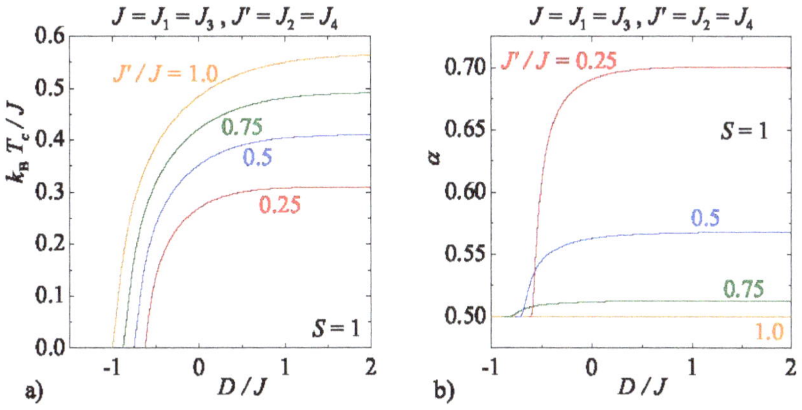

Figure 7. (**a**) The critical temperature of the model C as a function of the uniaxial single-ion anisotropy by considering the spin size $S = 1$ and several values of the interaction anisotropy J'/J; and (**b**) the respective changes of the critical exponent α for the specific heat along the critical lines displayed in Figure 7a.

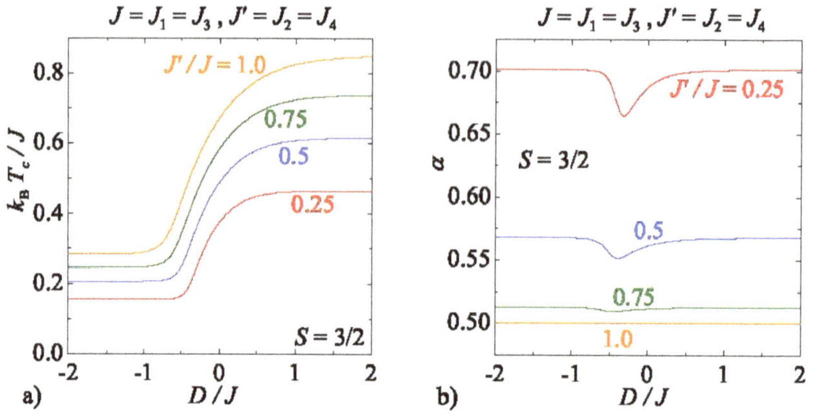

Figure 8. (**a**) The critical temperature of the model C as a function of the uniaxial single-ion anisotropy by considering the spin size $S = 3/2$ and several values of the interaction anisotropy J'/J; and (**b**) the respective changes of the critical exponent α for the specific heat along the critical lines displayed in Figure 8a.

3.4. Model D ($J \equiv J_1 = J_2, J' \equiv J_3 = J_4$)

Finally, we will perform a comprehensive analysis of critical behaviour of the model D with the triplet interactions $J \equiv J_1 = J_2$, $J' \equiv J_3 = J_4$, which are pairwise equal to each other within adjacent triangles of elementary square plaquettes. In this particular case one gets from Equation (5) three different Boltzmann's weights given by:

$$\omega_1 = \sum_{n=-S}^{S} \exp(\beta D n^2) \cosh\left[\frac{\beta n}{2}(J + J')\right], \qquad \omega_3 = \omega_5 = \sum_{n=-S}^{S} \exp(\beta D n^2),$$
$$\omega_7 = \sum_{n=-S}^{S} \exp(\beta D n^2) \cosh\left[\frac{\beta n}{2}(J - J')\right].$$

It is quite apparent that the Boltzmann's weights in Equation (19) satisfy the inequality $\omega_1 \geq \omega_7 \geq \omega_3$, which affords from Equation (7) the critical condition $\omega_1 = 2\omega_3 + \omega_7$. After substituting explicit form of the Boltzmann's weights (19) into the relevant critical condition one obtains the completely identical critical condition as given by Equation (17) for the model C. This result would suggest that the model D has exactly the same phase boundaries as the model C. However, it should be emphasized that the weak-universal critical behaviour of the model D will be characterized by different critical exponents:

$$\tan\left(\frac{\mu}{2}\right) = \sqrt{\frac{\omega_1}{\omega_7}} \quad \Rightarrow \quad \alpha = \alpha' = 2 - \frac{\pi}{\mu}, \quad \beta = \frac{\pi}{16\mu}, \quad \gamma = \gamma' = \frac{7\pi}{8\mu} \quad \nu = \nu' = \frac{\pi}{2\mu}, \tag{19}$$

because the definition of the parameter μ governing changes of the critical exponents is different (cf. Equations (18) and (19)).

For the sake of comparison, the critical temperature and critical exponent α are plotted in Figures 9 and 10 for the model D by assuming two different spin values $S = 1$ and $3/2$, respectively. It has already been argued that the phase boundaries of the model D coincide with the critical lines of the model C, so let us only comment the respective behaviour of the critical exponent α for the specific heat. Figure 9b would suggest that the critical exponent α for the model D with the integer-valued spin $S = 1$ falls down monotonically with increasing of the uniaxial single-ion anisotropy, whereas the interaction anisotropy generally reinforces the relevant decline. This behaviour is in sharp contrast with what has been previously reported for the model C, where exactly opposite tendency has been revealed (cf. Figures 7b and 9b). The reduction of the critical exponent α due to the interaction anisotropy has also been detected for the model D with the half-odd-integer spin $S = 3/2$, but this time the critical exponent α shows a peculiar nonmonotonous dependence on the uniaxial single-ion anisotropy with a round maximum. This finding is repeatedly in marked contrast with what has been previously found for the model C (c.f. Figures 8b and 10b). It could be thus concluded that the models C and D display remarkably different weak-universal critical behaviour of the critical exponents even though their phase boundaries are in a perfect coincidence.

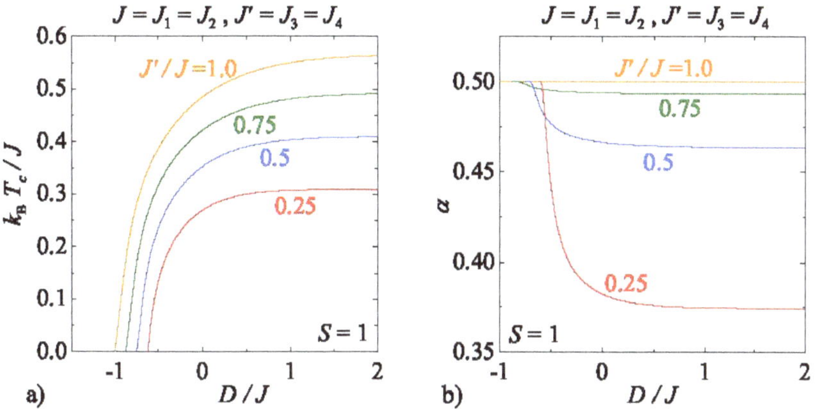

Figure 9. (**a**) The critical temperature of the model D as a function of the uniaxial single-ion anisotropy by considering the spin size $S = 1$ and several values of the interaction anisotropy J'/J; and (**b**) the respective changes of the critical exponent α for the specific heat along the critical lines displayed in Figure 9a.

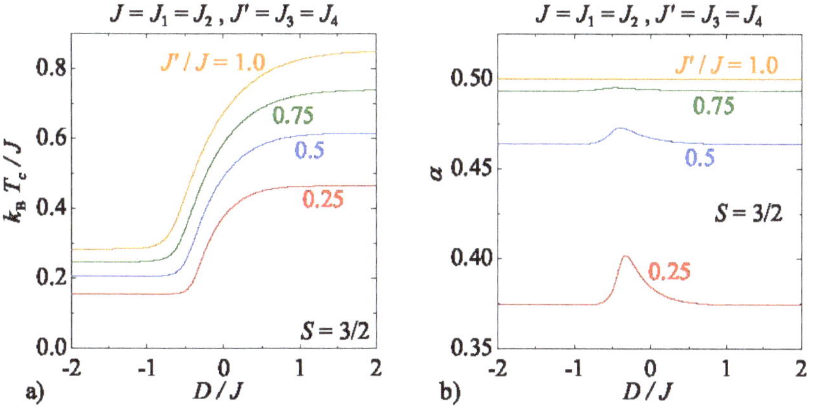

Figure 10. (**a**) The critical temperature of the model D as a function of the uniaxial single-ion anisotropy by considering the spin size $S = 3/2$ and several values of the interaction anisotropy J'/J; and (**b**) the respective changes of the critical exponent α for the specific heat along the critical lines displayed in Figure 10a.

4. Conclusions

The mixed spin-1/2 and spin-S Ising model with four different triplet interactions on a centered square lattice has been exactly solved by establishing a rigorous mapping equivalence with the corresponding zero-field eight-vertex model on a dual square lattice. A rigorous proof of the aforementioned exact mapping equivalence has been afforded by two independent ways, exploiting either the graph-theoretical formulation [10,11] or the spin representation [16,17] of the zero-field eight-vertex model, the latter one obtained after adapting of the generalized star-square mapping transformation [13–15]. It should be mentioned that the range of applicability of the present exactly solved mixed-spin Ising model with the triplet interactions goes beyond the scope of magnetic systems, because three-body interactions might play a crucial role in determining thermodynamic behavior of other complex physical systems such fluids.

The primary goal of the present work was to examine an influence of the interaction anisotropy, the uniaxial single-ion anisotropy and the spin parity upon phase transitions and critical phenomena. It has been shown that the considered model exhibits a strong-universal critical behaviour with constant critical exponents, which are independent of the uniaxial single-ion anisotropy as well as the spin parity when considering either the isotropic model A with four equal triplet interactions or the anisotropic model B with one triplet interaction differing from the other three. On the other hand, it has also been evidenced that the models C and D with the triplet interactions, which are pairwise equal to each other, exhibit a weak-universal critical behaviour characterized by continuously varying critical exponents. Under these circumstances, the relevant critical exponents are changing along the critical lines in dependence on a relative strength of the triplet interactions as well as the uniaxial single-ion anisotropy. Although the critical boundaries of the models C and D are completely the same, different interaction anisotropy of both models is responsible for a qualitatively different weak-universal behaviour of the critical exponents. Besides, it also turns out that the mixed-spin systems with the integer-valued spins S exhibit very different variations of the critical exponents in comparison with their counterparts with the half-odd-integer spins S. A more detailed study of the order parameter and other important thermodynamic quantities, especially in a vicinity of the critical points, is left as a challenging task for our future work.

Acknowledgments: This work was financially supported by the grant of The Ministry of Education, Science, Research and Sport of the Slovak Republic under the contract No. VEGA 1/0331/15 and by the grant of the Slovak Research and Development Agency under the contract No. APVV-16-0186.

Conflicts of Interest: The author declares no conflict of interest.

References

1. Baxter, R.J. *Exactly Solved Models in Statistical Mechanics*; Academic: New York, NY, USA, 1982.
2. Hintermann, A.; Merlini, D. Exact solution of a two dimensional Ising model with pure 3 spin interactions. *Phys. Lett.* **1972**, *41*, 208–210.
3. Baxter, R.J.; Wu, F.Y. Exact Solution of an Ising Model with Three-Spin Interactions on a Triangular Lattice. *Phys. Rev. Lett.* **1973**, *31*, 1294–1297.
4. Baxter, R.J.; Wu, F.Y. Ising Model on a Triangular Lattice with Three-spin Interactions. I. The Eigenvalue Equation. *Aust. J. Phys.* **1974**, *27*, 357–368.
5. Baxter, R.J. Ising Model on a Triangular Lattice with Three-spin Interactions. II. Free Energy and Correlation Length. *Aust. J. Phys.* **1974**, *27*, 369–382.
6. Wood, D.W. A free-fermion model, and the solution of an Ising model with pure triplet interactions. *J. Phys. C Solid State Phys.* **1973**, *6*, L135–L138.
7. Liu, L.L.; Stanley, H.E. Exact solution in an external magnetic field of Ising models with three-spin interactions. *Phys. Rev. B* **1974**, *10*, 2958–2961.
8. Barry, J.H.; Wu, F.Y. Exact Solutions for a 4-Spin-interaction Ising Model on the $d = 3$ Pyrochlore Lattice. *Int. J. Mod. Phys.* **1989**, *3*, 1247–1275.

9. Urumov, V. Examples of Spin and External Field Dependent Critical Exponents. In *Ordering in Two Dimensions*; Sinha, S.K., Ed.; Elsevier: New York, NY, USA, 1980; pp. 361–364.
10. Baxter, R.J. Eight-Vertex Model in Lattice Statistics. *Phys. Rev. Lett.* **1971**, *26*, 832–833.
11. Baxter, R.J. Partition function of the Eight-Vertex lattice model. *Ann. Phys.* **1972**, *70*, 193–228.
12. Suzuki, M. New Universality of Critical Exponents. *Prog. Theor. Phys.* **1974**, *51*, 1992–1993.
13. Fisher, M.E. Transformations of Ising Models. *Phys. Rev.* **1959**, *113*, 969–981.
14. Rojas, O.; Valverde, J.S.; de Souza, S.M. Generalized transformation for decorated spin models. *Physica A* **2009**, *388*, 1419–1430.
15. Strečka, J. Generalized algebraic transformations and exactly solvable classical-quantum models. *Phys. Lett. A* **2010**, *374*, 3718–3722.
16. Wu, F.Y. Ising Model with Four-Spin Interactions. *Phys. Rev. B* **1971**, *4*, 2312–2314.
17. Kadanoff, L.P.; Wegner, R.J. Some Critical Properties of the Eight-Vertex Model. *Phys. Rev. B* **1971**, *4*, 3989–3993.

entropy

MDPI

Article

Anomalous Statistics of Bose-Einstein Condensate in an Interacting Gas: An Effect of the Trap's Form and Boundary Conditions in the Thermodynamic Limit

Sergey Tarasov [1,*], Vladimir Kocharovsky [1,2] and Vitaly Kocharovsky [3]

1 Institute of Applied Physics, Russian Academy of Science, Nizhny Novgorod 603950, Russia; kochar@appl.sci-nnov.ru
2 Department of the Advanced School of General and Applied Physics, Lobachevsky State University, Nizhny Novgorod 603950, Russia
3 Department of Physics and Astronomy, Texas A&M University, College Station, TX 77843-4242, USA; vkochar@physics.tamu.edu
* Correspondence: serge.tar@gmail.com; Tel.: +7-906-353-7253

Received: 31 December 2017; Accepted: 24 February 2018; Published: 27 February 2018

Abstract: We analytically calculate the statistics of Bose-Einstein condensate (BEC) fluctuations in an interacting gas trapped in a three-dimensional cubic or rectangular box with the Dirichlet, fused or periodic boundary conditions within the mean-field Bogoliubov and Thomas-Fermi approximations. We study a mesoscopic system of a finite number of trapped particles and its thermodynamic limit. We find that the BEC fluctuations, first, are anomalously large and non-Gaussian and, second, depend on the trap's form and boundary conditions. Remarkably, these effects persist with increasing interparticle interaction and even in the thermodynamic limit—only the mean BEC occupation, not BEC fluctuations, becomes independent on the trap's form and boundary conditions.

Keywords: Bose-Einstein condensation; statistics of Bose-Einstein condensate; Bogoliubov coupling; interacting Bose gas; mesoscopic system

1. Critical Phenomena and Statistics of Anomalous Fluctuations in a Bose-Einstein Phase Transition

Any second order phase transition in a many-body system occurs near its critical point via development of the critical phenomena and anomalously large fluctuations of an order parameter [1]. Finding their microscopic theory and analytical description constitutes one of the major and still unsolved problem in theoretical physics. A phase transition to the Bose-Einstein condensed state in the interacting Bose gas of N particles confined in an optical/magnetic trap is one of the phenomena that have been the most intensively studied in the last two decades (for a review, see [2] and references therein).

In the present paper we analyze the statistics of the number of condensed particles at the temperature T well outside a narrow critical region around the critical temperature T_c, namely, at $T \ll T_c$ where a well-known mean-field approximation based on the Gross-Pitaevskii and Bogoliubov-de Gennes equations is valid. (In fact, such a mean-field approximation works reasonably well even up to the temperature $T_c/2$, that is for all temperatures in the interval $T < T_c/2$, as has been verified, for example, by the experiments on the dynamics and decay of the condensate excitations [2–6].) At somewhat higher temperatures, but still outside the critical region, one could employ a more complicated theory (beyond the mean field) that consistently accounts for a deeper thermal depletion of spatially inhomogeneous condensate and various many-body effects of particle interactions via an approximate solution to the Dyson equation for the Green's and self-energy functions of the standard quantum-field-theory method [3,4]. Even higher temperatures at the wings of critical region, which are still far from the central part of critical region, could be accessed by means of a phenomenological renormalization-group theory (for a review, see [7] and references therein). Finally, the BEC phase transition in the weakly interacting Bose gas at any temperatures outside and inside the entire critical region can be rigorously described by the general equations for the order parameter, self-energy and Green's functions within a recently developed microscopic theory of phase transitions [8–10].

The study of the BEC statistics has a long and rich history (see, for example, an arbitrary sample of relatively recent papers [11–25] and references therein.) However, nobody succeeded yet in deriving the BEC statistics in the interacting gas for the temperature interval $T_c/2 < T < T_c$ and the entire critical region neither by means of the standard quantum-field-theory method, renormalization-group theory or microscopic theory nor by means of the other known approaches (discussed, e.g., in [6]) which are more involved than the Bogoliubov approximation.

An analytical solution for the BEC statistics at any temperature, including the entire critical region, was found recently [15,26–30] only for the ideal, non-interacting gas confined in a trap of arbitrary geometry and dimensions. In particular, it provides a universal scaling of all thermodynamic quantities and a universal form of the BEC occupation probability distribution over the entire critical region in the thermodynamic limit of macroscopically large systems. This universal statistics is either non-Gaussian of an anomalously large variance or Gaussian, depending on the trap geometry, and clearly manifests itself in the mesoscopic systems starting from the number of particles N of the order of a thousand.

An essential novelty of the work [15] and the subsequent papers [26–30] was a formulation (and a solution) of the very problem of finding the entire probability distribution and its characteristic function for the BEC statistics as opposed to the calculation of just the first two statistical moments, the mean value and the variance, discussed in the preceding works (see, e.g., [11,13,14,18,23]). It was done mainly for the ideal Bose gas, with only one exception. Namely, in the paper [15] we put forward and solved this problem within the Bogoliubov approximation for a weakly interacting gas confined in a three-dimensional (3D) box trap with the periodic boundary conditions. Such a trap has a purely homogeneous ground state and a homogeneous condensate wave function.

In the present paper we continue this line of research. Namely, we find the probability distribution and the characteristic function, including all moments and cumulants, for the BEC statistics in a weakly interacting dilute gas confined in a 3D rectangular box of dimensions L_x, L_y, L_z with the Dirichlet (zeroth) or so-called fused boundary conditions (see (28) and thereinafter). In the Dirichlet's case a one-particle ground state is strongly inhomogeneous $\propto \sin(\pi x/L_x)\sin(\pi y/L_y)\sin(\pi z/L_z)$ and one has to solve the Gross-Pitaevskii and Bogoliubov-de Gennes equations for the inhomogeneous condensate and quasiparticle wave functions (see, e.g., [2,3,5–7,31,32]). For simplicity's sake, we do this within the Thomas-Fermi approximation assuming that the healing length ξ is much less than the box dimensions:

$$\xi \ll L_x, L_y, L_z; \qquad \xi = \sqrt{\frac{V}{8\pi N_0 a}}\ , \quad g = \frac{4\pi\hbar^2 a}{M}\ . \tag{1}$$

Here N_0 is the mean number of particles in the condensate, M the mass of a particle, $V = L_xL_yL_z$ the volume of a trap, $a > 0$ the s-wave scattering length determining the interaction constant g. If the first-order Born approximation for the two-body repulsive interaction with a potential $U(\mathbf{r_1} - \mathbf{r_2})$ is applicable, one can think of the interaction constant as of the $\mathbf{k} = 0$ component, $g = U_0$, of the Fourier transform $U_\mathbf{k} = \int U(\mathbf{r})\exp(i\mathbf{kr})d^3\mathbf{r}$ of the two-body potential and model the latter by a Dirac delta-function $U(\mathbf{r}) = U_0\delta(\mathbf{r})$. The diluteness of the gas means its low density $N/V \ll a^{-3}$, i.e., a small value of the gas parameter $Na^3/V \ll 1$. This assumption of the dilute gas together with the Thomas-Fermi condition (1) implies the following inequalities on the admissible values of the condensate gas parameter n_0a^3:

$$\frac{1}{\sqrt{8\pi}}\frac{a}{L} \ll \sqrt{n_0 a^3} \ll 1 \qquad \text{for } L = L_x, L_y, L_z; \qquad n_0 = \frac{N_0}{V}. \tag{2}$$

Then, possessing the BEC statistics for the mesoscopic system of N interacting particles in the box traps of different rectangular forms and different, periodic, fused or Dirichlet, boundary conditions, we compare their statistics and conclude on the effect of the trap's form and boundary conditions on them. In particular, we calculate the asymptotics of the BEC statistics and the effect of the trap's form and boundary conditions in the thermodynamic limit $N \to \infty$, $V \to \infty$ with a concentration of particles N/V being kept constant.

The paper is organized as following. In Section 2 we formulate the basic model of the BEC statistics in a weakly interacting gas confined in a flat trap within the mean-field theory of the Gross-Pitaevskii equation for the condensate and Bogoliubov—de Gennes equations for the quasiparticles, both being simplified by means of the Thomas-Fermi approximation. In Section 3 we present the general analytical formulas for the characteristic function and all cumulants of the noncondensate occupation probability distribution for a mesoscopic system of N particles in a weakly interacting gas in the case when the Bogoliubov coupling squeezes (forces to coherently correlate the particles' creation and annihilation operators) the contribution to the noncondensate occupation separately either from each quasiparticle or from each pair of symmetric counter-propagating quasiparticles. In Section 4 we derive the explicit analytical formulas for the asymptotics of the obtained characteristic function and all cumulants of the noncondensate occupation probability distribution in the thermodynamic limit. In Section 5 we apply these results for BEC statistics to the case of the cubic or rectangular box trap with the periodic, fused or Dirichlet boundary conditions in the regime of thermally-dominated fluctuations. Section 6 contains the discussion and implications of the obtained results for the statistical physics of the BEC phase transition. The main conclusions are stated in Section 7.

2. Quasiparticles: The Spectrum and Eigenstates Within the Gross-Pitaevskii and Bogoliubov—De Gennes Equations in the Thomas-Fermi Approximation

In the thermal equilibrium the mesoscopic system of N interacting particles is described by the statistical operator (density matrix)

$$\hat{\rho} = e^{-\hat{H}/T} \prod_j (1 - e^{-\varepsilon_j/T}), \quad \hat{H} = \sum_j \varepsilon_j \hat{b}_j^\dagger \hat{b}_j, \tag{3}$$

where the Hamiltonian $\hat{H}$ is given by the Bogoliubov approximation as the sum over the dressed by the condensate and many-body interaction quasiparticles with the eigenenergies ε_j and creation (annihilation) operators $\hat{b}_j^\dagger$ ($\hat{b}_j$) enumerated by the index $j = 1, 2, \ldots$. The field operator of the particles

$$\hat{\psi}(\mathbf{r}) = \sqrt{N_0}\phi(\mathbf{r}) + \hat{\theta}(\mathbf{r}), \quad \text{where} \quad \hat{\theta} = \sum_j \left[u_j(\mathbf{r})\hat{b}_j + v_j^*(\mathbf{r})\hat{b}_j^\dagger \right], \quad [\hat{b}_j, \hat{b}_{j'}^\dagger] = \delta_{j,j'}, \tag{4}$$

includes the c-number (classical) condensate wave function $\sqrt{N_0}\phi(\mathbf{r})$ and the field operator $\hat{\theta}(\mathbf{r})$ of the noncondensate particles. The operators $\hat{b}_j^\dagger$ and $\hat{b}_j$ obey the canonical commutation relations.

The profile of the condensate wave function ϕ (which is chosen to be real) obeys the Gross-Pitaevskii equation [2]

$$\left(-\frac{\hbar^2}{2M}\Delta_{\mathbf{r}} + U_{trap}(\mathbf{r}) + gN_0\phi^2(\mathbf{r}) - \mu \right) \phi(\mathbf{r}) = 0, \qquad \int \phi^2(\mathbf{r}) d^3\mathbf{r} = 1, \tag{5}$$

where $U_{trap}(\mathbf{r})$ is the external potential of the trap and μ the chemical potential. The eigenenergies and eigenstates $(u_j, v_j)^T$ of quasiparticles are determined by the Bogoliubov-de Gennes equations [2,31,32]

$$\left(-\frac{\hbar^2}{2M}\Delta_{\mathbf{r}} + U_{trap}(\mathbf{r}) + 2gN_0\phi(\mathbf{r})^2 - \mu \right) \begin{pmatrix} u_j \\ v_j \end{pmatrix} + gN_0\phi^2(\mathbf{r}) \begin{pmatrix} v_j \\ u_j \end{pmatrix} = \varepsilon_j \begin{pmatrix} +u_j \\ -v_j \end{pmatrix} \tag{6}$$

with the boundary conditions specified by the trap setup. Here $\Delta_\mathbf{r}$ is the Laplace operator.

The operator of the number of noncondensed particles that determines the statistics of the condensate depletion can be represented via the creation and annihilation operators of quasiparticles:

$$\hat{N}_{ex} \equiv \int \hat{\theta}^{+}\hat{\theta} d^3\mathbf{r} = \sum_{j,k} \left(\hat{b}_j^{+}\hat{b}_k \int u_j^* u_k d^3\mathbf{r} + \hat{b}_j\hat{b}_k \int u_j v_k d^3\mathbf{r} + \hat{b}_j^{+}\hat{b}_k^{+} \int u_j^* v_k^* d^3\mathbf{r} + \hat{b}_j\hat{b}_k^{+} \int v_j v_k^* d^3\mathbf{r} \right). \quad (7)$$

The subscript "ex" stands for "excitations". Since the total number of particles in the trap is conserved, $N = const$, the condensate occupation statistics is determined by the complimentary operator $\hat{N}_0 = N - \hat{N}_{ex}$. The parameter N_0, that enters the Equations (5) and (6) and plays a part of the mean field in this Bogoliubov approach, is the mean condensate occupation and should be found from the self-consistency equation

$$N_0 = N - \langle \hat{N}_{ex} \rangle, \qquad \langle ... \rangle = \mathrm{Tr}(...\hat{\rho}). \quad (8)$$

The angle brackets denote averaging over the quantum ensemble given by the statistical operator (3).

In fact, the particle-number-conservation constraint $\hat{N}_0 + \hat{N}_{ex} = N$ defines the canonical or microcanonical (along with the energy constraint $\hat{H} = E = const$) statistical ensembles and is responsible for the very existence of the BEC phase transition as a process leading to a macroscopic occupation of a single quantum state [1,2,9,31]. Thus, a rigorous microscopic description of the BEC requires an introduction of the constraint Hilbert space by an appropriate truncation of the original Fock space of the system of N particles. The most natural and straightforward way of doing so is to express all physical quantities via the particle-number-conserving creation and annihilation operators of the canonical quasiparticles which describe the transitions between the ground and excited one-particle states. Such operators were first introduced in the famous Holstein-Primakoff representation [33] (worked out also by Schwinger [34]) and are known in the theory of BEC as the Girardeau-Arnowitt operators [35,36]. The similar "number-conserving" approaches were widely discussed in the literature – for example, see [37] introducing the ladder operators similar to the Girardeau-Arnowitt ones (but conserving the particle number only approximately, [36]), [38] involving the modified ladder operators to establish the U(1)-symmetry model for calculating the dynamics of BEC, [39] formulating the number-conserving finite temperature theory which takes into account the non-quadratic terms of Hamiltonian, [15] describing the BEC statistics in the Bogoliubov approximation via the Girardeau-Arnowitt operators, [16,17] describing the higher-order cumulants of the BEC statistics in an ideal gas at zero temperature, [40,41] presenting the description of the coupled dynamics of the condensate and noncondensate fractions, [24] calculating the correlations in a Bose gas via introduction of the c-number constants for the Girardeau-Arnowitt operators and the so-called "coherent" energy levels which are highly-occupied and mostly form the condensate mode of the interacting gas, [19] where the number-conserving approach was used for a comparatative analysis of the generalized BEC in the anysotropic boxes within the canonical and grand-canonical ensembles, etc.

While the particle number conservation is crucially important for the correct description of BEC in the critical region [9,26–29], the constraint $\hat{N}_0 + \hat{N}_{ex} = N$ is less important for the well-developed condensate phase at low temperatures $T \ll T_c$ considered in the present paper. Here the role of the constraint is reduced mainly to the self-consistency Equation (8) for the mean condensate occupation and has only a minor effect on fluctuations. In fact, the particle number conservation does not affect directly and cannot eliminate the strong effect of squeezing of quasiparticle contributions to the condensate depletion in Equation (7) via the Bogoliubov couplings.

The Bogoliubov-de Gennes system of equations is not a standard realization of the Sturm–Liouville problem, and there is no usual orthogonality of the partial u's and v's components of quasiparticle wave functions in the general case [31,32]. This fact is crucial for the existence of the nonzero Bogoliubov couplings, that is the nonzero integrals playing a part of coefficients in the right hand side of Equation (7). Actually, the sensitivity of the BEC statistics to the trap's form and boundary conditions strongly involves the statistics' dependence on the spectrum of quasiparticles, while the aforementioned non-orthogonality of the partial components of eigenfunctions u_j and v_j determines the squeezing of quasiparticle contributions to the condensate depletion in Equation (7) via the Bogoliubov couplings. At the same time, the eigenvectors $(u_j, v_j)^T$ of the quasiparticle wave functions are orthogonal [31] in the sense of the difference-component scalar product:

$$\int (u_i u_j^* - v_i v_j^*) dV = \delta_{i,j}, \quad \int (u_i v_j - u_j v_i) dV = 0, \quad \int (u_i \phi - v_i \phi^*) dV = 0.$$

Thus, finding the statistics of the condensate depletion operator $\hat{N}_{ex}$ over the quantum statistical ensemble $\hat{\rho}$ is a problem, far from being trivial, of diagonalizing an infinite matrix (7) with mostly nonzero entries defined by the overlapping integrals of quasiparticle states. However, there are cases when only a small part of the overlapping integrals is significant. One of such cases is the case of the flat traps discussed below.

In the present paper we simplify calculations by employing the Thomas-Fermi approximation in which one ignores the kinetic energy term $-(\hbar^2/(2M))\Delta_{\mathbf{r}}$ in the Gross-Pitaevskii Equation (5) and obtain the analytical solution for the condensate wave function:

$$\sqrt{N_0}\phi(\mathbf{r}) \approx \sqrt{\frac{\mu - U_{trap}(\mathbf{r})}{g}} \quad \text{if} \quad U_{trap}(\mathbf{r}) < \mu \tag{9}$$

and zero elsewhere. It is known that this approximation works quite well if the interparticle interaction is strong enough for the condition (1) to be fulfilled [2,32]. Moreover, we restrict analysis to the case of flat traps with a constant value (taken to be zero) of trapping potential inside the entire trap's volume and the infinite potential walls at the trap's borders. According to the Thomas-Fermi approximation (9), it amounts to the homogeneous profile of the condensate, $\phi(\mathbf{r}) = V^{-1/2}$, and $\mu = gn_0$. In this case the solution to the Bogoliubov-de Gennes Equations (6) for the quasiparticle states acquires the form

$$u_j(\mathbf{r}) = \frac{f_j(\mathbf{r})}{\sqrt{1 - A_j^2}}, \qquad v_j(\mathbf{r}) = \frac{A_j f_j(\mathbf{r})}{\sqrt{1 - A_j^2}} \tag{10}$$

with the spatial profile of both components $u_j(\mathbf{r})$ and $v_j(\mathbf{r})$ given by the same function $f_j(\mathbf{r})$ which is the solution to the one-particle Shrödinger equation of the empty (without condensate) trap:

$$-\frac{\hbar^2}{2M}\Delta_{\mathbf{r}} f_j(\mathbf{r}) = \epsilon_j^{(0)} f_j(\mathbf{r}). \tag{11}$$

The Shrödinger Equation (11) establishes the orthogonal basis of the one-particle wave functions $\{f_j(\mathbf{r})\}$ and the corresponding spectrum of the bare energies: $\epsilon_1^{(0)} \le \epsilon_2^{(0)} \le \epsilon_3^{(0)} \le \dots$

We assume that the latter equation is equipped with the same boundary conditions as were imposed on the one-particle wave function at the borders of the original (without condensate) trap. This means that the quasiparticles are mostly determined by the Bogoliubov-de Gennes equations applied to the central part of the almost homogeneous condensate which demonstrates a significant density drop only in the relatively thin, of the order of the healing length $\xi \ll L_x, L_y, L_z$, inhomogeneous boundary layers. At the same time, the boundary conditions imposed at the borders of the trap are important and strongly influence the quasiparticles in the following ways. First, the high-energy quasiparticles, which have the quasi-classical (WKB) properties, propagate adiabatically (freely) through the boundary layers, that is, they carry a full information about the boundaries and fully "feel" the boundary conditions. Second, the low-energy quasiparticles, the wavelengths of which are of the order of the trap's dimension, also know about boundaries since even in the central part of the volume these quasiparticles must obey the symmetry of the system which is determined by the boundary conditions. For example, switching between the Dirichlet and periodic boundary conditions, roughly speaking, enables or disables a half of all "wavevector components" (quantum numbers) in the spectrum of eigenstates in each spatial direction. Since the BEC statistics is determined mainly by the large groups of excited states and the trending terms of energy spectra, we assume that the corrections to eigenenergies $\epsilon_j^{(0)}$ caused by the inhomogenety of the condensate in the boundary layers via varying coefficients in the Bogoliubov-de Gennes Equations (6) do not strongly affect the BEC statistics in the asymptotics of vanishing parameter $\xi / V^{1/3} \to 0$. Thus, although the boundary condensate layers are formally omitted in the Thomas-Fermi approximate solution (9), they are actually present in the basic model formulated above. They are responsible for a transition from the homogeneous value of the condensate profile $\phi(\mathbf{r}) = V^{-1/2}$ in the trap's volume to the value (for example, zero in the Dirichlet's case) required by the original boundary conditions at the trap's borders and also determine the symmetry of the whole system.

The model (10) and (11), based on the Thomas-Fermi approximation, of the Bogoliubov-de Gennes Equations (6) describes the quasiparticles with the spectrum of eigenenergies ε_j and Bogoliubov couplings determined by the coupling parameter A_j given by the following equations:

$$\varepsilon_j = \sqrt{\left(\epsilon_j^{(0)}\right)^2 + 2gn_0\epsilon_j^{(0)}}, \qquad A_j = \frac{\varepsilon_j - \epsilon_j^{(0)} - gn_0}{gn_0}, \tag{12}$$

where the interaction coupling energy gn_0 is proportional to the condensate density $n_0 = N_0/V$. Similar approaches are known in the literature on the BEC in the inhomogeneous systems and were used by many authors (see, e.g., [32,42]) for the calculation of quantum depletion and other parameters of BEC.

It is worth noting that the formulated above model for the BEC statistics (10)–(12) is very basic and poses a series of further questions which go beyond the scope of the present paper. They include, for example,

(i) the problem of finding the actual spectrum and Bogoliubov couplings of quasiparticles from the solutions of the exact Gross-Pitaevskii and Bogoliubov-de Gennes Equations (5) and (6) beyond the Thomas-Fermi approximate model (10)–(12) as well as more specific questions, like
(ii) the problem of the possibly missed or excessive quasiparticles in the model (10)–(12),
(iii) the problem of an accurate transfer of the actual trap's boundary conditions into the equivalent boundary conditions for the approximate Shrödinger Equation (11) for quasiparticles at the borders of the main body of the quasi-homogeneous condensate through the inhomogeneous condensate boundary layer of thickness $\xi \ll L_x, L_y, L_z$ (for example, in terms of the impedance boundary conditions dictated by the exact solution to the Gross-Pitaevskii Equation (5)).

A possible approach to these problems, based on the second-order perturbation theory, can be found in [42] where the Bogoliubov analysis of the weakly inhomogeneous Bose system as well as the results for the corresponding deformation of the condensate and for the expectation value of the quantum depletion were presented. In particular, a suppression of the quantum depletion induced by a weak lattice was predicted. (The effect is probably related to the quasiparticle selection and corresponding limitation of quasiparticle spectrum by means of the Bragg scattering on the lattice.)

We'll discuss these problems and effects elsewhere. Below we focus on the technique for the analysis of BEC statistics in the interacting gas within the basic model (10)–(12) and, in particular, on a possible effect of the trap's form and boundary conditions on the statistics of condensate depletion in the thermodynamic limit.

3. The Characteristic Function and Cumulants of the BEC Depletion Probability Distribution for a Mesoscopic System of Weakly Interacting Particles

Here we describe a technique for evaluating the statistics of the total number of noncondensed particles. The statistics is derived in terms of a Fourier transform of its characteristic function which is described via the cumulants and generating cumulants related to the BEC depletion operator $\hat{N}_{ex}$ in Equation (7) and calculated analytically.

We consider a mesoscopic system with any finite number N of trapped particles and dimensions of the trap L_x, L_y, L_z. Let us introduce dimensionless energy variables:

$$\lambda_j^{(0)} \equiv \epsilon_j^{(0)}/\epsilon_1^{(0)}, \qquad \lambda_j \equiv \varepsilon_j/\epsilon_1^{(0)}, \qquad \Delta \equiv gn_0/\epsilon_1^{(0)}, \qquad \alpha \equiv \epsilon_1^{(0)}/T,$$
$$\text{so that} \qquad \epsilon_j^{(0)}/T = \alpha\lambda_j^{(0)}, \qquad \varepsilon_j/T = \alpha\lambda_j, \qquad gn_0/T = \alpha\Delta. \tag{13}$$

The dimensionless spectra $\{\lambda_j^{(0)}\}$ and $\{\lambda_j\}$ represent the bare one-particle and renormalized by condensate quasiparticle energies, respectively. The strength of the interparticle interaction is characterized by the parameter $\Delta = gn_0/\epsilon_1^{(0)}$ which is the ratio of the interaction coupling energy to the energy gap in the bare one-particle spectrum. The applicability condition of the Thomas-Fermi approximation (1) means that one has $\Delta \gg 1$. The dimensionless parameter α characterizes the size of the system, so that the thermodynamic limit of a macroscopically large system means $\alpha \to 0$. The statistical operator (3) and Bogoliubov quasiparticle eigenenergies ε_j and coupling coefficients A_j of Equation (12) acquire the following forms with an explicit appearance of the asymptotic parameter α:

$$\hat{\rho} = e^{-\alpha \sum_j \lambda_j \hat{b}_j^+ \hat{b}_j} \prod_j (1 - e^{-\alpha\lambda_j}), \qquad \lambda_j = \sqrt{\left(\lambda_j^{(0)}\right)^2 + 2\lambda_j^{(0)}\Delta}, \qquad A_j = \frac{\lambda_j - \lambda_j^{(0)} - \Delta}{\Delta}. \tag{14}$$

The present analysis implies the Bogoliubov approximation which is valid at $T \ll T_c$ (or maybe even up to $T < T_c/2$), that is, anyway far outside the critical region. Hence, the condensate is well developed and the particle number constraint of the canonical ensemble $N = \hat{N}_0 + \hat{N}_{ex}$ does not play any significant role for the occupation numbers of the quasiparticle states and different blocks of quasiparticles (which are, in fact, pairs of quasiparticles in the case of the box trap), which have mutually nonzero overlapping integrals in the excited particle occupation operator (7), produce independent contributions to the noncondensate occupation. This means that the characteristic function is the product of the partial characteristic functions, corresponding to the occupation contributions from such blocks of quasiparticles, and the probability distribution $\rho(n)$ of the condensate depletion operator $\hat{N}_{ex}$ could be effectively obtained via the Fourier transform of this characteristic function

$$\Theta(u) \equiv \langle e^{iu\hat{N}_{ex}} \rangle, \qquad \rho(n) = \frac{1}{2\pi} \int_{-\pi}^{\pi} e^{-iun} \Theta(u) du. \tag{15}$$

Thus, it remains to calculate the partial characteristic function for the noncondensate occupation from each separate block of quasiparticles contributing to the noncondensate with mutually coherent correlations, or squeezing, due to Bogoliubov couplings in Equations (7) and (10). This is straightforward to do either by the algebraic method suggested in [15] or by the developed later, equivalent and more general method of the Wigner transform [43]. The result can be written as the product of contributions from each quasiparticle j as follows

$$\Theta(u) = \prod_j \sqrt{\frac{z(A_j) - 1}{z(A_j) - e^{iu}} \times \frac{z(-A_j) - 1}{z(-A_j) - e^{iu}}}, \qquad z(A_j) = \frac{A_j - e^{\alpha\lambda_j}}{A_j e^{\alpha\lambda_j} - 1}. \tag{16}$$

The corresponding generating cumulants (which are defined via the following derivatives of the logarithm of the characteristic function) have a symmetric transparent form:

$$\tilde{\kappa}_m \equiv \left.\frac{\partial^m \ln \Theta(u)}{\partial (e^{iu}-1)^m}\right|_{u=0} = \frac{\Gamma(m)}{2}\sum_j \left[\frac{1}{(z(A_j)-1)^m} + \frac{1}{(z(-A_j)-1)^m}\right], \tag{17}$$

$\Gamma(m)$ denotes the Euler gamma functions. Here the product and the sum run over all quasiparticles of the considered basic model (10)–(12). These formulas hold for any homogeneously condensed mesoscopic system, including both the case of the real-valued eigenfunction f_j (when Bogoliubov coupling squeezes the contribution to noncondensate occupation from each quasiparticle separately and the corresponding matrix blocks in the noncondensate occupation operator (7) are 2×2 blocks) and the case of the complex-valued eigenfunction f_j (when Bogoliubov coupling squeezes the contribution to noncondensate occupation from each pair of symmetric counter-propagating quasiparticles and the corresponding matrix blocks in the noncondensate occupation operator (7) are 4×4 blocks) as well as even the case when both real- and complex-valued eigenfunctions are simultaneously present among the quasiparticles.

The presented generating cumulants are very useful for the analysis of the BEC statistics. The ordinary cumulants $\kappa_m \equiv d^m \ln \Theta(u)/d(iu)^m|_{u=0}$ and the initial moments $\alpha_m \equiv \langle \hat{N}_{ex}^m \rangle$ immediately follow from the generating cumulants (for the other general relations see, e.g., [15,26,44]):

$$\kappa_r = \sum_{m=1}^{r} \sigma_r^{(m)} \tilde{\kappa}_m, \qquad \alpha_m = \sum_{r=1}^{m} \sum^{(m,r)} (m; a_1, ..., a_m)' \kappa_1^{a_1} ... \kappa_m^{a_m}.$$

Here $\sigma_r^{(m)}$ are the Stirling numbers of the second kind [44], $(m; a_1, ..., a_m)' = m!/[(1!)^{a_1} a_1! ... (m!)^{a_m} a_m!]$ is a multinomial coefficient, and the sum $\sum^{(m,r)}$ runs over the non-negative integers $a_1, ..., a_r$ which satisfy the following two conditions: $a_1 + 2a_2 + ... + r a_r = r$ and $a_1 + ... + a_r = m$. In particular, the first four cumulants constitute themselves the four most important characteristics of the shape of the probability density function – the mean value, the variance (square of the standard deviation), the asymmetry and the excess:

$$\begin{aligned}
&\text{Mean} = \langle \hat{N}_{ex} \rangle = \kappa_1 = \tilde{\kappa}_1; \\
&\text{Variance} = \sigma^2 = \kappa_2 = \tilde{\kappa}_2 + \tilde{\kappa}_1; \\
&\text{Asymmetry} = \kappa_3 = \tilde{\kappa}_3 + 3\tilde{\kappa}_2 + \tilde{\kappa}_1; \\
&\text{Excess} = \kappa_4 + 3\kappa_2^2, \qquad \kappa_4 = \tilde{\kappa}_4 + 6\tilde{\kappa}_3 + 7\tilde{\kappa}_2 + \tilde{\kappa}_1.
\end{aligned} \tag{18}$$

The meaning and the main advantage of the cumulants are in their direct and explicit characterization of the non-Gaussian properties of statistics as well as its asymptotics at the wings of the probability density function. In particular, the cumulants of the Gaussian statistics are exactly zero for all orders higher than the second order. Thus, the simple analytical expressions (17) for the generating cumulants yield the analytical results for any other statistical parameters which one could be interesting in.

The Equation (17) for the first cumulant, $m = 1$, has a special meaning. It is a nonlinear self-consistency Equation (8) for determining the BEC order parameter that is the BEC mean occupation:

$$N_0 = N - \sum_j \frac{1 + A_j^2}{1 - A_j^2} \frac{1}{e^{\alpha \lambda_j} - 1} - \sum_j \frac{A_j^2}{1 - A_j^2}. \tag{19}$$

This is because the total number of particles in the trap N is fixed, so that the occupation of condensate is determined by the number of excited particles, $N_0 = N - \langle \hat{N}_{ex} \rangle$, and, at the same time, the mean number of excited particles $\langle \hat{N}_{ex} \rangle$ is determined via the Bogoliubov-de Gennes quasiparticles' spectrum λ_j and couplings A_j which are governed by the interaction coupling energy gN_0/V. Formally, the self-consistency equation has a nontrivial solution not only in the whole region of validity of the Bogoliubov approximation at low enough temperatures, say, $T \ll T_c$, but also at the temperatures much closer to the critical temperature T_c (see, for example, [15]). The solution to the self-consistency equation could be found as a power series in the asymptotic parameter α via the Mellin transform technique discussed, for example, in [27,28].

When the self-consistent value of the condensate occupation N_0 has been found, the Equations (16) and (17) analytically yields the BEC statistics of the mesoscopic system. This statistics is non-Gaussian as is clearly seen from the formulas for the higher-order cumulants which establish nonzero values.

4. Asymptotics of the BEC Statistics in the Thermodynamic Limit

Let us now calculate the thermodynamic limit of the BEC statistics (15) and (16) obtained above for the basic model (10)–(12). The crucial fact is that this BEC statistics, in general, is non-Gaussian and depends on the trap's form and boundary conditions not only for a mesoscopic system, but even for a macroscopically large system in the thermodynamic limit.

The cumulant analysis provides the most convenient and transparent description of the BEC statistics. First of all, let us find the asymptotic scaling of the cumulants in the limit $\alpha \ll 1, \Delta \gg 1$ via a direct evaluation of the sum in the Equation (17). The two terms in the summand forming the generating cumulant $\tilde{\kappa}_m$ in the Equation (17) could be written in the following forms:

$$\begin{aligned} \frac{1}{z(+A_j)-1} &= \frac{1}{\lambda_j^{(0)}} \times \frac{\lambda_j}{e^{\alpha\lambda_j}-1} - \frac{A_j}{1+A_j}, \\ \frac{1}{z(-A_j)-1} &= \frac{1}{\lambda_j^{(0)}+2\Delta} \times \frac{\lambda_j}{e^{\alpha\lambda_j}-1} + \frac{A_j}{1-A_j}. \end{aligned} \tag{20}$$

They are the combinations of the two summands which have a different origin and different dependencies on the two large parameters, $\alpha^{-1} \gg 1$ and $\Delta \gg 1$.

The first terms in the right hand sides describe the effects of a nonzero temperature on the BEC statistics associated with the thermal depletion of BEC and strongly depend on the parameter α. They predictably vanish in the limit of very low temperatures $T \to 0$, or $\alpha\lambda_1 \gg 1$, but play a major role in the moderate temperature regime. It is important to note that these "thermal" terms curiously involve the bare particle energies $\lambda_j^{(0)}$, which happens due to the exact algebraic relations $(1+A_j)/(1-A_j) = \lambda_j^{(0)}/\lambda_j$ and $(1-A_j)/(1+A_j) = (\lambda_j^{(0)}+2\Delta)/\lambda_j$ valid for the Bogoliubov spectrum (12) and its dimensionless counterpart (14).

The second terms in the right hand sides are responsible for the effects associated with the quantum depletion. They are totally independent on α (and temperature) and depend only on the interaction parameter Δ which determines the Bogoliubov couplings A_j.

Taking into account the asymptotic behavior of the coupling coefficient A_j,

$$A_j \simeq -1 + \sqrt{2\lambda_j^{(0)}/\Delta} \quad \text{for } \lambda_j^{(0)} \ll \Delta, \qquad \text{and} \qquad A_j \simeq -\Delta/2\lambda_j^{(0)} \quad \text{for } \lambda_j^{(0)} \gg \Delta, \tag{21}$$

we immediately conclude that the term $\left(z(+A_j)-1\right)^{-1}$ in Equation (17) determines the main contributions from the states with low energies, $\lambda_j^{(0)} \ll \Delta$, since it has the resonant denominators both in the "quantum" and "thermal" parts. For the states with high energies, $\lambda_j^{(0)} \gg \Delta$, both terms in brackets in Equation (17) are equally important.

For the first generating cumulant these "thermal" and "quantum" summands are present as the separate contributions which is clearly illustrated by Equation (19). However, for all higher-order generating cumulants there is an interplay of these "thermal" and "quantum" contributions in accord with the following binomial expansion:

$$\tilde{\kappa}_m = \frac{\Gamma(m)}{2} \sum_j \sum_{l=0}^{m} C_m^l \left(\frac{(-A_j)^{m-l}}{\left(\lambda_j^{(0)}\right)^l (1+A_j)^{m-l}} + \frac{(A_j)^{m-l}}{\left(\lambda_j^{(0)}+2\Delta\right)^l (1-A_j)^{m-l}} \right) \left(\frac{\lambda_j}{e^{\alpha\lambda_j}-1} \right)^l, \quad (22)$$

where C_m^l stands for the binomial coefficient.

Let us consider separately the three significantly different cases corresponding to the three possible relations between the two large parameters, $\alpha^{-1} \gg 1$ and $\Delta \gg 1$. The case of a very large interaction energy and low temperatures, $gn_0 \gg T, \epsilon_1^{(0)}$, corresponds to a very cold, somewhat degenerate interacting gas with the negligible thermal effects. The opposite case of relatively high temperatures, $T \gg gn_0 \gg \epsilon_1^{(0)}$, corresponds to an interacting gas with thermally dominated fluctuations. In the intermediate case the interplay between the thermal and quantum effects should be taken into account. The following analysis of the cumulants' scaling requires a comparison of the contributions from various individual states (quasiparticles) to the noncondensate occupation and is closely related to the applicability of the central limit theorem to the corresponding random variable (7) which is the sum of many random squeezed contributions from the quasiparticles.

In the case of very low temperatures, the probability distribution of the noncondensate occupation is determined solely by the quantum effects, and all Bose thermal exponents in Equation (20) are negligibly small. The cumulants are scaled in terms of the large parameter $\Delta \gg 1$. The individual contribution of a low-energy quasiparticle (with an almost linear, or acoustic, Bogoliubov-de Gennes spectrum λ_j) to the generating cumulant $\tilde{\kappa}_m$ has the order of $\left(\Delta/\lambda_j^{(0)}\right)^{m/2}$ which easily follows from the Equations (17) and (20). The bare energy $\lambda_j^{(0)}$ has a quadric dependence on the quantum numbers. Thus, the first two cumulants, $\langle \hat{N}_{ex} \rangle$ and σ^2, can be approximated by the well-convergent integrals over all states $\{j\}$ that yields the scaling $\langle \hat{N}_{ex} \rangle \sim \sigma^2 \sim \Delta^{3/2}$. The corresponding sums are much larger than the individual contributions since a plenty of states contribute considerably. The higher-order cumulants cannot be approximated in the same way since the corresponding integrals are divergent at low energies. It means that only a small fraction of low-energy states makes the dominant contribution to the value of higher-order cumulants which has the same order of magnitude as the individual summands. In a result, the higher-order cumulants scale as $\kappa_m \sim \Delta^{m/2}$ for $m \geq 4$. The third cumulant has a little bit different scaling $\kappa_3 \sim \Delta^{3/2} \ln \Delta$ which corresponds to the logarithmic divergence of the approximating integral and means that a number of high-energy states make a small, but not totally negligible contributions.

These calculations show that in the thermodynamic limit of the macroscopically large system, $V \to \infty$ at a constant condensate density, $N_0/V \simeq$ const, the total noncondensate occupation in a quantum-dominated regime experiences the standard thermodynamic fluctuations with the standard deviation $\sigma \sim \sqrt{\langle \hat{N}_{ex} \rangle}$ and has the normal, Gaussian probability distribution with vanishing normalized higher-order cumulants: $\kappa_m/\sigma^m \to 0$. Exactly the same situation holds for the statistics of the noncondensate occupation in an ideal Bose gas trapped in the three-dimensional harmonic potential [27,28] when the largest parameter in the system is α^{-1}.

In the opposite case of thermally-dominated fluctuations, the thermal depletion is much larger than the quantum depletion. In this case the main contribution to the generating cumulants immediately follows from the binomial expansion (17) if one keeps only the largest terms:

$$\tilde{\kappa}_m = \frac{\Gamma(m)}{2} \sum_j \left[\frac{1}{\left(\lambda_j^{(0)}\right)^m} + \frac{1}{\left(\lambda_j^{(0)}+2\Delta\right)^m} \right] \left(\frac{\lambda_j}{e^{\alpha\lambda_j}-1} \right)^m . \quad (23)$$

(Let us stress again that the brackets in this expression involve the bare energy spectrum.) The individual "thermal" contribution of a quasiparticle to the generating cumulant $\tilde{\kappa}_m$ has the order of $\left(\alpha\lambda_j^{(0)}\right)^{-m}$ and demonstrates the dependence on $\lambda_j^{(0)}$ that is significantly steeper than the one revealed for the "quantum" contribution. Thus, for any 3D flat trap the sum could be obtained via the integral over all states $\{j\}$ only for the first cumulant $\tilde{\kappa}_1$, which immediately yields the following result

$$\langle \hat{N}_{ex} \rangle \simeq V\,\zeta\left(\frac{3}{2}\right)\left(\frac{MT}{2\pi\hbar^2}\right)^{3/2} \sim \alpha^{-3/2}, \tag{24}$$

where $\zeta(j) = \sum_{l=1}^{\infty} l^{-j}$ is a Riemann zeta function. It is worth noting that the contribution of each quasiparticle j is on the order of α^{-1}, but a huge number of the contributing states makes the collective result much larger, namely, $\langle \hat{N}_{ex} \rangle \sim \alpha^{-3/2}$.

At the same time, for all higher-order cumulants including the second one (which gives the variance of the distribution) the corresponding integral is divergent at low energies. It means, that the main contribution to the value of $\tilde{\kappa}_m$ for $m \geq 2$ comes from a small enough number of the lowest energy quasiparticles j, and the discrete structure of the energy spectrum should be taken into account. Employing the series expansion for the Boltzmann exponent $e^{\alpha\lambda_j} - 1 \simeq \alpha\lambda_j$, we obtain

$$\tilde{\kappa}_m = \frac{\Gamma(m)\alpha^{-m}}{2}\sum_j\left[\frac{1}{\lambda_j^{(0)m}} + \frac{1}{(\lambda_j^{(0)} + 2\Delta)^m}\right]. \tag{25}$$

The Equation (25) means that the total value of the sum is of the same order as the values of the individual contributions. The latter result is similar to the one obtained in [27,28] for the ideal Bose gas. The only difference is that now, for the interacting gas, it involves the two spectral sums, $\sum_j \frac{\Gamma(t)}{\lambda_j^{(0)t}}$ and $\sum_j \frac{\Gamma(t)}{(\lambda_j^{(0)}+2\Delta)^t}$, instead of a single spectral sum $\sum_j \frac{\Gamma(t)}{(\lambda_j-\lambda_0)^t}$ that was employed in the BEC statistics of the ideal gas with the ground state energy λ_0. The Equation (25) for $m = 2$ immediately reveals the scaling of the variance, $\sigma^2 \sim \tilde{\kappa}_2 \sim \alpha^{-2}$. Thus, the system is characterized by the anomalously large variance [11,12,14,15,18,25], $\sigma^2 \sim \langle \hat{N}_{ex} \rangle^{4/3} \gg \langle \hat{N}_{ex} \rangle$, as is clear from the Equations (24) and (25).

This, partly heuristic method of finding the cumulants' asymptotics could be verified by a more rigorous method involving the Mellin transform and employed previously for the ideal gas in [27,28]. The latter method yields even the smaller next order terms of asymptotics omitted in (25). However, here we skip discussion of these terms and other details of the asymptotics. Let us just note that the generating cumulant $\tilde{\kappa}_m$ and the ordinary cumulant κ_m have the same main term in the asymptotics.

It is convenient to introduce the normalized random variable $x = (N_{ex} - \langle \hat{N}_{ex} \rangle)/\sigma$ which characterizes the statistics of the number of noncondensed particles, has the zero mean value and the unity variance. The cumulants and characteristic function of this normalized variable could be easily obtained by applying the scaling and employing the trap function $S(t, u)$ discussed in detail in [27]. The crucial point of the calculation is that the scaling of the variance $\sigma^2 \sim \alpha^{-2}$ is related to the scaling of the higher-order cumulants $\kappa_m \sim \alpha^{-m}$ in quite a special way. Thus, in the thermodynamic limit $\alpha \to 0$ the simple calculations of the scaled higher-order cumulants $\kappa_m^{(x)}$ of the normalized variable x yield the following explicit analytical result:

$$\begin{gathered}\kappa_1^{(x)} = 0; \quad \kappa_2^{(x)} = 1; \quad \kappa_m^{(x)} \equiv \frac{\kappa_m}{\sigma^m} = \frac{\tilde{S}(m)}{\tilde{S}(2)^{m/2}} \quad \text{for} \quad m > 2, \\ \tilde{S}(t) = \frac{\Gamma(t)}{2}\sum_j\left[\frac{1}{\left(\lambda_j^{(0)}\right)^t} + \frac{1}{\left(\lambda_j^{(0)} + 2\Delta\right)^t}\right], \; \tilde{S}(t,u) = \frac{\Gamma(t)}{2}\sum_j\left[\frac{1}{\left(\lambda_j^{(0)} - iu\right)^t} + \frac{1}{\left(\lambda_j^{(0)} + 2\Delta - iu\right)^t}\right],\end{gathered} \tag{26}$$

where we employed the trap function $\tilde{S}(t)$ and the extended trap function $\tilde{S}(t,u)$ determined by the spectrum $\{\lambda_j\}$. It is immediate to conclude that in the thermodynamic limit $\alpha \to 0$ the scaled higher-order cumulants tend to the nonzero constants determined by the trap's form (geometry) and boundary conditions as well as by the strength of the interparticle interaction.

The probability density function ρ_x of the normalized random variable $x = (N_{ex} - \langle \hat{N}_{ex} \rangle)/\sigma$ can be written straightforwardly in the form of the Fourier transform:

$$\rho_x = \frac{1}{2\pi} \int_{-\infty}^{\infty} e^{-iux} \Theta^{(x)}(u) du, \qquad \Theta^{(x)}(u) = \tilde{S}\left(0, \frac{u}{\sqrt{\tilde{S}(2)}}\right) - \tilde{S}(0) - \frac{iu}{\sqrt{\tilde{S}(2)}} \tilde{S}(1). \tag{27}$$

The characteristic function $\Theta^{(x)}(u)$ in Equation (27) is a well-defined function due to an exact cancellation of all singular terms (see the Appendix in [27]) and appears to be non-Gaussian. Thus, within the basic model (10)–(12), the probability distribution $\rho(n)$ of the noncondensate occupation (15) and, hence, the probability distribution of the complimentary condensate occupation are significantly non-Gaussian for any number N of interacting particles and any dimensions of the trap, and the difference with the Gaussian statistics is significant.

In the intermediate case, both thermal and quantum contributions to the cumulants, which are fused in the binomial formula (22), are important and should be taken into account. The Gaussian or non-Gaussian behavior of the BEC statistics is closely related to the way the second cumulant, or the variance, $\kappa_2 \equiv \sigma^2$, is accumulated. Its "thermal" contribution scales as α^{-2}, and its "quantum" contribution scales as $\Delta^{3/2}$. They are of the same order, that is $\sigma \sim \alpha^{-1} \sim \Delta^{3/4}$, when $\alpha\Delta^{3/4} \sim 1$. At the same condition, all higher-order cumulants κ_m, $m \geq 3$, are mostly determined by the thermal contribution which is of the order of α^{-m}. (The quantum contributions are of the same importance only for $\alpha\Delta^{1/2} \sim 1$ which corresponds to a stronger interparticle interaction.) Hence, the scaled higher-order cumulants $\kappa_m^{(x)}$ tend in the thermodynamic limit to some nonzero constants which implies the non-Gaussian statistics. However, the difference from the Gaussian statistics is somewhat suppressed compared to the thermally-dominated regime due to a larger value of the variance. This non-Gaussian distribution is characterized by the standard deviation $\sigma \sim \sqrt{\langle \hat{N}_{ex} \rangle}$ which is not anomalously large yet. It happens so because the first cumulant, $\kappa_1 = \langle \hat{N}_{ex} \rangle$, is mostly determined by the quantum contribution. The thermal contribution to $\langle \hat{N}_{ex} \rangle$ becomes of the same order as the quantum one only for $\alpha\Delta \sim 1$ which corresponds to the higher temperatures.

In summary, the non-Gaussian statistics of the total noncondensate occupation clearly appears in the 3D flat traps starting from $\alpha\Delta^{3/4} \sim 1$ and remains such for all smaller values of α, that is, in other words, starting from the temperature $T \sim \sqrt[4]{(gn_0)^3\epsilon_1^{(0)}}$ and for the higher temperatures. The quantum-dominated and thermally-dominated regimes correspond to the large and small values of the same dimensionless parameter, $\alpha\Delta^{3/4} \gg 1$ and $\alpha\Delta^{3/4} \ll 1$, respectively.

The same conclusions are true for the large enough mesoscopic systems (that is, when α is small, but finite) since the relative corrections to the leading asymptotic terms discussed above become much less than 1 when the thermodynamic-limit parameter α and the inverse particle-interaction parameter Δ^{-1} become much less than unity: $\alpha \ll 1$, $\Delta^{-1} \ll 1$. Thus, the scaled statistics of the BEC in the mesoscopic systems is quite similar to the thermodynamic-limit statistics.

5. Statistics of the BEC Depletion in a Weakly Interacting Gas Trapped in the Cubic or Rectangular Box with the Periodic, Dirichlet or Fused Boundary Conditions in the Regime of Thermally-Dominated Fluctuations

The most curious and nontrivial regime among the ones described above is the regime of thermally-dominated fluctuations where the central limit theorem is not applicable and the BEC statistics is non-Gaussian. In the present section we consider this regime closely and present the detailed analysis of the corresponding probability density function in the thermodynamic limit.

For simplicity's sake, we employ mostly graphics and skip the long analytical calculations. For the mesoscopic, but relatively large systems (that is, when $\alpha \ll 1$, say, $\alpha \sim 10^{-2}$) the scaled BEC statistics is close enough to the one in the thermodynamic limit and could be easily obtained, including all finite-size corrections, on the basis of the mesoscopic values of cumulants given by Equations (17).

We start with some general properties of the probability density function ρ_n or its scaled counterpart ρ_x, (27), described by the basic model (10)–(12), common for all three-dimensional cubic or rectangular box traps, no matter what are the boundary conditions. The leading term of the asymptotics of the characteristic function at the large values of the argument is given by a power law $\Theta_x(u) \sim (-iu)^{3/2}$ as could be shown in the same way as it was done in [27,28] for the case of an ideal gas. It means that the actual distribution ρ_x is described by an exponential slope $\rho_x \sim e^{-x}$ for the positive arguments x, i.e., for $n > \langle \hat{N}_{ex} \rangle$, and by much steeper, super-exponential slope $\rho_x \sim e^{-|x|^3}$ for the negative arguments x, i.e., for $n < \langle \hat{N}_{ex} \rangle$. The form of both wings of the probability distribution is quite different from the Gaussian law.

The origin of the strongly non-Gaussian statistics is closely related to an inapplicability of the central limit theorem. The point is that the statistical distribution and the cumulants of the noncondensate occupation are dominated by the anomalously large contributing occupations (random variables) of a relatively small number of the lowest energy quasiparticles. The inapplicability of the central limit theorem also means that the shape of the probability distribution could be affected by a modification of a relatively small fraction of the quasiparticle energy spectrum, for example, caused by some perturbation of the trap's form or boundary conditions.

Let us illustrate this general conclusion by a straightforward comparison of the probability distributions of the total noncondensate occupation for three, in general, anisotropic box traps of dimensions $L_x \leq L_y \leq L_z$ with the periodic, Dirichlet or fused boundary conditions, respectively. According to the basic model (10)–(12), these traps possess the bare spectra $\alpha\lambda_{\mathbf{j}}^{(0)} \equiv \epsilon_{\mathbf{j}}^{(0)}/T$ enumerated by the integer vector $\mathbf{j} = (j_x, j_y, j_z)$, namely,

$$\begin{aligned}
\alpha^{(p)} &= \frac{2\hbar^2\pi^2}{ML_z^2T}, \quad \lambda_{\mathbf{j}}^{(0)(p)} = \frac{L_z^2}{L_x^2}j_x^2 + \frac{L_z^2}{L_y^2}j_y^2 + j_z^2, \quad j_{x,y,z} = \pm 0, \pm 1, \pm 2, \ldots, \text{ excluding } \mathbf{j} = (0,0,0) \\
\alpha^{(D)} &= \frac{\hbar^2\pi^2}{2ML_z^2T}, \quad \lambda_{\mathbf{j}}^{(0)(D)} = \frac{L_z^2}{L_x^2}j_x^2 + \frac{L_z^2}{L_y^2}j_y^2 + j_z^2, \quad j_{x,y,z} = +1, +2, +3, \ldots, \\
\alpha^{(f)} &= \frac{\hbar^2\pi^2}{2ML_z^2T}, \quad \lambda_{\mathbf{j}}^{(0)(f)} = \frac{L_z^2}{L_x^2}j_x^2 + \frac{L_z^2}{L_y^2}j_y^2 + j_z^2, \quad j_{x,y,z} = 0, +1, +2, \ldots, \text{ excluding } \mathbf{j} = (0,0,0).
\end{aligned} \tag{28}$$

Here and hereafter we denote the quantities specific to the periodic, Dirichlet or fused boundary conditions by the superscripts (p), (D), or (f), respectively. Note that the parameters $\alpha^{(p)}$ for the box with the periodic boundary conditions is 4 times larger than the Dirichlet's or fused trap's $\alpha^{(D)}$ and $\alpha^{(f)}$ because for this box the lowest-energy nontrivial one-particle state should demonstrate a full wavelength spatial variance along the trap, while the Dirichlet (or any impedance) boundary conditions allow also a half wavelength one-particle states. The periodic boundary conditions are introduced mainly for the convenience of theoretical calculations. In the actual experiments, usually the Dirichlet boundary conditions are more relevant. Since the true effective boundary conditions for the Shrödinger Equation (11) are not fully understood yet, we model them not only by the classical Dirichlet (zero) boundary conditions which prohibit the zero quantum values for any of the quantum numbers j_x, j_y, j_z, but also by the so-called fused boundary conditions which allow for the zero quantum numbers and combine some but not all features of the periodic and Dirichlet boundary conditions. Including such zero quantum numbers in the quasiparticle energy spectrum has a direct physical meaning as allowing for the entire planes, like $(0, j_z, i_z)$, or lines, like $(0, 0, j_z)$, of quasiparticles more closely related to the state $\mathbf{j} = (0,0,0)$ corresponding to the trivial, zero-energy Goldstone mode $u(\mathbf{r}) = -v(\mathbf{r}) = \phi(\mathbf{r})$. This mode corresponds to an arbitrary phase of the condensate wave function and, therefore, does not contribute to the noncondensate occupation. However, it is known that the Goldstone mode should always be present among the solutions of the Bogoliubov-de Gennes equations [18,32,45]. This fact suggests that the quasiparticles with the entire lines or planes of the zero quantum numbers (see Equation (28)) could also be present in the system. Note also that the most significant difference between the three spectra in Equation (28) is the difference in the ranges (28) in which the quantum numbers (j_x, j_y, j_z) vary.

First, let us elaborate, by means of the analytical result in Equations (27) and (28), on the values of the cumulants and their dependence on the interaction strength parameter Δ (see Equation (13)) for the case of the cubic box traps, $L_x = L_y = L_z = L$.

The box with the periodic boundary conditions provides 8 times larger number of quasiparticles compared to the Dirichlet's one because the integer vectors $\mathbf{j} = (j_x, j_y, j_z)$ enumerating the quasiparticles for the box with periodic boundary conditions fill all 8 octants of the three-dimensional integer lattice compared to the only one positive octant filled by the integer vectors enumerating the quasiparticles for the Dirichlet's box. However, since the energies (and, hence, the parameter $\alpha^{(p)}$) for the periodic boundary conditions are 4 times larger and the leading term in the asymptotics of the first cumulant $\kappa_1 \sim \alpha^{-3/2}$ scales as the 3/2 power of α^{-1}, the thermodynamic limit of the mean noncondensate occupation is the same, no matter which boundary conditions are applied.

The eightfold shortage of the number of quasiparticles is not the only important thing which happens as one alters the boundary conditions. Another significant fact is that the change of the boundary conditions disproportionately adds or removes all quasiparticles with the zero quantum numbers. That group of quasiparticles is quite large since in the sphere of quantum numbers (j_x, j_y, j_z) with energies no larger than Λ, $\lambda_j \le \Lambda$, there are $\sim \Lambda^{2/3}$ quasiparticles with the zero quantum numbers. This observation explains a pronounced effect of the boundary conditions on all higher-order cumulants which appear to be noticably different for the different traps. Numerically, the variance and asymmetry in a cubic trap, $L_x : L_y : L_z = 1:1:1$, are as follows

$$\begin{aligned} \kappa_2^{(p)} : \kappa_2^{(D)} : \kappa_2^{(f)} &= \frac{\tilde{S}_2^{(p)}}{16} : \tilde{S}_2^{(D)} : \tilde{S}_2^{(f)} = 0.57 : 0.35 : 2.62, \\ \kappa_3^{(p)} : \kappa_3^{(D)} : \kappa_3^{(f)} &= \frac{\tilde{S}_3^{(p)}}{64} : \tilde{S}_3^{(D)} : \tilde{S}_3^{(f)} = 0.13 : 0.066 : 3.56, \end{aligned} \tag{29}$$

where the calculations were done for the interaction strength $\Delta^{(D,f)} \equiv gn_0/\epsilon_1^{(0)(D,f)} \equiv 2MgN_0/(\hbar^2\pi^2 L) = 80$ in the case of the boxes with the Dirichlet or fused boundary conditions (see Equation (28)) that corresponds to $\Delta^{(p)} \equiv gn_0/\epsilon_1^{(0)(p)} \equiv MgN_0/(2\hbar^2\pi^2 L) = 20$ in the case of the box with the periodic boundary conditions.

The values of the cumulants significantly depend on the parameter Δ which characterizes the strength of interparticle interaction. However, even in the limit of very strong interaction $\Delta \to \infty$ (i.e., $gn_0 \gg \epsilon_1^{(0)}$) the higher-order cumulants keep nonzero values, so that the probability distribution remains non-Gaussian. The point is that with increasing Δ the contribution of a spectral sum with the shifted energies $\lambda_j + 2\Delta$ in Equation (26) decreases. As a result, the values of "the amplitudes" $\tilde{S}(m)$ which determine the asymptotic values of the cumulants κ_m in the thermodynamic limit tend to a half of their values at $\Delta = 0$. (Note that the zero value of the interaction strength parameter, $\Delta = 0$, corresponds to a limit of the ideal gas that is admissible within the basic model (10)–(12) only for the trap with the periodic boundary conditions where the condensate is strictly homogeneous. The moderate or small values of $\Delta \sim 1$ or $\ll 1$ have only a formal meaning for the traps with the other boundary conditions since for the inhomogeneous condensate the Thomas-Fermi approximation (1) requires $\Delta \gg 1$.) This decrease of the higher-order cumulants κ_m can be described analytically via the following asymptotics of their "amplitudes" (the universal numbers)

$$\tilde{S}(m) \to \frac{1}{2}\left[\tilde{S}(m)\Big|_{\Delta=0} + \frac{k\,\pi^{3/2} L_x L_y}{L_z^2} \times \frac{\Gamma(m-3/2)}{(2\Delta)^{m-3/2}} \right] \quad \text{at} \quad \Delta \gg 1 \tag{30}$$

with the coefficient k equal to $1/8$ or 1 for the Dirichlet or periodic boundary conditions, respectively. We see that "the amplitudes" $\tilde{S}(m)$ of the cumulants of the orders $m = 3, 4, 5, \ldots$ decrease by almost two times and reach their ultimate minimum values already at the moderate interaction strength, $MgN_0/(2\hbar^2\pi^2 L) \sim 5$, which corresponds to $\Delta \sim 5$ and $\Delta \sim 20$ for periodic and Dirichlet boundary conditions, respectively.

Importantly, the situation is opposite for the variance κ_2 and the corresponding value of its "amplitude" $\tilde{S}(2)$ which decreases very slowly, namely, as $1/\sqrt{2\Delta}$. This is demonstrated in Figure 1a: Even at $\Delta = 50$ the variance does not quite reach its ultimate minimum value. Note that for the case of the periodic boundary conditions at small Δ the variance tends exactly to the value known for the ideal Bose gas as it should be since in this case the condensate is homogeneous for any strength of repulsive interaction and, hence, the basic model is valid for any values of Δ. The situation is different for the case of the Dirichlet boundary conditions where the formal value of κ_2 at $\Delta = 0$ is 30% less than the true value of the ideal-gas variance.

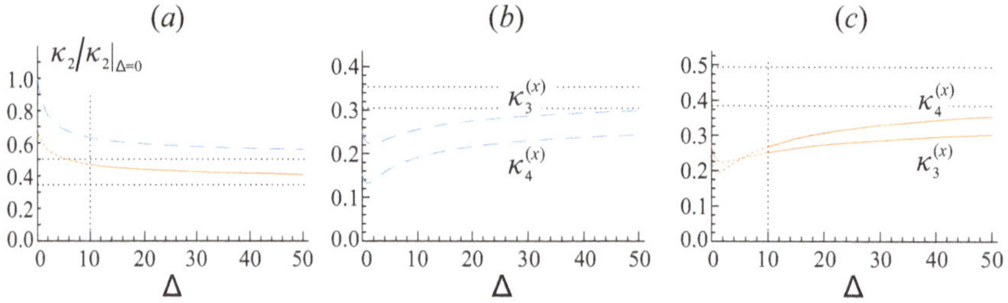

Figure 1. The dependences of cumulants on the interaction strength $\Delta \equiv gn_0/\epsilon_1^{(0)}$ in the thermodynamic limit of a very large system ($\alpha \to 0$): (**a**) the variance κ_2 for a cubic trap with the periodic (a dashed blue line) and Dirichlet (a solid orange line) boundary conditions normalized to its (hypothetical) value $\kappa_2|_{\Delta=0}$ at zero interaction strength; (**b**) the normalized cumulants $\kappa_3^{(x)}$ and $\kappa_4^{(x)}$ for a cubic trap with the periodic boundary conditions and their asymptotic values at $\Delta \to \infty$; (**c**) the normalized cumulants $\kappa_3^{(x)}$ and $\kappa_4^{(x)}$ for a cubic trap with the Dirichlet boundary conditions and their asymptotic values at $\Delta \to \infty$. The curves for the Dirichlet's trap are dotted in the region $\Delta < 10$ where the Thomas-Fermi approximation is expected to be inaccurate.

The evolution of the normalized asymptotic cumulants $\kappa_m^{(x)}$ shown in Figure 1b,c in accord with the analytical results (27) and (28) clearly follows from the evolution of their "amplitudes"—the universal numbers $\tilde{S}(m)$. Namely, for small enough values of Δ (which has a physical meaning only for the case of the periodic boundary conditions) both $\tilde{S}(m)$ and $\tilde{S}(2)$ decrease with an increase of Δ. However, the larger the exponent in the spectral sum, the faster it goes down. As a result, the normalized cumulant $\kappa_m^{(x)}$, which is equal to the ratio $\tilde{S}(m)/\tilde{S}(2)^{m/2}$, decreases. For the moderate values of Δ the values of $\tilde{S}(m)$ at $m = 3, 4, 5...$ are already halved, but the value of $\tilde{S}(2)$ still is decreasing slowly. Therefore, the normalized cumulant slowly grows up to the level of $2^{m/2-1}\kappa_m^{(x)}|_{\Delta=0}$. In fact, that means that even for a large value $\Delta \sim$ 100–500 one should not totally neglect the second shifted spectral sum of the terms containing $\lambda_j + 2\Delta$ in Equation (26). However, we'll see below that the governed by these cumulants dependence of the probability distribution function on the value of the interaction strength parameter Δ is not very fast and is driven almost entirely by the Δ-dependence of the variance.

The result in Equation (30) demonstrates that the variance and asymmetry calculated for the traps with different boundary conditions are not the same. Therefore, in the general case the scaled asymptotic distributions for the corresponding traps do not coincide as well. It is also worth to note that changing the boundary conditions is not the only way to affect the higher-order cumulants. For example, the squeezing of a cubic trap to a prolonged or shortened rectangular box results into changing the asymptotic cumulants and significantly affects the scaled probability distributions.

Both of these effects are illustrated in Figure 2. More general deformation of the trap geometry, for example, into some kind of a cylindrical form also would alter the noncondensate occupation statistics.

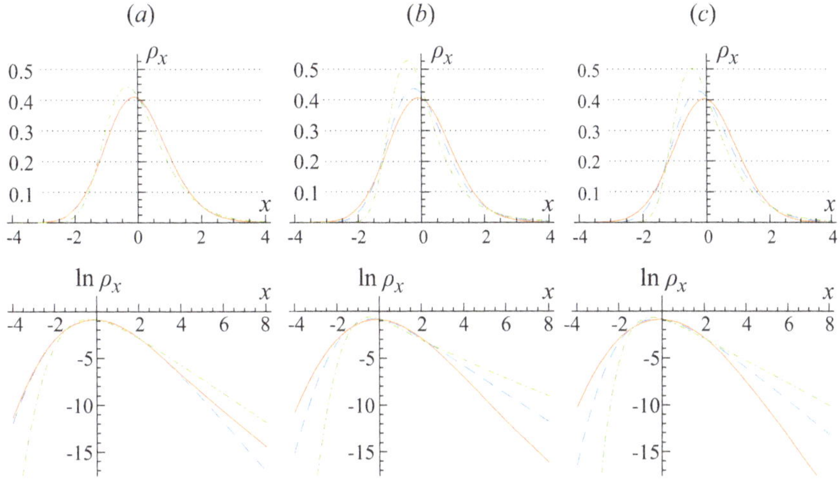

Figure 2. The normalized asymptotic distribution ρ_x of the scaled total number of noncondensed particles, $x = (n - \langle \hat{N}_{ex} \rangle)/\sigma$, plotted in the thermodynamic limit ($\alpha \to 0$) for different rectangular traps in the linear (up) and log (down) scales. The aspect ratio of a trap $L_x : L_y : L_z$ is (**a**) $1 : 1 : 1$; (**b**) $1 : 1 : 2$; and (**c**) $0.33 : 1 : 1$. The dashed blue, solid orange and dot-dashed green curves are for the periodic, Dirichlet and fused boundary conditions, respectively. The interaction strength is $\Delta^{(D,f)} \equiv 2MgN_0/(\hbar^2\pi^2 L) = 80$ in the case of the boxes with the Dirichlet or fused boundary conditions (see (28)) that corresponds to $\Delta^{(p)} \equiv MgN_0/(2\hbar^2\pi^2 L) = 20$ in the case of the box with the periodic boundary conditions.

An effect of the trap's form (geometry) on the probability distribution of the noncondensate occupation can be explicitly shown via dependence of the asymptotics of the characteristic function $\Theta^{(x)}(u)$ at the large values of its arguments, $|u| \gg 1$, on the dimensions L_x, L_y, L_z of the anisotropic box trap. Indeed, we find that this asymptotics

$$\ln \Theta^{(x)}(u) \sim F\left(\frac{-iu}{\sqrt{\tilde{S}(2)}}\right) + F\left(\frac{-iu}{\sqrt{\tilde{S}(2)}} + 2\Delta\right), \quad \text{where}$$
$$F(y) = \frac{2\sqrt{\pi} R_{3/2}\, y^{3/2}}{3} + \frac{R_1\, y \ln y}{2} + \frac{y}{2}\left(\tilde{S}^{(0)}(1) - R_1 + \gamma R_1\right) + \dots, \tag{31}$$

depends on the ratios $L_x : L_y : L_z$ via the R's dimensionless coefficients given by the following formulas

$$R_{3/2}^{(p)} = \frac{\pi^{3/2} L_x L_y}{L_z^2}, \qquad R_1^{(p)} = 0,$$
$$R_{3/2}^{(D)} = \frac{\pi^{3/2} L_x L_y}{8\, L_z^2}, \qquad R_1^{(D)} = -\frac{\pi}{8}\left(\frac{L_x L_y}{L_z^2} + \frac{L_x}{L_z} + \frac{L_y}{L_z}\right), \tag{32}$$
$$R_{3/2}^{(f)} = \frac{\pi^{3/2} L_x L_y}{8\, L_z^2}, \qquad R_1^{(f)} = +\frac{\pi}{8}\left(\frac{L_x L_y}{L_z^2} + \frac{L_x}{L_z} + \frac{L_y}{L_z}\right),$$

for the periodic, Dirichlet and fused boundary conditions, respectively (see Equation (28)). The coefficient $\tilde{S}^{(0)}(1)$ entering function $F(y)$ in Equation (31) is a regularized (if it is singular) value of the trap function $\tilde{S}(t)$ at $t = 1$, and $\gamma \simeq 0.577$ is the Euler–Mascheroni constant.

The change of the rectangular box form via the dimensions' ratios $L_z : L_y : L_z$ affects "the variance amplitude" $\tilde{S}_2$ which is the characteristic scale of the variable u. (This is also clearly seen from the expression for the $\Theta^{(x)}(u)$ in Equation (27)). A simple analysis of the Fourier integral (27) shows that the quantity $(\tilde{S}_2)^{-1/2}$ is the characteristic scale of the argument x of the scaled asymptotic distribution ρ_x. The dimensions' ratios also determine the coefficients in the asymptotics (31) of $\Theta^{(x)}(u)$ at the large values of the argument. All these coefficients, starting from the next-to-leading term coefficient R_1, strongly depend on the dimensions' ratios. Only the coefficient in front of the leading term could not be changed by means of the anisotropic squeezing or stretching of the box form if the volume of the trap is kept constant. (One has to take into account that the eightfold difference in Equation (32) between the coefficient $R^{(p)}_{3/2}$ and the coefficients $R^{(D)}_{3/2}$, $R^{(f)}_{3/2}$ for the periodic and the other two boundary conditions is exactly compensated by the difference in the definitions of $\alpha^{(p)}$ and $\alpha^{(D)}$, $\alpha^{(f)}$, see Equation (28)). The increasing anisotropy of the trap significantly decreases "the amplitude" $\tilde{S}_2$ of the anomalously large variance

$$\sigma^2 \simeq \frac{\tilde{S}_2}{(R_{3/2})^{4/3}\ \zeta^{4/3}(3/2)} \langle \hat{N}_{ex} \rangle^{4/3} \tag{33}$$

since the leading term in the asymptotics of the mean value $\langle \hat{N}_{ex} \rangle$ does not feel the anisotropy and remains almost constant. Figure 3 demonstrates this effect for a very large, thermodynamic-limit rectangular box trap with the Dirichlet boundary conditions: the variance decreases by 20% when any dimension of the box becomes about four times longer (or shorter) than the other dimensions.

The result in the Equation (33) explicitly proves that the fluctuations of the condensate depletion in the weakly interacting gas are anomalously large compared to the standard thermodynamic fluctuations. The thermodynamic fluctuations would have considerably smaller variance, $\sigma^2 \sim \langle \hat{N}_{ex} \rangle$, proportional only to the first power of the mean value, while the variance (33) of the true noncondensate occupation is proportional to the larger power, $\sigma^2 \sim \langle \hat{N}_{ex} \rangle^{4/3}$, of the mean noncondensate occupation $\langle \hat{N}_{ex} \rangle \gg 1$. Thus, the basic model (10)–(12) is in agreement with this known fact [14,15,18].

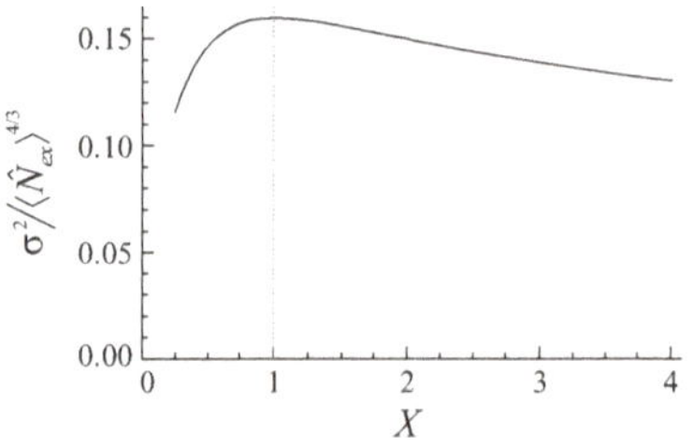

Figure 3. The thermodynamic-limit ($\alpha \to 0$) asymptotics of the anomalously large variance σ^2, Equation (33), of the condensate depletion fluctuations in the interacting gas trapped in the rectangular box with the Dirichlet boundary conditions vs the box trap anisotropy characterized by the dimensions' ratios $L_x : L_y : L_z = 1 : 1 : X$, where $0.25 < X < 4$. The interaction strength is set to be $\Delta^{(D)} \equiv gn_0/\epsilon_1^{(0)(D)} = 80$.

The main effect of the change of boundary conditions is that a large group of states with one zero quantum number appears to be added or removed disproportionately. As a result, the next-to-leading logarithmic term appears in the characteristic function (31) that leads to the logarithmic shift of an argument in the exponent, which could make the left-wing decay of the distribution considerably faster or slower. However, for the cases considered in Figure 2 this effect is not a major one since in these cases the variance values $\tilde{S}_2$ differ so significantly (which, in fact, is not the general case!) that the most important effect of the trap's form on the noncondensate occupation probability distribution consists in altering the characteristic scale of the noncondensate occupation random variable x.

The change of the boundary conditions also considerably affects the lowest energy quasiparticles which determine the right-wing asymptotics of the noncondensate occupation probability distribution. The smaller is the lowest energy and the higher is the corresponding degeneracy, the weaker exponential decay is realized for the larger than average noncondensate occupations $x > 0$ (that is, for $n > \langle \hat{N}_{ex} \rangle$).

It is worth noting that a relatively pronounced effect of the change of the boundary conditions in the rectangular trap from the periodic to Dirichlet ones calculated above cannot be directly observed in the experiments since the periodic boundary conditions constitute an auxiliary theoretical model. Only some partial analogues of such switching of boundary conditions could be realized in a real system, for example, in a toroidal trap with an azimuthal symmetry and, hence, the periodic boundary condition along the azimuthal direction via cutting the torus by a high, almost vertical potential wall. This or similar cylindrically symmetric setups which are more relevant to the actual experiments will be discussed elsewhere. In the present paper we just draw attention to such effect of boundary conditions on the BEC statistics.

The overall picture of the effects of the trap's form and boundary conditions on the noncondensate occupation statistics for a weakly interacting gas within the basic model (10)–(12) is similar to the one known for an ideal gas. (For example, statistics of BEC in an ideal gas trapped in the anisotropic box or slabs was discussed in [19,21,22,27,46].) As is illustrated in Figure 2, a typical change in the thermodynamic-limit asymptotics of the noncondensate occupation probability distribution due to a change of the trap's form or boundary conditions is about 10% or so. Note that this is already normalized and centered probability distribution, so that the change cannot be eliminated by some kind of mesoscopic effects or perturbations.

Of course, the leading terms in the thermodynamic-limit asymptotics for the mean value $\langle \hat{N}_{ex} \rangle$ are the same for the different trap's forms and boundary conditions. This happens since a very large number of the quasiparticles contribute to $\langle \hat{N}_{ex} \rangle$ and the individual contributions are not really important. The non-Gaussian distribution appears to be a robust property of the system—the trends of asymptotics could not be changed without changing the index in the power law for the density of states, and modification of a small number of quasiparticles hardly affects the behavior of the probability distribution. The effect of boundary conditions on the distributions is mostly governed by the large enough group of energies with one zero quantum number. The effect of the trap's form deformation disproportionately affects the whole energy spectrum of quasiparticles and, hence, is significant. The same mechanisms determine the BEC statistics in an ideal gas as is shown in [27,28].

A further analysis beyond the basic model (10)–(12) could reveal the other cases when a change of the trap affects the BEC statistics significantly. For example, the effect of a change of the trapping potential in the entire volume of the trap, say, by means of imposing an additional weak potential lattice considered in [42], at moderate (but still less than critical, $T \ll T_c$) temperatures could be analyzed by the perturbation method of [42]. Unfortunately, the authors of [42] discussed only the case of the quantum-dominated regime at very close to zero temperatures when the BEC statistics is Gaussian in the bulk limit. Moreover, the ref. [42] is devoted to the calculation of the mean value (i.e., only the first cumulant) of quantum depletion, while we are interesting in the entire statistics (i.e., all higher-order cumulants) of condensate depletion and, mostly, in the opposite case of the thermally-dominated fluctuations when the statistics is significantly non-Gaussian even for the macroscopically large system. Note, however, that the dependence of the low-temperature limit of the mean condensate depletion on the large interaction parameter Δ, calculated in [42], is definitely in agreement with the "quantum" contribution to the first cumulant calculated in the present paper from Equation (25).It is possible, in principle, to calculate the characteristic function of the BEC statistics starting from Equations (29) and (31) of [42]. We expect that in the result one will obtain the regimes of the quantum-dominated, Gaussian and thermally-dominated, non-Gaussian fluctuations similar to the ones described in the present paper. Thus, these calculations could reveal the dependence of BEC statistics on the lattice perturbation of the trapping potential and provide a basis for the design of an actual experiment on its measurement.

6. Discussion: Why the BEC Statistics Remains Non-Gaussian and Dependent on the Trap's Form and Boundary Conditions Even in the Thermodynamic Limit and Upon Increasing Interaction?

The proposed basic model (10)–(12) of the BEC statistics is relatively simple. It takes into account the main physical mechanism of an influence of the interparticle interaction on the noncondensate (and complimentary condensate) occupation and its fluctuations. The point is that the noncondensate occupation is a random variable equal to the sum of many bare-particle excited-state occupations created by the independently fluctuating Bogoliubov-de Gennes quasiparticles, as is clearly seen from the Equations (16) and (17). The interparticle interaction, mediated by the condensate, forces, via the Bogoliubov coupling, each quasiparticle or pair of counter-propagating quasiparticles to populate the bare-particle excited states in a squeezed, or coherently correlated mode. This is exactly what the basic model describes via the Equations (7) and (10).

Thus, on the one hand the basic model admits the nontrivial analysis presented in the Sections 3–5 and, in particular, the analytical solution for its thermodynamic-limit asymptotics. On the other hand, it allows one to take into account the effect of the interparticle interaction in the Bose gas and, in particular, the renormalization (dressing) of the quasiparticles and trap by a well-developed condensate fraction. As a result, this basic model yields a series of interesting conclusions about the probability distribution, moments and cumulants of BEC fluctuations as the functions of the interaction strength, temperature, the number of trapped particles, the dimension of the system, the form (geometry) of the trap and the imposed boundary conditions.

First of all, the noncondensate occupation fluctuations in an interacting gas are anomalously large compared to the standard thermodynamic fluctuations as is proven by the result in the Equation (33).

The second, but very important fact is that the BEC statistics turns out to be strongly non-Gaussian, as it was in the case of the non-interacting, ideal gas [15,26–30], and remains so with an increase of the interparticle interaction. The origin of this non-Gaussianness is, of course, in an inapplicability of the central limit theorem. Indeed, if the contribution of quasiparticles to the noncondensate occupation decreases slowly with an increase of their quantum numbers j_x, j_y, j_z, a large number of quasiparticles contributes significantly to the statistics, the central limit theorem works and one finally obtains the Gaussian distribution. In the opposite case of rapidly decreasing contributions the group of quasiparticles which mainly determines the statistics is not large enough and the central limit theorem is not applicable. Hence, the resulting statistics is non-Gaussian. This is exactly the picture behind the analysis of the values of cumulants and the characteristic function as the sums of the accumulating contributions from the lattice of quasiparticle states presented in the Sections 3–5.

As is known for the case of an ideal Bose gas [27,28], the three-dimensional systems with the quadratic spectrum, $\lambda_{\mathbf{j}} \sim j_x^2 + j_y^2 + j_z^2$, and the linear spectrum, $\lambda_{\mathbf{j}} \sim j_x + j_y + j_z$, constitute the examples of the "rapidly growing" and "slowly growing" spectra, respectively. The crucial nontrivial fact about the BEC statistics in a weakly interacting gas is as follows. The Bogoliubov-de Gennes quasiparticles have a "slowly growing" spectrum (12), ε_j, of an almost linear dependence of eigenenergies on the quantum numbers vector $\mathbf{j}$ starting from the zero energy until the interaction coupling energy gn_0. Nevertheless, the statistics of the noncondensate occupation (consisting of the excited bare particle occupations) is accumulated (due to the squeezing caused by Bogoliubov coupling) in accord with the bare energies $\epsilon_j^{(0)}$ (see the first sum in the square brackets of Equation (25)). But the bare energies form the quadratic spectrum and, hence, grow fast enough to disable the central limit theorem!

Next, we find that in the region of parameters where the thermal depletion is much larger than the quantum depletion the noncondensate occupation probability distribution converges in the thermodynamic limit to the special function determined by the trap's form and boundary conditions. Within the basic model, the effect of the trap's form and boundary conditions on the probability distribution could exist even in the thermodynamic limit. Moreover, any change of the spectrum, which strongly affects (or even excludes, as it happens when the boundary conditions are changed from the periodic to Dirichlet ones) a large enough group of quasiparticles, results into the probability distribution altering. This is the reason why the system geometry and dimensionality are very important for the BEC statistics. In particular, a proper choice of the trap's geometry and a change of dimensionality from $D = 3$ value to the lower, quasi-1D or quasi-2D values, for which the non-Gaussian BEC statistics is known [27] to be more pronounced and achievable starting from the lower exponent $p = 2D/(4 - D)$ of the power-law trapping potential $U_{trap} \sim |r|^p$, can increase the observable effect on the condensate depletion statistics.

The other mechanism of the influence of the trap's form and boundary conditions on the BEC statistics in the interacting gas could be seen from the analysis of a mesoscopic system when the condensate boundary layers near the borders of the trap occupy only a small fraction of the trap's volume in the Thomas-Fermi regime. The remaining, main part of the trap's volume contains an almost homogeneous condensate and determines the main part of the structure and Bogoliubov couplings of all low energy excitations. The contributions from these excitations dominate the values of all higher-order generating cumulants $\tilde{\kappa}_m$, $m \geq 3$, and, hence, are responsible for the non-Gaussian behavior of the BEC statistics. At the same time, the second generating cumulant, i.e., the variance, $\sigma^2 \sim \tilde{\kappa}_2$ could contain an appreciable contribution from the high energy, $\varepsilon_j \sim$ or $> gn_0$, excitations the wavelengths of which are shorter than the healing length. The latter implies a validity of the WKB (quasi-classical) approximation for the solution to the Bogoliubov-de Gennes Equations (6) for these high energy quasiparticles even in the boundary layer of the healing-length thickness. It means that the high energy excitations propagate adiabatically (as in the geometrical optics approximation) through the boundary layer and, hence, fully feel the boundary conditions of the trap. As a result of this interplay between the different properties of the low and high energy excitations and their different roles in the BEC statistics, the scaled higher-order cumulants in Equation (26) and, hence, the statistics of condensate fluctuations in Equation (27) in the interacting gas depend on the trap's geometry (form) and boundary conditions and are non-Gaussian. A very similar difference in the dependences of the second-order and higher-order cumulants exists also in the case of the varying interparticle interaction g, i.e., the parameter Δ in Equation (13), as is discussed in Section 5 and in Figure 1.

Of course, there are many effects which are not included in the formulated above basic model of the BEC statistics such as (a) the inhomogeneity of the condensate, in particular, within the boundary layers of the healing-length thickness near the borders of the trap, (b) the effect of such inhomogeneities on the modification of the quasiparticle energy spectrum, Bogoliubov couplings and squeezing of the noncondensate occupation, (c) the thermal and many-body effects in the range of temperatures from $T \sim T_c/2$ to the critical temperature $T = T_c$ which are beyond the Bogoliubov approximation, etc. They require much more involved theoretical analysis and can considerably modify the BEC statistics in the interacting gas. Moreover, in order to describe correctly the full range of parameters and, in particular, an evolution of the BEC statistics in its transition from a small mesoscopic system to a macroscopic system or from an ideal gas with zero interaction $g = 0$ to an interacting gas in the Thomas-Fermi regime as well as in order to get numerically accurate parameters of the BEC statistics one has to improve the basic model by the accurate account and solutions for these effects.

So, the intriguing problems of whether or not the nature allows the surface's geometry (form) and boundary conditions of a macroscopically large trap to influence the BEC statistics and how the BEC statistics evolves with increasing trap's size and interparticle interaction remain open. A similarly general question is whether or not the BEC statistics in the real macroscopic systems remains non-Gaussian in the thermodynamic limit. A universal mechanism leading to the Gaussian statistics in the many-body thermodynamic systems is well known. It is the central limit theorem valid for the sum of a large system of independent random variables in which there are no strongly dominating small subsystems. The considered systems significantly differ from the usual thermodynamic systems since the spontaneous symmetry breaking creates a well-developed condensate, or a coherent macroscopic wave function, which spreads over the entire volume of the system from border to border and coherently transfers the information about the trap's form (geometry) and boundary conditions over the whole system. This could be the basis of the general mechanism keeping the BEC statistics non-Gaussian and dependent on the trap's form and boundary conditions in the thermodynamic limit. Independently on the answers to these questions, the basic model presented in this paper can be employed as a useful constructive basis for the search, comparison and classification of various mechanisms affecting the BEC statistics.

Certainly, the ultimate answers to all these questions should be given by the experimental studies of the BEC statistics in the interacting gas. They constitute a promising field of research but require very accurate measurements and a very stable confinement of BEC with the precisely controllable parameters. A recent progress in the experimental techniques suggests that the BEC statistics is already becoming accessible experimentally (see, e.g., [47]). Among the recent experiments, especially promising are the experiments of the Hadzibabic's group [48–50] with an almost flat cylindrical trap where the thermally-dominated regime seems to be accessible. For example, the setup of [49] in the thermally-dominated regime of condensate depletion at $T \sim T_c/3$ constitutes a macroscopically large system since the parameter of the thermodynamic limit in Equation (13), $\alpha = \hbar^2\pi^2/(2ML^2T)$, is indeed very small, $\alpha \sim 0.003$. Hence, the mesoscopic effects should be relatively weak and the described above anomalous thermodynamic-limit BEC statistics could be revealed experimentally. Another interesting example of the uniform BEC with the almost flat potential inside the box trap and vertical "walls" at the borders was demonstrated in the experiments of the Raizen's group [51–53]. Other promising experimental setups include a two-dimensional flat trap of Dalibard's group [54], the circular waveguides for BEC trapping [55] and the system which "paints" an arbitrary border of the trap by rapidly moving laser beams [56].

7. Conclusions

We formulated and solved the basic model (10)–(12) of the BEC statistics in a weakly interacting gas confined in a flat trap within the Thomas-Fermi approximation of the mean-field theory of Gross-Pitaevskii and Bogoliubov-de Gennes. This model yields the non-Gaussian condensate occupation statistics which is different from also non-Gaussian ideal-gas BEC statistics (for the same trap) but shares with it a similar dependence on the trap's form and boundary conditions. Such a dependence does not vanish with the increase of the interparticle interaction, the number of trapped particles and the volume of the system, that is in the thermodynamic limit.

Acknowledgments: The research was supported by RFBI (grant 16-32-00471), the program of fundamental research of the DPS RAS IV.2.6 "Fundamental Basis and Experimental Realization of Promising Semiconductor Lasers for Industry and Technologies" (Section 4) and the Dynasty Foundation. We are grateful to the organizing committee of the SigmaPhi2017 Conference on Statistical Physics for covering the costs of the publication.

Author Contributions: All coauthors, Sergey Tarasov, Vladimir Kocharovsky and Vitaly Kocharovsky contributed equally to deriving the results and writing the paper.

Conflicts of Interest: The authors declare no conflict of interest.

References

1. Landau, L.D.; Lifshitz, E.M. *Statistical Physics, Part 1*; Pergamon: Oxford, UK, 1981.
2. Pitaevskii, L.; Stringary, S. *Bose-Einstein Condensation*; Clarendone: Oxford, UK, 2003.
3. Fedichev, P.O.; Shlyapnikov, G.V. Finite-temperature perturbation theory for a spatially inhomogeneous Bose-condensed gas. *Phys. Rev. A* **1998**, *58*, 3146–3158.
4. Shi, H.; Griffin, A. Finite-temperature excitations in a dilute Bose-condensed gas. *Phys. Rep.* **1998**, *304*, 1–87.
5. Zagrebnov, V.A.; Bru, J.B. The Bogoliubov model of weakly imperfect Bose gas. *Phys. Rep.* **2001**, *350*, 291–434.

6. Proukakis, N.P.; Jackson, B. Finite-temperature models of Bose–Einstein condensation. *J. Phys. B At. Mol. Opt. Phys.* **2008**, *41*, 203002.
7. Andersen, J.O. Theory of the weakly interacting Bose gas. *Rev. Mod. Phys.* **2004**, *76*, 599–639.
8. Kocharovsky, V.V.; Kocharovsky, Vl.V. Microscopic theory of a phase transition in a critical region: Bose–Einstein condensation in an interacting gas. *Phys. Lett. A* **2015**, *379*, 466–470.
9. Kocharovsky, V.V.; Kocharovsky, Vl.V. Microscopic theory of phase transitions in a critical region. *Phys. Scr.* **2015**, *90*, 108002.
10. Kocharovsky, V.V.; Kocharovsky, Vl.V. Exact general solution to the three-dimensional Ising model and a self-consistency equation for the nearest-neighbors' correlations. *arXiv* **2016**, arXiv:1510.07327v3.
11. Ziff, R.M.; Uhlenbeck, G.E.; Kac, M. The ideal Bose-Einstein gas, revisited. *Phys. Rep.* **1977**, *32*, 169–248.
12. Holthaus, M.; Kalinowski, E.; Kirsten, K. Condensate fluctuations in trapped Bose gases: Canonical vs. microcanonical ensemble. *Ann. Phys.* **1998**, *270*, 198–230.
13. Grossmann, S.; Holthaus, M. Maxwell's Demon at work: Two types of Bose condensate fluctuations in power-law traps. *Opt. Express* **1997**, *1*, 262–271.
14. Giorgini, S.; Pitaevskii, L.P.; Stringari, S. Anomalous Fluctuations of the Condensate in Interacting Bose Gases. *Phys. Rev. Lett.* **1998**, *80*, 5040–5043.
15. Kocharovsky, V.V.; Kocharovsky, Vl.V.; Scully, M.O. Condensation of N bosons. III. Analytical results for all higher moments of condensate fluctuations in interacting and ideal dilute Bose gases via the canonical ensemble quasiparticle formulation. *Phys. Rev. A* **2000**, *61*, 053606.
16. Boers, D.; Holthaus, M. Canonical statistics of occupation numbers for ideal and weakly interacting Bose-Einstein condensates. In *Dynamics and Thermodynamics of Systems with Long-Range Interactions*; Springer: Berlin, Germany, 2002.
17. Weiss, C.; Holthaus, M. Asymptotics of the number partitioning distribution. *Europhys. Lett.* **2002**, *59*, 486.
18. Zwerger, W. Anomalous Fluctuations in Phases with a Broken Continuous Symmetry. *Phys. Rev. Lett.* **2004**, *92*, 027203.
19. Pule, J.V.; Zagrebnov, V.A. The canonical perfect Bose gas in Casimir boxes. *J. Math. Phys.* **2004**, *45*, 3565–3583.
20. Idziaszek, Z. Microcanonical fluctuations of the condensate in weakly interacting Bose gases. *Phys. Rev. A* **2005**, *71*, 053604.
21. Martin, P.A.; Zagrebnov, V.A. The Casimir effect for the Bose-gas in slabs. *Europhys. Lett.* **2005**, *73*, 15.
22. Toms, D.J. Statistical mechanics of an ideal Bose gas in a confined geometry. *J. Phys. A Math. Gen.* **2006**, *39*, 713–722.
23. Idziaszek, Z.; Gajda, M.; Rzążewski, K. Fluctuations of a weakly interacting Bose-Einstein condensate. *Europhys. Lett.* **2009**, *86*, 10002.
24. Wright, T.M.; Proukakis, N.P.; Davis, M.J. Many-body physics in the classical-field description of a degenerate Bose gas. *Phys. Rev. A* **2011**, *84*, 023608.
25. Chatterjee, S.; Diaconis, P. Fluctuations of the Bose–Einstein condensate. *J. Phys. A Math. Theor.* **2014**, *47*, 085201.
26. Kocharovsky, V.V.; Kocharovsky, Vl.V. Analytical theory of mesoscopic Bose-Einstein condensation in an ideal gas. *Phys. Rev. A* **2010**, *81*, 033615.
27. Tarasov, S.V.; Kocharovsky, Vl.V.; Kocharovsky, V.V. Universal scaling in the statistics and thermodynamics of a Bose-Einstein condensation of an ideal gas in an arbitrary trap. *Phys. Rev. A* **2014**, *90*, 033605.
28. Tarasov, S.V.; Kocharovsky, Vl.V.; Kocharovsky, V.V. Universal fine structure of the specific heat at the critical λ-point for an ideal Bose gas in an arbitrary trap. *J. Phys. A Math. Theor.* **2014**, *47*, 415003.
29. Tarasov, S.V.; Kocharovsky, Vl.V.; Kocharovsky, V.V. Grand Canonical Versus Canonical Ensemble: Universal Structure of Statistics and Thermodynamics in a Critical Region of Bose–Einstein Condensation of an Ideal Gas in Arbitrary Trap. *J. Stat. Phys.* **2015**, *161*, 942–964.
30. Kocharovsky, V.V.; Kocharovsky, Vl.V.; Tarasov, S.V. Bose–Einstein Condensation in Mesoscopic Systems: The Self-Similar Structure of the Critical Region and the Nonequivalence of the Canonical and Grand Canonical Ensembles. *JETP Lett.* **2016**, *103*, 62–75.
31. Leggett, A.J. Bose-Einstein condensation in the alkali gases: Some fundamental concepts. *Rev. Mod. Phys.* **2001**, *73*, 307–356.
32. Pethick, C.J.; Smith, H. *Bose-Einstein Condensation in Dilute Gases*; Cambridge University Press: Cambridge, UK, 2002.

33. Holstein, T.; Primakoff, H. Field dependence of the intrinsic domain magnetization of a ferromagnet. *Phys. Rev.* **1940**, *58*, 1098–1113.
34. Schwinger, J. *Quantum Theory of Angular Momentum*; Academic Press: New York, NY, USA, 1965.
35. Girardeau, M.; Arnowitt, R. Theory of many-boson systems: Pair theory. *Phys. Rev.* **1959**, *113*, 755.
36. Girardeau, M. Comment on "Particle-number-conserving Bogoliubov method which demonstrates the validity of the time-dependent Gross-Pitaevskii equation for a highly condensed Bose gas". *Phys. Rev. A* **1997**, *58*, 775.
37. Gardiner, C.W. Particle-number-conserving Bogoliubov method which demonstrates the validity of the time-dependent Gross-Pitaevskii equation for a highly condensed Bose gas. *Phys. Rev. A* **1997**, *56*, 1414.
38. Castin, Y.; Dum, R. Low-temperature Bose-Einstein condensates in time-dependent traps: Beyond the U(1) symmetry-breaking approach. *Phys. Rev. A* **1998**, *57*, 3008.
39. Morgan, S.A. A gapless theory of Bose-Einstein condensation in dilute gases at finite temperature. *J. Phys. B At. Mol. Opt. Phys.* **2000**, *33*, 3847–3893.
40. Gardiner, S.A.; Morgan, S.A. Number-conserving approach to a minimal self-consistent treatment of condensate and noncondensate dynamics in a degenerate Bose gas. *Phys. Rev. A* **2007**, *75*, 043621.
41. Billam, T.P.; Mason, P.; Gardiner, S.A. Second-order number-conserving description of nonequilibrium dynamics in finite-temperature Bose-Einstein condensates. *Phys. Rev. A* **2013**, *87*, 033628.
42. Müller, C.A.; Gaul, C. Condensate deformation and quantum depletion of Bose–Einstein condensates in external potentials. *New J. Phys.* **2012**, *14*, 075025.
43. Englert, B.-G.; Fulling, S.A.; Pilloff, M.D. Statistics of dressed modes in a thermal state. *Optics Commun.* **2002**, *208*, 139–144.
44. *Handbook of Mathematical Functions*; Abramowitz, M., Stegun, I.A., Eds.; Dover: New York, NY, USA, 1972.
45. Okumura, M.; Yamanaka, Y. Unitarily inequivalent vacua in Bose–Einstein condensation of trapped gases. *Physica A* **2006**, *365*, 429–445.
46. Kirsten, K.; Toms, D.J. Bose-Einstein condensation under external conditions. *Phys. Lett. A.* **1998**, *243*, 137–141.
47. Perrin, A.; Bücker, R.; Manz, S.; Betz, T.; Koller, C.; Plisson, T.; Schumm, T.; Schmiedmayer, J. Hanbury Brown and Twiss correlations across the Bose–Einstein condensation threshold. *Nat. Phys.* **2012**, *8*, 195–198.
48. Gaunt, A.L.; Schmidutz, T.F.; Gotlibovych, I.; Smith, R.P.; Hadzibabic, Z. Bose-Einstein Condensation of Atoms in a Uniform Potential. *Phys. Rev. Lett.* **2013**, *110*, 200406.
49. Lopes, R.; Eigen, C.; Navon, N.; Clement, D.; Smith, R.P.; Hadzibabic, Z. Quantum Depletion of a Homogeneous Bose-Einstein Condensate. *Phys. Rev. Lett.* **2017**, *119*, 190404.
50. Lopes, R.; Eigen, C.; Barker, A.; Viebahn, K.G.; Robert-de-Saint-Vincent, M.; Navon, N.; Hadzibabic, Z.; Smith, R.P. Quasiparticle energy in a strongly interacting homogeneous Bose-Einstein condensate. *Phys. Rev. Lett.* **2017**, *118*, 210401.
51. Meyrath, T.P.; Schreck, F.; Hanssen, J.L.; Chuu, C.-S.; Raizen, M.G. Bose-Einstein condensate in a box. *Phys. Rev. A* **2005**, *71*, 041604(R).
52. Dudarev, A.M.; Raizen, M.G.; Niu, Q. Quantum Many-Body Culling: Production of a Definite Number of Ground-State Atoms in a Bose-Einstein Condensate. *Phys. Rev. Lett.* **2007**, *98*, 063001.
53. Pons, M.; del Campo, A.; Muga, J.G.; Raizen, M.G. Preparation of atomic Fock states by trap reduction. *Phys. Rev. A* **2009**, *79*, 033629.
54. Chomaz, L.; Corman, L.; Bienaimé, T.; Desbuquois, R.; Weitenberg, C.; Nascimbène, S.; Beugnon, J.; Dalibard, J. Emergence of coherence via transverse condensation in a uniform quasi-two-dimensional Bose gas. *Nat. Commun.* **2015**, *6*, 6162.
55. Gupta, S.; Murch, K.W.; Moore, K.L.; Purdy, T.P.; Stamper-Kurn, D.M. Bose-Einstein condensation in a circular waveguide. *Phys. Rev. Lett.* **2005**, *95*, 143201.
56. Henderson, K.; Ryu, C.; MacCormick, C.; Boshier, M.G. Experimental demonstration of painting arbitrary and dynamic potentials for Bose–Einstein condensates. *New J. Phys.* **2009**, *11*, 043030.

entropy

MDPI

Article

Study on Bifurcation and Dual Solutions in Natural Convection in a Horizontal Annulus with Rotating Inner Cylinder Using Thermal Immersed Boundary-Lattice Boltzmann Method

Yikun Wei [1,2], Zhengdao Wang [3,*,†], Yuehong Qian [4,*,†] and Wenjing Guo [5]

1 Faculty of Mechanical Engineering & Automation, Zhejiang Sci-Tech University, Hangzhou 310018, China; yikunwei@zstu.edu.cn
2 State-Province Joint Engineering Lab of Fluid Transmission System Technology, Hangzhou 310018, China
3 Shanghai Institute of Applied Mathematics and Mechanics, Shanghai University, Shanghai 200072, China
4 School of Mathematical Sciences, Soochow University, Suzhou 215006, China
5 Basic Courses Department, Shandong University of Science and Technology, Taian 271019, China; wjguo1983@126.com
* Correspondence: dao1210@shu.edu.cn (Z.W.); yuehongqian@suda.edu.cn (Y.Q.)
† These authors contributed equally to this work.

Received: 22 July 2018; Accepted: 17 September 2018; Published: 25 September 2018

Abstract: A numerical investigation has been carried out to understand the mechanism of the rotation effect on bifurcation and dual solutions in natural convection within a horizontal annulus. A thermal immersed boundary-lattice Boltzmann method was used to resolve the annular flow domain covered by a Cartesian mesh. The Rayleigh number based on the gap width is fixed at 10^4. The rotation effect on the natural convection is analyzed by streamlines, isotherms, phase portrait and bifurcation diagram. Our results manifest the existence of three convection patterns in a horizontal annulus with rotating inner cylinder which affect the heat transfer in different ways, and the linear speed (U_i^*) determines the proportion of each convection. Comparison of average *Nusselt* number versus linear speed for the inner cylinder indicates the existence of the three different mechanisms which drive the convection in a rotation system. The convection pattern caused by rotation reduces the heat transfer efficiency. Our results in phase portraits also reveal the differences among different convection patterns.

Keywords: natural convection; bifurcation; horizontal annulus; thermal IB-LBM; rotation

1. Introduction

Natural convection in a horizontal annulus is a useful subject due to its importance in theoretical study and in many engineering applications, such as thermal energy storage systems, cooling systems in electronic components, brake assembly and bearing fittings. Considering its simple geometry, this system has become a classical model and has been well studied by many researchers. Early in 1966, Grigull and Hauf [1] investigated natural convection in a horizontal annulus experimentally. In 1969, Powe et al. [2] designed an experiment and delineated the convection with different types of flow depending on the Grashof number and radius ratio. Kuehn and Goldstein [3] studied the natural convection in concentric cylinders numerically and experimentally. Nguyen et al. [4] presented the flow patterns and heat transfer rates in terms of the radius ratio, the Rayleigh number and inversion parameter. Dyko et al. [5] presented a three-dimensional numerical and experimental study for flow between two horizontal coaxial cylinders at Rayleigh numbers approaching and exceeding the critical values. The instability and bifurcation phenomena in the natural convection in a horizontal annulus have also attracted the interest of many researchers. Yoo et al. [6–8] studied the effect of the Prandtl number on the bifurcation phenomenon of natural convection in an annulus with a narrow or wide gap. Results show that natural convection with high Prandtl number in a narrow annulus has bifurcation solutions. Petrone et al. [9,10] analyzed the bifurcation phenomena theoretically through a bifurcation diagram by numerical simulation. Luo et al. [11] reported the eccentricity effect on the dual solutions. Hu et al. [12] demonstrated three solutions in the wide gap annulus with a constant heat flux at wall. Zhang et al. [13] investigated the natural convection in a circular enclosure partitioned by a thin flat plate. Natural convection in other types of enclosures has also been studied by many other researchers [14,15].

It has been confirmed that the rotation effect plays a significant role in natural convection in an annulus in applications such as the heat transfer in bearing system, i.e., in the work of Lopez et al. [16] which studied the rotation effect on convection in a vertical annulus. For a square enclosure within which there is a rotating circular cylinder, Zhang et al. [17] found that the heat transfer can be conduction-dominated or mixed convection-dominated depending on the rotating speed. In this paper, we focus on the rotation effect on the natural convection in a horizontal annulus. The mechanism of the rotation effect is quantified by linear speed of rotational inner cylinder, and is presented and analyzed by the streamlines and isotherms at different dimensionless linear speeds. We also use the bifurcation diagram of rotation system to discuss the bifurcation solutions of natural convection in a horizontal annulus with stationary inner and outer cylinders. The present study of the rotation effect on natural convection provides a thorough insight in understanding the heat transfer in the bearing system problem.

All simulations in this study are carried out using a thermal immersed boundary-lattice Boltzmann method (IB-LBM). The lattice Boltzmann method (LBM) has been widely adopted in computational fluid dynamics for its convenience in treating curved boundaries and performing parallel computation [12,14–16,18–30]. It exhibits great convenience in heat and mass transfer [6–8,12]. The IB-LBM has achieved great success in dealing with moving boundaries with complex geometry since 2004 since Feng et al. [21] first introduced the IBM into LBM. Since then, a lot of efforts have been made to improve the accuracy of this method [22,24–26]. Feng et al. [22] used the idea of direct force to calculate the boundary force. Niu et al. [24] considered the moment exchange on the boundary. Wu et al. [25,26] proposed an implicit velocity correction-based model. For thermal fluids, researchers made efforts to deal with the thermal boundary conditions [12,28–30]. Hu et al. [12] developed a thermal IB-LBM to simulate both the temperature Dirichlet boundary condition and the Neumann boundary condition. Wu et al. discussed the accuracy of thermal IB-LBM [30]. Recently, the thermal lattice Boltzmann method with curving boundary treatment becomes popular in simulating natural convection in a horizontal annulus [11,12,31,32].

The remainder of this paper is organized as follows. Section 2 describes the scope of simulations and briefly introduces the numerical method used in this paper. In Section 3, numerical results are compared with experiments. In Section 4, numerical results and the effect of rotation on the natural convection are presented. We summarize our findings and conclude in Section 5.

2. Numerical Model and Method

In this section, the configuration of natural convection in a horizontal annulus and a treatment for thermal flow with curved boundaries are introduced. We follow the idea of the coupled lattice BGK model (CLBGK) [21] to treat thermal flow field. The IB-LBM is used to treat the boundary conditions.

2.1. Numerical Model

The configuration of a two-dimensional natural convection is shown in Figure 1. R_i and R_o denote the radius of inner and outer cylinders, respectively. The flow is driven by gravity G_α. The governing equations of the convection are [23]:

$$\begin{array}{ll} \partial_t \rho + \partial_\alpha(\rho u_\alpha) = 0 & (a) \\ \partial_t(\rho u_\alpha) + \partial_\beta(\rho u_\alpha u_\beta) = G_\alpha + F_\alpha - \partial_\alpha p + \partial_\beta\left[\mu(\partial_\alpha u_\beta + \partial_\beta u_\alpha)\right] & (b) \\ \partial_t T + u_\alpha \partial_\alpha T = \kappa \partial_\alpha^2 T + \Phi & (c) \end{array} \quad (1)$$

where ρ is the fluid density, u_α is velocity, G_α is the buoyancy, F_α is the body force, μ is the dynamic viscosity, T is the temperature, κ is thermal diffusivity and Φ is the heat source term. The inner and outer cylinder surfaces are maintained at different uniform temperatures T_h and $T_c(T_h > T_c)$, respectively. With the Boussinesq approximation [21], the fluid density is assumed to be a linear function of the temperature and the buoyancy can be written as: $G_\alpha = -g_\alpha \beta(T - T_0)$ and β is the coefficient of thermal expansion and T_0 is the average temperature of fluid.

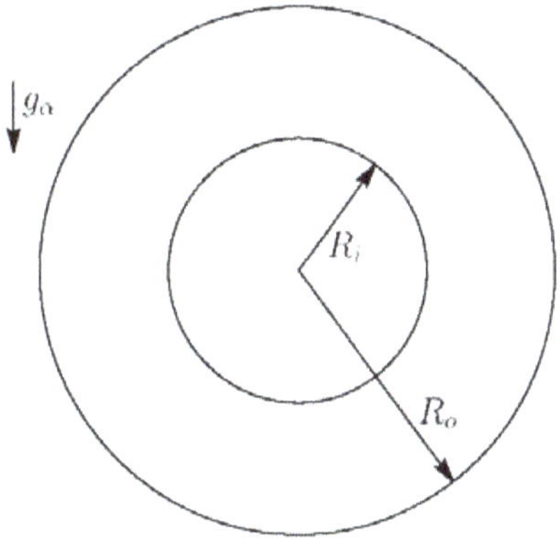

Figure 1. Configuration of natural convection in a horizontal annulus.

Some dimensionless variables in natural convection are introduced as follows. The Mach number (Ma):

$$Ma = \frac{\sqrt{|g_\alpha|\beta \Delta T L}}{c_s}, \quad (2)$$

where $\Delta T = T_h - T_c$ is temperature difference between hot wall and cold wall, L is the characteristic length across the temperature field which is $R_o - R_i$ for the convection in a horizontal annulus. The Mach number used in present work is fixed at 0.01 to ensure that the flow is incompressible.

The Rayleigh number (Ra):

$$Ra = \frac{|g_\alpha|\beta \Delta T L^3}{\nu \kappa} = \frac{(MaLc_s)^2 Pr}{\nu^2}, \quad (3)$$

where Pr is the Prandtl number defined as $Pr = \nu/\kappa$.

The *Nusselt* number (Nu):

$$Nu = \frac{|q_\alpha|}{Q_c} = \frac{|u_\alpha T - \kappa \partial_\alpha T|}{Q_c}, \quad (4)$$

where $|q_\alpha|$ is the quantity of local heat flux and Q_c is the average heat flux resulting from pure conduction.

2.2. Thermal IB-LBM

To solve the governing equations, the CLBGK model [23] is used in this paper. The general philosophy is to solve the velocity field and the temperature field using two independent lattice BGK equations, respectively, and then to combine them into one coupled model for the Boussinesq incompressible flow.

The lattice BGK equation with external force term reads:

$$f_i(x_\alpha + c_{i\alpha}\delta_t, t + \delta_t) - f_i(x_\alpha, t) = -\omega_f\left[f_i(x_\alpha, t) - f_i^{(eq)}(x_\alpha, t)\right] + F_i\delta_t, \tag{5}$$

where f_i is the density distribution function, $f_i^{(eq)}$ is the equilibrium distribution function, $c_{i\alpha}$ is the ith discretized velocity vector and ω_f is the relaxation parameter. For the D2Q9 model [18], the equilibrium function is:

$$f_i^{(eq)}(x_\alpha, t) = w_i\rho\left[1 + \frac{c_{i\alpha}v_\alpha}{c_s^2} + \frac{v_\alpha v_\beta}{2c_s^2}\left(\frac{c_{i\alpha}c_{i\beta}}{c_s^2} - \delta_{\alpha\beta}\right)\right], \tag{6}$$

where sound speed c_s and the weight coefficient w_i are:

$$c_s = \frac{1}{\sqrt{3}}\frac{\delta_x}{\delta_t}, \quad w_i = \begin{cases} 4/9 & i = 0 \\ 1/9 & i = 1 \sim 4 \\ 1/36 & i = 5 \sim 8 \end{cases}. \tag{7}$$

Using Chapman-Enskog expansion till the second order of approximation and considering the conserved quantity $\rho = \sum f_i$ and $\rho u_\alpha = \sum f_i c_{i\alpha} + F_\alpha\delta_t/2$, the relation between the relaxation parameter ω_f and kinematic viscosity ν can be obtained as:

$$\nu = c_s^2\left(\frac{1}{\omega_f} - \frac{1}{2}\right)\delta_t. \tag{8}$$

The force scheme proposed by Buick [27] is applied in this work:

$$F_i = w_i\left(1 - \frac{\omega_f}{2}\right)\frac{G_\alpha c_{i\alpha}}{c_s^2}, \tag{9}$$

where G_α is the external force.

Coupling temperature field to flow field, the well-known Boussinesq approximation is often used in the study of natural convection. With the Boussinesq approximation, the gravity can be rewritten as:

$$G_y = -\rho|g_\alpha|[1 - \beta(T - T_0)]. \tag{10}$$

To simplify this formula, T_0 is chosen as $-1/\beta$ and the gravity is:

$$G_y = \rho|g_\alpha|\beta T. \tag{11}$$

For the flow field, non-slip boundary condition is considered in all simulations. The direct force IB-LBM proposed by Feng [22] is applied to realize the dynamical boundary condition. The discretion of the derivative of time in above equation is related to the desired boundary condition (Equation (12)). This method is also termed as direct forcing model by Wu [25] penalty method by Feng et al. [21] and direct forcing method by Feng et al. 2005 [22]. They proposed a direct force term to realize the no-slip boundary condition as formed below.

$$F_\alpha = \rho(\partial_t u_\alpha + u_\beta\partial_\beta u_\alpha) - \mu\partial_\beta^2 u_\alpha + \partial_\alpha p, \tag{12}$$

where $\partial_t u_\alpha = \frac{u_\alpha^D - u_\alpha}{\delta_t}$ and u_α^D is the desired value of velocity on the boundary.

Based on their method, the force term used in a thermal IB-LBM has the following formula:

$$F_\alpha = \rho(\partial_t u_\alpha + u_\alpha \partial_\beta u_\beta) - \partial_\beta^2 u_\alpha + \partial_\alpha p - \rho \left| g_\alpha \right| \beta T e_{y\alpha}, \tag{13}$$

where $e_{y\alpha}$ is the unit vector in the y direction.

To simulate the thermal field, we follow the idea of CLBGK proposed by Guo [23] and use a 9-discrete-velocity model with the distribution function of temperature g_i.

$$g_i(x_\alpha + c_{i\alpha}\delta_t, t + \delta_t) - g_i(x_\alpha, t) = -\omega_g \left[g_i(x_\alpha, t) - g_i^{(eq)}(x_\alpha, t) \right] + \Phi_i \delta_t \tag{14}$$

where Φ_i is discretized heat source term which is related to heat source Φ. The thermal equilibrium function is:

$$g_i^{(eq)}(x_\alpha, t) = w_i T \left[1 + \frac{c_{i\alpha} v_\alpha}{c_s^2} + \frac{v_\alpha v_\beta}{2c_s^2} \left(\frac{c_{i\alpha} c_{i\beta}}{c_s^2} - \delta_{\alpha\beta} \right) \right]. \tag{15}$$

To compare with Equation (1) (c), the Chapman-Enskog expansion is also used to decide the relation between ω_g and κ, which is:

$$\kappa = c_s^2 \left(\frac{1}{\omega_g} - \frac{1}{2} \right) \delta_t. \tag{16}$$

Following the idea of previous direct force model and considering $u_\alpha \partial_\alpha T = 0$ at boundary, the isothermal boundary condition is realized by the following modified heat source:

$$\Phi = \partial_t T - \kappa \partial_\alpha^2 T \tag{17}$$

where $\partial_t T$ is calculated by $\partial_t T = \frac{T^D - T}{\delta_t}$ and T^D is the desired value of temperature on the boundary.

Since LBM deals with Eulerian flow points and boundaries are represented by Lagrangian points, Delta function is used to exchange the message between Eulerian points and Lagrangian points and has the following formula:

$$\begin{gathered} D(x_\alpha - x_{0\alpha}) = \delta(x - x_0)\delta(y - y_0) \\ \delta(r - r_0) = \begin{cases} \frac{1}{4}\left[1 + \cos\left(\frac{\pi |r|}{2}\right)\right] & |r| \leq 2 \\ 0 & |r| > 2 \end{cases} . \end{gathered} \tag{18}$$

The messages from both velocity field and thermal field are exchanged by Delta function mentioned in the above section.

3. Verification

The accuracy of the present method in natural convection in an annulus is verified. To qualify the results, the average *Nusselt* number ($\overline{Nu}$)is defined by the following equation:

$$\overline{Nu} = \frac{\overline{q}_n}{Q_c} = \frac{\int q_n \mathrm{d}s}{Q_c S} \tag{19}$$

where S is the area of the flow filed. For pure conduction, the exact solution of the temperature field in an annulus is:

$$\frac{T - T_c}{\Delta T} = \frac{\ln \frac{R_o}{r}}{\ln R_\beta} \tag{20}$$

where radius ratio $R_\beta = \frac{R_o}{R_i}$. Considering the definition of the average *Nusselt* number, its formula in an annulus yields:

$$\overline{Nu} = \frac{\int q_n ds}{Q_c S} = \gamma \frac{L \int q_r r dr d\theta}{\kappa \Delta T \pi S} = \gamma \frac{\int q_r r dr d\theta}{\kappa \Delta T \pi (R_o + R_i)} \tag{21}$$

where the modified parameter γ is:

$$\gamma = \frac{\ln R_\beta (R_\beta + 1)}{2(R_\beta - 1)}. \tag{22}$$

The relation between γ and R_β is shown in Figure 2.

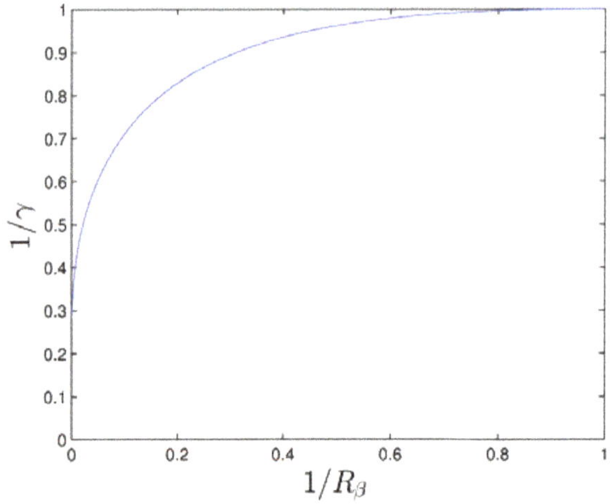

Figure 2. The distribution of the modified parameter γ versus the radius ratio R_β.

The grid independence tests of present thermal IB-LBM for natural convection in horizontal annuli are conducted. The grid sizes of 128 × 128, 256 × 256 and 512 × 512(*l.u.*) are tested for 4 different Rayleigh numbers, namely, $Ra = 10^2, 10^3, 10^4$ and 2×10^4 to verify the grid independence and the accuracy of the boundary treatment by comparing with the results from experiments [3]. Table 1 presents the results of average *Nusselt* number. It is shown that numerical results of the present method are in excellent agreement with experimental results. The relative errors between grid size of 256 × 256(*l.u.*) and extrapolated results are only 1.39%, 1.02%, 1.29% and 1.17% for $Ra = 10^2, 10^3, 10^4$ and 5×10^4, respectively. Results show that the grid size of 256 × 256(*l.u.*) which is used in following studies is fine enough.

Table 1. The grid independence test and comparison for the average Nusselt number. $R_\beta = 2.6$ and $Pr = 0.7$.

Ra	Grid Size				Experiment [3]
	128 × 128	256 × 256	512 × 512	Extrapolation	
1×10^2	1.034	1.016	1.008	1.002	1.000~1.002
1×10^3	1.101	1.092	1.087	1.081	1.081~1.084
1×10^4	1.955	1.977	1.989	2.003	2.005~2.010
2×10^4	2.346	2.364	2.375	2.392	2.394~2.405

4. Results and Discussion

In this section, the rotation effect on bifurcations of natural convection is analyzed with $Pr = 0.7$ and $Ra = 10^4$ in an annulus with its radius ratio $R_\beta = 1.5$. The streamlines, isotherms and phase portrait of average *Nusselt* number will be further analyzed. In order to study the rotation effect of the inner cylinder, the linear speed of the inner cylinder is non-dimensionalized by the characteristic speed $U_0 = \sqrt{|g_\alpha|\beta\Delta TL}$. As a matter of convenience, the *Nusselt* number Nu is used to represent the average *Nusselt* number in the flow field $\overline{Nu}$ in the following section.

4.1. Bifurcation Solutions in Stationary Concentric Cylinders

It has been found by many researchers that the bifurcation phenomenon appears in the natural convection in a horizontal annulus [7,11,12,31]. Different initial conditions will lead to different structures of convection. The schematic of different initial conditions used in our paper is shown in Figure 3. With different temperature distributions at Γ_u, we obtained three different structure of convection. From isotherms, different numbers of hot peaks can be found in different states. In Figure 4c, for instance, one can find three hot peaks at (0.2, 0.65), (0.5, 0.8) and (0.8, 0.65), respectively. A number of hot peaks are used to distinguish these different states in this paper. In this subsection, the bifurcation phenomena are simulated, which mainly verify the accuracy of the thermal IB-LBM treatment and obtain the initial conditions for following rotating cases.

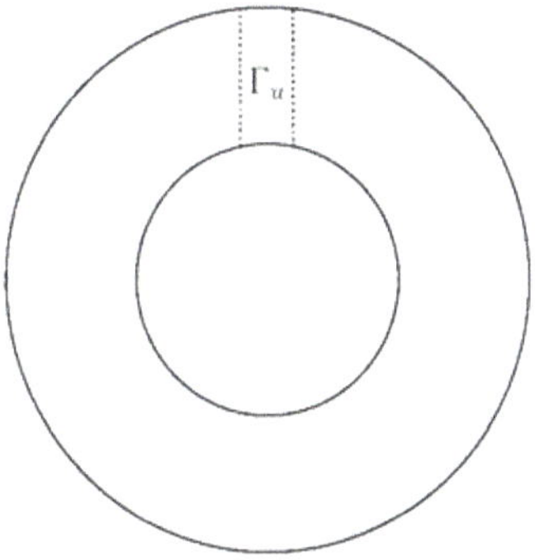

Figure 3. Schematic of initial condition of bifurcation solutions.

The bifurcation solutions of natural convection in a horizontal annulus with stationary inner and outer cylinders are presented. We first use zero velocity and low temperature for all flow fields as zero initial condition and obtained 1-peak state. Then 2-peak state is obtained by using the final state of 1-peak state by removing internal energy step as its initial condition. Finally, 3-peak state is obtained by using the final state of 2-peak state with an injecting internal energy step as its initial condition. The treatment of removing/injecting internal energy step in dealing with the temperature field at top region Γ_u is shown in Figure 3. When implementing removing internal energy step, temperature at top region Γ_u is set as the lowest temperature T_c, compulsively. The injecting internal energy step is to set the temperature at top region the highest temperature T_h, compulsively.

Figure 4 shows the results of streamlines and isotherms at $Ra = 10^4$, $Pr = 0.7$, $R_\beta = 1.5$. For 1-peak-state which is shown in Figure 4a, it is observed with zero initialization. It is seen that a pair of symmetrical vortices are formed in the annulus because of the buoyancy and one hot peak in the isotherms appears at the top region. According to the formula of local *Nusselt* number (Equation (4)), it is larger at top region than the remained region. After removing internal energy step, another solution with two hot peaks of this system is observed in Figure 4b. It is seen that a new pair of vortices appear at the top region of the annulus and velocity at middle of the top region turns from upwards to downwards. Injecting internal energy to Γ_u region at 2-peak state, another bifurcation consists of three hot peaks branches. 3-peak-state is presented in Figure 4c. A new pair of vortices appears at the top region. It is further observed that new vortex pair stretches and compresses the previous vortex pair and the previous vortex pair turns into two small vortices which beside the newly appeared large vortex pair. One also sees that 3-peak-state is very unstable from Figure 5c. This also suggests that with a small perturbation (inner cylinder rotate clockwise with its dimensionless linear speed exceeds 0.03), the flow state turns from 3-peak-state to 2-peak-state. It is further found that new pairs of vortices constantly appear at the top region of the annulus and the *Nusselt* number gradually increases with increasing number of hot peaks at the same Ra, Pr and R_β, which are 1.579, 1.791 and 1.888, respectively.

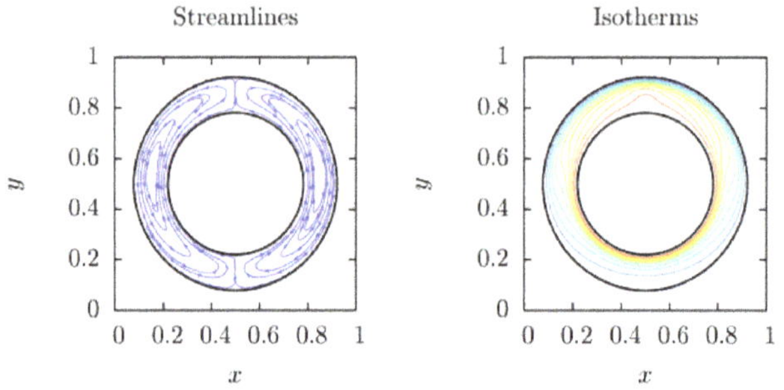

(**a**) 1-Peak-State

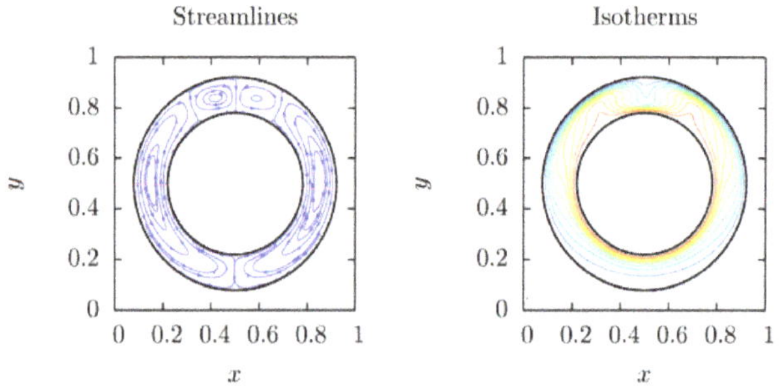

(**b**) 2-Peak-State

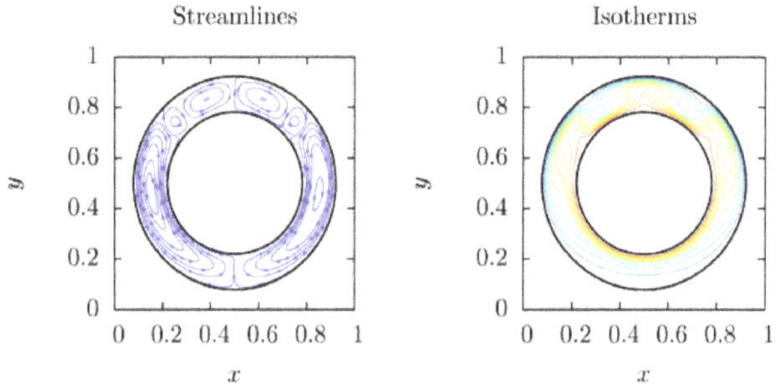

(**c**) 3-Peak-State

Figure 4. Different states at the same Ra, Pr and R_β. $Ra = 10^4$, $Pr = 0.7$, $R_\beta = 1.5$.

The phase portrait of *Nusselt* number is shown in Figure 5. In order to show the detail in the phase portrait, final phase trajectory of *Nusselt* number are zoomed in and presented in the inset of Figure 5. From Figure 5a,b, the phase portraits indicate that the *Nusselt* number increases with increasing hot peaks. Since the configuration is symmetric, the phase portraits in both Figure 5a,b indicate that the 2-peak-state and 3-peak-state are both stable. However, as discussed above, 3-peak-state will be unstable when the symmetry of the system breaks. Figure 5c shows the phase trajectory of the progress when a small perturbation is introduced into 3-peak-state. It is seen that any perturbation which breaks the symmetry of the system makes one of two previous small vortices compressed, the average *Nusselt* number decreases, and the compressed vortex finally disappears. Two large vortices beside the disappeared small vortex then merged into one vortex and the system turns to 2-peak-state. One also sees that the average *Nusselt* number of the newly formed 2-peak-state is smaller than that of the equilibrium system, vortices then resize under the buoyancy and the average *Nusselt* number increases till the system reaches equilibrium state.

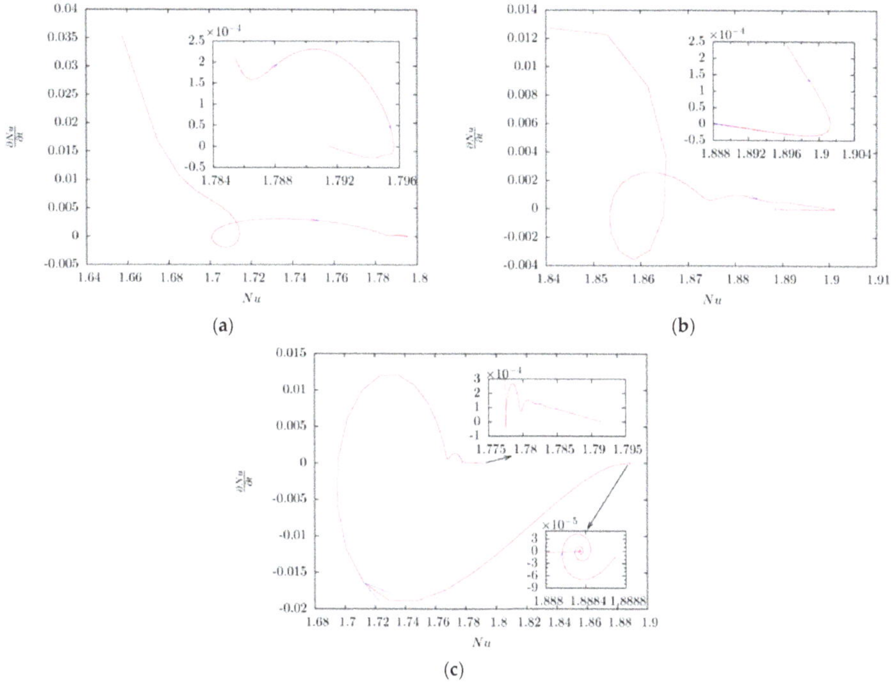

Figure 5. The phase portrait of Nusselt number. (**a**) Flow state turns from 1-peak-state to 2-peak-state after removing energy step; (**b**) Flow state turns from 2-peak-state to 3-peak-state after injecting energy step; (**c**) Flow state turns from 3-peak-state back to 2-peak-state after a small perturbation introduced into the system.

4.2. Rotation Effect on Natural Convection

The effect of linear speed of rotational inner cylinder is further analyzed in this subsection. Since 3-peak-state is an unstable system for rotating inner cylinder, 1-peak-state and 2-peak-state are the only two bifurcation systems discussed in this subsection. The inner cylinder rotates clockwise and its dimensionless linear speed (U_i^*) is ranging from 0 to 4. Figures 6a and 7a show the results obtained from stationary inner cylinder of 1-peak-state and 2-peak-state, respectively. The linear speed is increasingly considered after that.

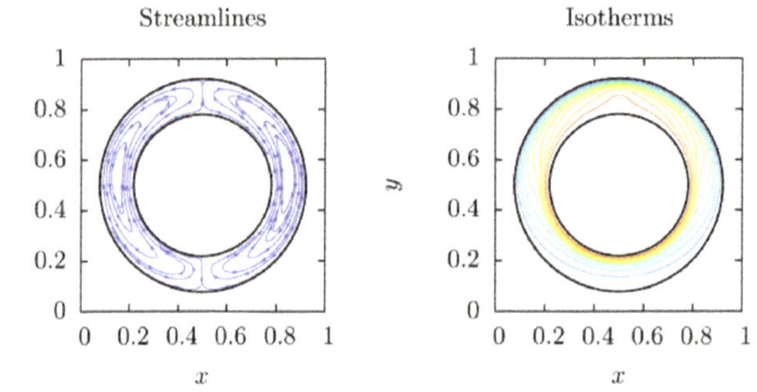

(**a**) $U_i^* = 0$

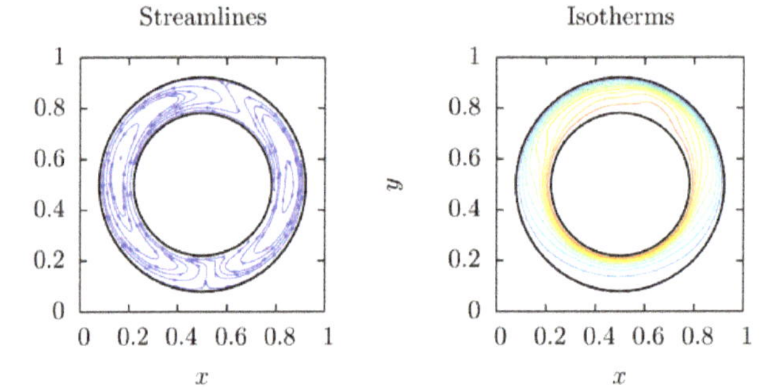

(**b**) $U_i^* = 0.5$

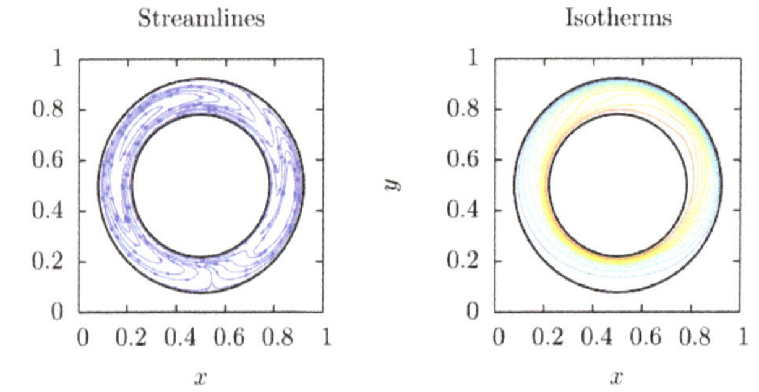

(**c**) $U_i^* = 2.0$

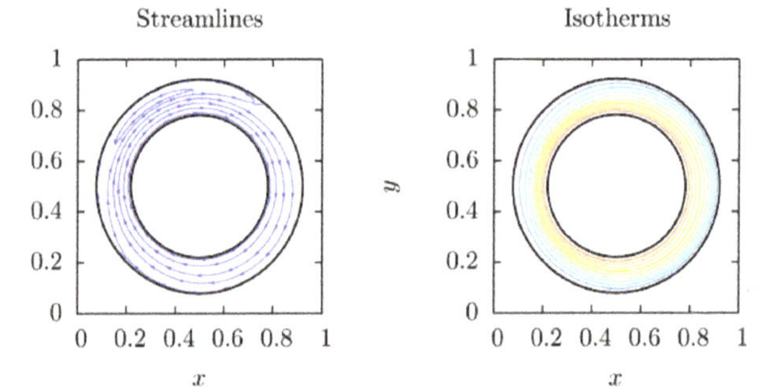

(**d**) $U_i^* = 4.0$

Figure 6. The streamlines and isotherms of rotation effect on 1-peak-state natural convection for $0 \leq U_i^* \leq 4$.

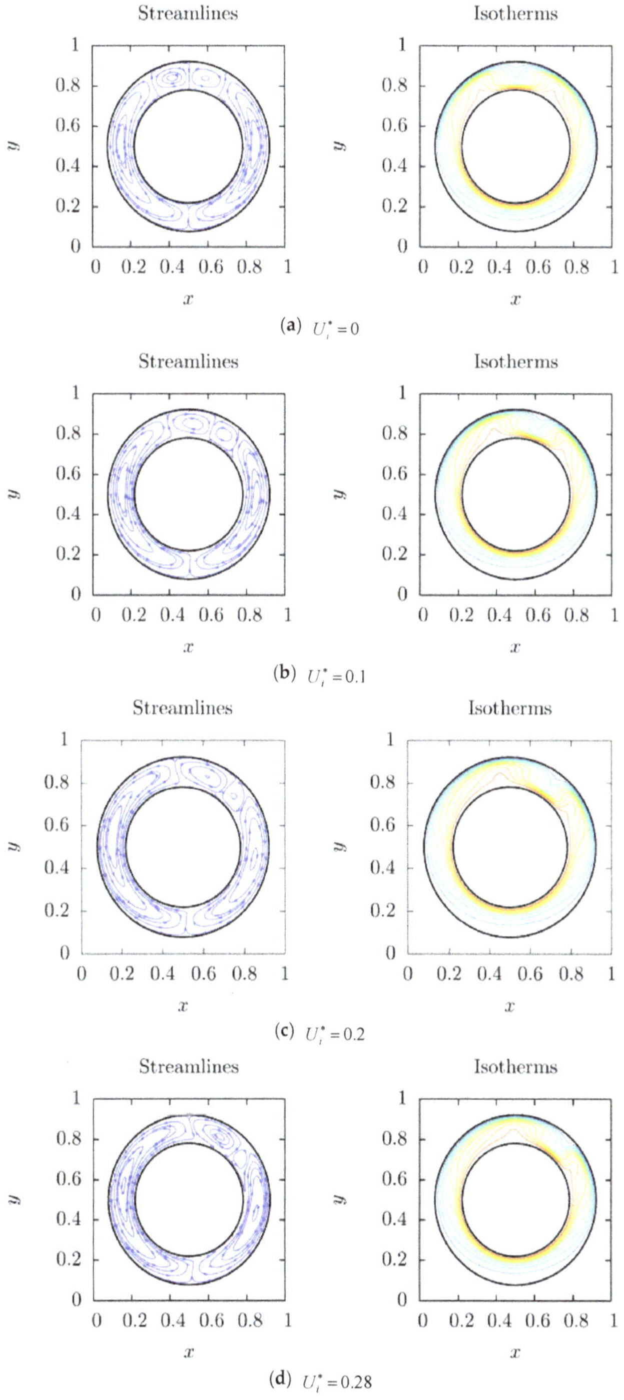

Figure 7. The streamlines and isotherms of rotation effect on 1-peak-state natural convection $0 \leq U_i^* \leq 0.28$.

It is seen that for 1-peak-state, two symmetrical vortices are formed in a stationary annulus as shown in Figure 6a. With increasing linear speed, left vortex is stretched and the right one is compressed, the hot peak moves clockwise and the width of hot peak increases as shown in Figure 6b. One also sees that the separated part of convection at top region deviates from the middle line more obviously than that at bottom region. Left vortex wedges into the right one at top region and compresses the right vortex. The right vortex then approaches the outer cylinder. From Figure 6c, three main convection appear in the annulus. It is obtained that the first convection is the left vortex and the second one is the right vortex. This suggests that the third convection starts from right part of bottom region near the right vortex, clings to the hot inner cylinder, streams around the right vortex and finally back to the right part of bottom region. From Figure 6d, one also sees that the first convection, the left vortex is compressed and is located in the top region, the second convection, and the right main vortex disappears, the third convection is developed into a main convection in the annulus when the dimensionless linear speed is faster than $U_i^* = 4.0$. This mainly implies that the disappearance of the second convection leads to a rapid decrease of the *Nusselt* number after the critical number of dimensionless linear speed, which is between 3 and 4.

For 2-peak-state, a stable range of $0 \leq U_i^* \leq 0.28$ is considered, in which 2-peak-state is stable. The streamlines and isotherms of rotation effect on 2-peak-state are shown in Figure 7, where four dimensionless speed, namely, U_i^*=0, 0.1, 0.2, 0.28 are considered. As shown in Figure 7a, one can see that a pair of main vortices and a pair of small vortices at top region emerge symmetrically in the annulus at stationary condition. From Figure 7b,c, it is seen that as linear speed of the inner cylinder increases, the symmetry of the flow are broken, the left vortices of both pairs are stretched and the right ones are compressed. It is also obtained that two hot peaks move clockwise and the width of left hot peak becomes larger than the right one. Figure 7d shows the result from highest linear speed before critical speed. It is seen that the compressed vortex of the small pair approaches the vertical position of the annulus. Amazingly, it is further found that the compressed small vortex disappears, and the stretched small vortex and the compressed main vortex merge with each other; the system then turns to 1-peak-state as the dimensionless linear speed is larger than a critical number.

Figure 8 illustrates the phase portrait of the progress of 2-peak-state turning to 1-peak-state. In order to show the detail in the phase portrait, final phase trajectory of *Nusselt* number are zoomed and presented in the inset of Figure 8. The steady 2-peak-state flow field with $U_i^* = 0.28$ is used as the initial condition of this simulation, the speed of inner cylinder is increased to 0.29. After the non-equilibrium progress of small vortex disappearing and two vortex merging with each other, as analyzed before. This also suggests that the *Nusselt* number becomes that of steady 1-peak-state with inner cylinder speed equals 0.29 gradually. It is further found that the critical number of dimensionless linear speed is between 0.28 and 0.29, the bifurcation phenomenon disappears and the steady state is 1-peak-state when the linear speed is higher than the critical speed.

As shown in Figure 9, both duel solutions can be obtained till $U_i^* = 0.28$. With the increment of linear speed, the rotation effect leads to the decrement of the average *Nusselt* number, and it is more obvious in 2-peak-state than that in 1-peak-state, which mainly illustrates that 2-peak-state is more sensitive than 1-peak-state. It is obtained that the average *Nusselt* number of 2-peak-state is larger than that of 1-peak-state, i.e., heat convection is more important in 2-peak-state than that in 1-peak-state. It is implied that when U_i^* is greater than 0.29, 2-peak-state is unstable. A local minimum value of average *Nusselt* number which appears at $U_i^* = 0.5$, indicating two different main mechanisms driving the convection. For $U_i^* \leq 0.5$, the system is over-damped for $U_i^* \geq 0.6$, the system is under-damped. Figure 9 shows the phase trajectories of cases at $U_i^* = 0.3$ and $U_i^* = 0.9$ as examples of these two systems. This also suggests that the third convection seems leading to the behavior of under-damped trajectories while the first and the second convection seems leading to the behavior of over-damped trajectories. It is further found that a small increase occurs for the *Nusselt* number, when the third convection is taken control, in which this increment results in the local minimum value of average *Nusselt* number at around $U_i^* = 0.5$.

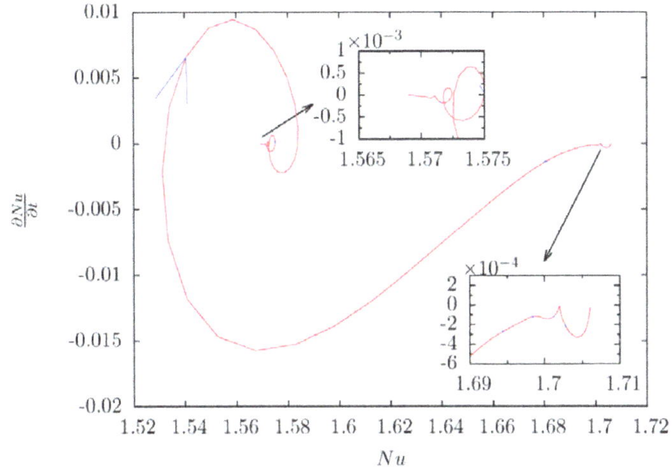

Figure 8. The phase portrait of 2-peak-state turns to 1-peak-state.

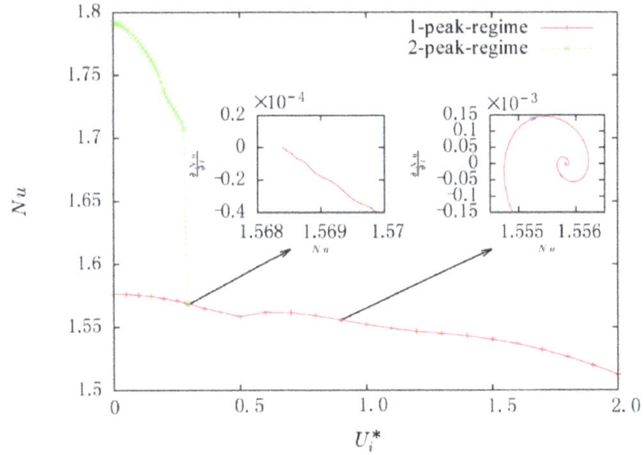

Figure 9. The bifurcation diagram of natural convection in horizontal concentric cylinders with rotating inner cylinder.

5. Conclusions

In this paper, numerical simulations have been performed to demonstrate the existence of bifurcation phenomena in natural convection in a horizontal annulus to investigate the rotation effect on it by using the thermal IB-LBM. Several conclusions can be summarized as follows.

In the first place, it is validated that numerical results of the present method are well consistent with experimental results. Our results manifest that 1-peak-state is the most stable system, 2-peak-state is more sensitive than 1-peak-state and 3-peak-state is the most sensitive system which will breakdown by a very small perturbation.

After that it is found that the rotation effect on 2-peak-state is more obvious than 1-peak-state, since 2-peak-state is more sensitive than 1-peak-state. It is noted that in the range of the present parameters, 2-peak-state is stable. When the linear speed of inner cylinder exceeds the critical speed, 2-peak-state breakdown and turns to 1-peak-state. From the flow pattern shown by streamlines, one may indicate that the right small vortex is compressed and finally disappears, then the left small vortex and the right main vortex merge with each other and the flow turns to 1-peak-state. Finally, three convections in this system which are affected by different linear speed of inner cylinder were discussed in detail. It is implied that the first and the second convections are contributed by the left and right main vortices, respectively. The third convection is a new vortex under the effect of rotation. With increasing linear speed, left main vortex is stretched and period of the first convection increases. Right main vortex is compressed and the scope of the second convection decreases. The third convection grows with the acceleration of rotation. It is further found that 2-peak-state is stable in the range of $0 \leq U_i^* \leq 0.28$, and when the linear speed of inner cylinder exceeds the critical speed, 2-peak-state breakdown and turns to 1-peak-state.

Author Contributions: Y.W. contributed significantly to analysis and manuscript preparation. Z.W. performed the data analyses and wrote the manuscript. Y.Q. contributed the concept of the study. W.G. helped perform the analysis with constructive discussions.

Funding: This work was funded by the National Natural Science Foundation of China (11872337, 11502237 and 51706205), Zhejiang Province Natural Science Foundation (LY18A020010), Zhejiang province public welfare technology application research project (2017C31075), the Young Researchers Foundation of Zhejiang Provincial Top Key Academic Discipline of Mechanical Engineering, and Dept. Social Security and Human Resources, Project of Ten Thousands Talents no. W01060069.

Acknowledgments: The referees' valuable suggestions are greatly helpful for us to improve our work. We appreciate them.

Conflicts of Interest: The authors declare no conflicts of interest.

References

1. Grigull, U.; Hauf, W. Natural convection in horizontal cylindrical annuli. In Proceedings of the 3rd International Heat Transfer Conference, Chicago, IL, USA, 7–12 August 1966; pp. 182–195.
2. Powe, R.E.; Carley, C.T.; Bishop, E.H. Free convective flow patterns in cylindrical annuli. *J. Heat Transf.* **1969**, *91*, 310. [CrossRef]
3. Kuehn, T.H.; Goldstein, R.J. An experimental and theoretical study of natural convection in the annulus between horizontal concentric cylinders. *J. Fluid Mech.* **1976**, *74*, 695–719. [CrossRef]
4. Nguyen, T.H.; Vasseur, P.; Robillard, L. Natural convection between horizontal concentric cylinders with density inversion of water for low Rayleigh numbers. *Int. J. Heat Mass Transf.* **1982**, *25*, 1559–1568. [CrossRef]
5. Dyko, M.P.; Vafai, K.; Mojtabi, A.K. A numerical and experimental investigation of stability of natural convective flows within a horizontal annulus. *J. Fluid Mech.* **1999**, *381*, 27–61. [CrossRef]
6. Yoo, J.S. Natural convection in a narrow horizontal cylindrical annulus: Pr = 0.3. *Int. J. Heat Mass Transf.* **1998**, *41*, 3055–3073. [CrossRef]
7. Yoo, J.S. Prandtl number effect on bifurcation and dual solutions in natural convection in a horizontal annulus. *Int. J. Heat Mass Transf.* **1999**, *42*, 3279–3290. [CrossRef]
8. Yoo, J.S. Transition and multiplicity of flows in natural convection in a narrow horizontal cylindrical annulus: Pr = 0.4. *Int. J. Heat Mass Transf.* **1999**, *42*, 709–722. [CrossRef]
9. Petrone, G.; Chénier, E.; Lauriat, G. Stability of free convection in air-filled horizontal annuli: Influence of the radius ratio. *Int. J. Heat Mass Transf.* **2004**, *47*, 3889–3907. [CrossRef]
10. Petrone, G.; Chénier, E.; Lauriat, G. Stability analysis of natural convective flows in horizontal annuli: Effects of the axial and radial aspect ratios. *Phys. Fluids* **2006**, *18*, 104107. [CrossRef]
11. Luo, K.; Yi, H.L.; Tan, H.P. Eccentricity effect on bifurcation and dual solutions in transient natural convection in a horizontal annulus. *Int. J. Therm. Sci.* **2014**, *89*, 283–293. [CrossRef]
12. Hu, Y.; Li, D.; Shu, S.; Niu, X.D. Study of multiple steady solutions for the 2D natural convection in a concentric horizontal annulus with a constant heat flux wall using immersed boundary-lattice Boltzmann method. *Int. J. Heat Mass Transf.* **2015**, *81*, 591–601. [CrossRef]
13. Zhang, W.; Wei, Y.K.; Chen, X.P.; Dou, H.-S.; Zhu, Z.C. Partitioning effect on natural convection in a circular enclosure with an asymmetrically placed inclined plate. *Int. Commun. Heat Mass Transf.* **2018**, *90*, 11–22. [CrossRef]

14. Wei, Y.K.; Dou, H.S.; Wang, Z.D.; Qian, Y.H. A novel two-dimensional coupled lattice Boltzmann model for incompressible flow in application of turbulence Rayleigh-Taylor instability. *Comput. Fluids* **2017**, *156*, 97–102. [CrossRef]
15. Wei, Y.K.; Dou, H.S.; Wang, Z.D.; Qian, Y.H.; Yan, W.W. Simulations of natural convection heat transfer in an enclosure at different Rayleigh number using lattice Boltzmann method. *Comput. Fluids* **2016**, *124*, 30–38. [CrossRef]
16. Lopez, J.M.; Marques, F.; Mercader, I.; Batiste, O. Onset of convection in a moderate aspect-ratio rotating cylinder: Eckhaus-Benjamin-Feir instability. *J. Fluid Mech.* **2007**, *590*, 187–208. [CrossRef]
17. Zhang, W.; Wei, Y.K.; Dou, H.-S.; Zhu, Z.C. Transient behaviors of mixed convection in a square enclosure with an inner impulsively rotating circular cylinder. *Int. Commun. Heat Mass Transf.* **2018**, in press.
18. Qian, Y.H.; d'Humières, D.; Lallemand, P. Lattice BGK models for Navier-Stokes equation. *Europhys. Lett.* **1992**, *17*, 479–484. [CrossRef]
19. Liang, H.; Shi, B.C.; Chai, Z.H. Lattice Boltzmann simulation of three-dimensional Rayleigh-Taylor instability. *Phys. Rev. E* **2016**, *93*, 033113–033119. [CrossRef] [PubMed]
20. Liang, H.; Xu, J.; Chen, J.; Wang, H.; Chai, Z.H.; Shi, B.C. Phase-field-based lattice Boltzmann modeling of large-density-ratio two-phase flows. *Phys. Rev. E* **2018**, *97*, 033309. [CrossRef] [PubMed]
21. Feng, Z.G.; Michaelides, E.E. The immersed boundary-lattice Boltzmann method for solving fluid-particles interaction problems. *J. Comput. Phys.* **2004**, *195*, 602–628. [CrossRef]
22. Succi, S. *The Lattice-Boltzmann Equation for Fluid Dynamics and Beyond*; Oxford University Press: Oxford, UK, 2001.
23. Guo, Z.L.; Shi, B.C.; Zheng, C.G. A coupled lattice BGK model for the Boussinesq equations. *Int. J. Numer. Meth. Fluids* **2002**, *39*, 325–342. [CrossRef]
24. Niu, X.D.; Shu, C.; Chew, Y.T.; Peng, Y. A momentum exchange-based immersed boundary-lattice boltzmann method for simulating incompressible viscous flows. *Phys. Lett. A* **2006**, *354*, 173–182. [CrossRef]
25. Wu, J.; Shu, C. Implicit velocity correction-based immersed boundary-lattice Boltzmann method and its applications. *J. Comput. Phys.* **2009**, *228*, 1963–1979. [CrossRef]
26. Wu, J.; Shu, C.; Zhang, Y.H. Simulation of incompressible viscous flows around moving objects by a variant of immersed boundary-lattice Boltzmann method. *Int. J. Numer. Meth. Fluids* **2010**, *62*, 327–354. [CrossRef]
27. Buick, J.M.; Greated, C.A. Gravity in lattice Boltzmann model. *Phys. Rev. E* **2000**, *61*, 5307–5320. [CrossRef]
28. Liu, C.H.; Lin, K.H.; Mai, H.C.; Lin, C.A. Thermal boundary conditions for thermal lattice Boltzmann simulations. *Compt. Math. Appl.* **2010**, *59*, 2178–2193. [CrossRef]
29. Le, G.; Zhang, J. Boundary slip from the immersed boundary lattice Boltzmann models. *Phys. Rev. E* **2009**, *79*, 026701. [CrossRef] [PubMed]
30. Wu, J.; Cheng, Y.; Miller, L.A. An iterative source correction based immersed boundary-latticeBoltzmann method for thermal flow simulations. *Int. J. Heat Mass Transf.* **2017**, *115*, 450–460. [CrossRef]
31. Luo, K.; Yi, H.L.; Tan, H.P. Radiation effects on bifurcation and dual solutions in transient natural convection in a horizontal annulus. *Int. J. Therm. Sci.* **2015**, *89*, 283–293. [CrossRef]
32. Amiri, D.A.; Nazari, M.; Kayhani, M.H.; Succi, S. Non-Newtonian unconfined flow and heat transfer over a heated cylinder using the direct-forcing immersed boundary-thermal lattice Boltzmann method. *Phys. Rev. E* **2014**, *89*, 053312. [CrossRef] [PubMed]

entropy

MDPI

Article

A Mathematical Realization of Entropy through Neutron Slowing Down

Barry Ganapol [1,*], Domiziano Mostacci [2] and Vincenzo Molinari [2]

1 Department of Aerospace and Mechanical engineering, University of Arizona, Tucson, AZ 85721, USA
2 Department of Industrial Engineering, University of Bologna, 40136 Bologna, Italy; domiziano.mostacci@unibo.it (D.M.); Vincenzo.molinari@unibo.it (V.M.)
* Correspondence: ganapol@cowboy.ame.arizona.edu

Received: 23 February 2018; Accepted: 20 March 2018; Published: 28 March 2018

Abstract: The slowing down equation for elastic scattering of neutrons in an infinite homogeneous medium is solved analytically by decomposing the neutron energy spectrum into collision intervals. Since scattering physically smooths energy distributions by redistributing neutron energy uniformly, it is informative to observe how mathematics accommodates the scattering process, which increases entropy through disorder.

Keywords: entropy; elastic scattering; neutron slowing down

---

1. Introduction

Neutron slowing down in an infinite homogeneous medium [1] is a classic problem in neutron transport theory. Neutrons (test particles) collide elastically with nuclei (field particles) and thereby lose energy to nuclear recoil. Thus, we have a common collisional process as described by a balance in energy phase space between a neutron source and neutrons scattering into and out of an infinitesimal energy increment, leading to the slowing down equation. One can analytically solve this equation for the neutron collision density distribution as it tends toward its equilibrium state. In addition, neutron loss is possible through radiative capture but will not be considered. Limiting our investigation to an infinite medium has naturally eliminated spatial and directional variation. While the slowing down equation is deterministic, it nevertheless describes the statistical scattering process, as illustrated by the associated mathematics.

In the following, we argue that the solution to the neutron slowing down equation characterizes the evolution of disorder associated with neutron–nucleus collisions. While it is not strictly correct to attribute disorder to entropy [2], in our case, starting from monoenergetic neutrons representing complete order, subsequent scattering creates disorder by uniformly redistributing neutron energy and recoil energy transfer to field particles. The nucleus scattering model conserves kinetic energy; however, it should be noted that the slowing down process assumes background nuclei are at rest. This considerably simplifies the scattering kernel and allows an analytical solution. Beginning with oscillations of the collision density in lethargy (logarithm of energy), called Placzek transients, neutron slowing down demonstrates increasing entropy with increasing lethargy. The oscillations originate from the discontinuity of derivatives submerged further into the solution at collision interval boundaries. As will be shown, the initially sharp discontinuity from the singular delta function source embeds itself in higher-order derivatives. Hence, with increasing lethargy, the solution becomes smoother, which is a tendency toward increased randomness and equilibrium. Therefore, neutron slowing down is a physical example of the mathematical representation of increasing disorder since one begins with a source of zero entropy (certainty), and, with an ever-increasing number of collision intervals, smoothing (uncertainty) of the energy distribution follows.

2. Solution

2.1. The Slowing Down Equation

The neutron slowing down equation in a purely scatter material in the fast neutron regime is

$$F(E) = \int_0^{E_0^+} dE' P(E' \to E) F(E') + \delta(E - E_0) \tag{1a}$$

for the collision (energy) density

$$F(E) \equiv \Sigma_s(E)\phi(E), \tag{1b}$$

where $\Sigma_s(E)$ is the scattering cross section and $\phi(E)$ is the neutron scalar flux. The neutron scatters elastically from a nucleus uniformly to the energy interval $\alpha E' \leq E \leq E'$ with probability of scattering into dE given by

$$P(E' \to E)dE = \frac{dE}{(1-\alpha)E'}; \tag{1c}$$

otherwise, the probability is zero. The scattering parameter is

$$\alpha \equiv \left(\frac{A-1}{A+1}\right)^2, \tag{1d}$$

where A is the mass number of the nucleus, and a monoenergetic source emits neutrons at energy E_0. Therefore, Equation (1a) becomes

$$F(E) = \frac{1}{1-\alpha} \int_E^{\min(E_0^+, E/\alpha)} \frac{dE'}{E'} F(E') + \delta(E - E_0). \tag{2}$$

Note that to include source neutrons, the upper limit in the scattering integral must come from just above E_0.

Change to Lethargy Variable

With the change of energy to the lethargy variable,

$$u \equiv \ln\left(\frac{E_0}{E}\right),$$

Equation (2) becomes

$$F(u) = \frac{1}{1-\alpha} \int_{\max(0^-,\, u-q)}^{u} du e^{\,(u-u')} F(u') + \delta(u), \tag{3a}$$

where

$$q \equiv \ln\left(\frac{1}{\alpha}\right); \tag{3b}$$

and with the further transformation

$$F(u) \equiv e^{-u} g(u), \tag{4a}$$

there results,

$$g(u) = \frac{1}{1-\alpha} \int_{\max(0^-, u-q)}^{u} du' g(u') + \delta(u). \tag{4b}$$

2.2. Determination of $g(u)$ by Collision Interval

A natural decomposition of lethargy into scattering collision intervals (n), shown in Figure 1, enables an explicit solution. The lethargy interval q is the maximum lethargy gain a neutron experiences after a single collision.

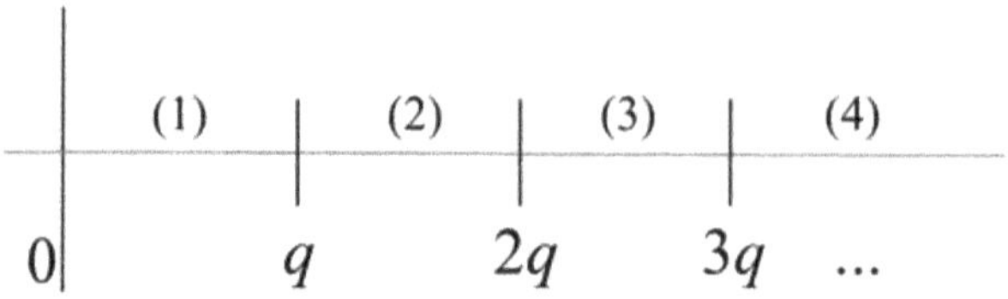

Figure 1. Scattering collision intervals (n = 1, 2, 3, 4).

2.2.1. Collision Interval (1)

In the first collision interval (1), Equation (4b) is

$$g_1(u) = \frac{1}{1-\alpha}\int_0^u du' g_1(u') + \delta(u). \tag{5a}$$

For

$$g(u) \equiv g_1(u),\ 0 \leq u \leq q, \tag{5b}$$

a convenient solution is

$$g_1(u) = g_0(u) + g_{1c}(u). \tag{6a}$$

The source, emitting uncollided neutrons, defines

$$g_0(u) \equiv \delta(u), \tag{6b}$$

and introducing Equation (6a) into Equation (5a) gives

$$g_{1c}(u) = \frac{1}{1-\alpha}\int_0^u du' g_{1c}(u') + \frac{1}{1-\alpha} \tag{6c}$$

for neutrons experiencing at least one collision. Therefore, upon differentiation

$$\frac{dg_{1c}(u)}{du} = \frac{1}{1-\alpha} g_{1c}(u), \tag{7a}$$

and solving, with initial condition

$$g_{1c}(0) = \frac{1}{1-\alpha}, \tag{7b}$$

from Equation (6c)

$$g_{1c}(u) = \frac{1}{1-\alpha} e^{u/(1-\alpha)}. \tag{7c}$$

The solution in interval (1) is

$$g_1(u) = \delta(u) + \frac{1}{1-\alpha} e^{u/(1-\alpha)}, \tag{8}$$

exhibiting the delta function source discontinuity at $u = 0$ with no disorder and source neutrons scattering to the end of interval (1).

2.2.2. Collision Interval (2)

For the second collision interval, $q \leq u \leq 2q$, Equation (4b) becomes

$$g(u) = \frac{1}{1-\alpha}\int_{u-q}^{u} du' g(u');$$

and, if

$$g(u) \equiv g_2(u),\ q \leq u \leq 2q, \tag{9a}$$

then

$$g_2(u) = \frac{1}{1-\alpha}\int_{u-q}^{q} du' g_1(u') + \frac{1}{1-\alpha}\int_{q}^{u} du' g_2(u'), \tag{9b}$$

where scattering from interval (1) contributes to interval (2). Differentiating gives

$$\frac{dg_2(u)}{du} = \frac{1}{1-\alpha}g_2(u) - \frac{1}{(1-\alpha)^2}e^{(u-q)/(1-\alpha)} - \frac{1}{1-\alpha}\delta(u-q). \tag{9c}$$

Before solving Equation (9c), we note the delta function source singularity, originally at $u = 0$, has moved to the derivative of $g_2(u)$ at $u = q$, and continues on to higher derivatives in subsequent collision intervals, as will be shown.

From Equation (9b),

$$g_2(q^+) = \lim_{\varepsilon\to 0} g_2(q+\varepsilon) = \frac{1}{1-\alpha}\int_{0^+}^{q} du' g_1(u') = \frac{1}{1-\alpha}\left[e^{q/(1-\alpha)} - 1\right]; \tag{9d}$$

and on solving Equation (9c) as a sum of the solution to the homogeneous equation and the particular solution gives

$$g_2(u) = \left[g_2(q^+) - \frac{(u-q)}{(1-\alpha)^2}\right]e^{(u-q)/(1-\alpha)} + \frac{1}{1-\alpha}[1-\Theta(u-q)], \tag{9e}$$

or, since from Equation (8),

$$g_1(0^+) = \frac{1}{1-\alpha},$$

$$g_2(u) = \left[g_2(q^+) - \frac{(u-q)}{(1-\alpha)}g_1(0^+)\right]e^{(u-q)/(1-\alpha)} + \frac{1}{1-\alpha}[1-\Theta(u-q)]. \tag{9f}$$

Though the last term vanishes in interval (2), it is theoretically necessary to give the delta function discontinuity in the derivative.

Note that since Equation (8) evaluated at q^- is

$$g_1(q^-) = \frac{1}{1-\alpha}e^{q/(1-\alpha)}, \tag{10a}$$

across the boundary of intervals (1) and (2), one observes a finite discontinuity in $g(u)$,

$$\Delta g_2(q) \equiv g_2(q^+) - g_1(q^-) = -\frac{1}{1-\alpha}; \tag{10b}$$

hence, the delta function in Equation (9c) at $u = q$ in the derivative of $g_2(u)$.

2.2.3. Collision Interval (3)

To establish a pattern, we continue to interval (3) with Equation (4b) for $n = 3$:

$$g_3(u) = \frac{1}{1-\alpha}\int_{u-q}^{2q} du' g_2(u') + \frac{1}{1-\alpha}\int_{2q}^{u} du' g_3(u'), \tag{11a}$$

where

$$g(u) \equiv g_3(u),\ 2q \le u \le 3q. \tag{11b}$$

On differentiation of Equation (11a):

$$\frac{dg_3(u)}{du} = \frac{1}{1-\alpha} g_3(u) - \frac{1}{1-\alpha} g_2(u-q) \tag{11c}$$

and solving

$$g_3(u) = g_3(2q^+)e^{(u-2q)/(1-\alpha)} - \frac{1}{1-\alpha}\int_{2q}^{u} du' e^{(u-u')/(1-\alpha)} g_2(u'-q).$$

After integration of the last term, we find

$$g_3(u) = \left[g_3(2q^+) - \frac{(u-2q)}{(1-\alpha)} g_2(q^+) + \frac{1}{2}\frac{(u-2q)^2}{(1-\alpha)^3}\right] e^{(u-2q)/(1-\alpha)}, \tag{11d}$$

which is also

$$g_3(u) = \left[g_3(2q^+) - \frac{(u-2q)}{(1-\alpha)} g_2(q^+) + \frac{1}{2}\frac{(u-2q)^2}{(1-\alpha)^2} g_1(0^+)\right] e^{(u-2q)/(1-\alpha)}. \tag{11e}$$

The initial condition, $g_3(2q^+)$, for interval (3) is approached from within the interval and is given by Equation (11a) as

$$g_3(2q^+) = \lim_{\varepsilon \to 0} g_3(2q+\varepsilon) = \frac{1}{1-\alpha}\int_{q}^{2q} du' g_2(u').$$

However, from Equation (9b), we also find

$$g_2(2q^-) = \frac{1}{1-\alpha}\int_{q}^{2q} du' g_2(u'), \tag{11f}$$

demonstrating the continuity of $g(2q)$ across intervals (2) and (3), and completing the solution for collision interval (3).

2.2.4. Collision Interval (n)

Continuity for $u > 2q^+$

In general, defining the solution $g_n(u)$ for interval (n),

$$g(u) = g_n(u),\ (n-1)q \le u \le nq, \tag{12a}$$

and partitioning the scattering integral in Equation (4b) into current and previous intervals gives

$$g_n(u) = \frac{1}{1-\alpha}\int_{u-q}^{(n-1)q} du' g_{n-1}(u') + \frac{1}{1-\alpha}\int_{(n-1)q}^{u} du' g_n(u'). \tag{12b}$$

The interpretation of this expression is shown in Figure 2, and, as already mentioned, includes a contribution to interval (n) from scatter in the previous interval.

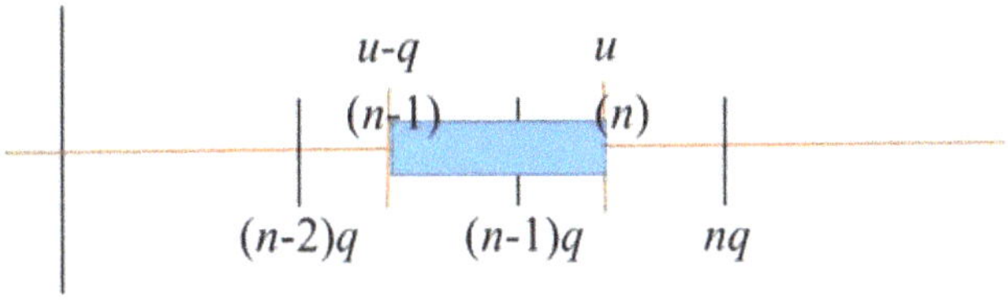

Figure 2. Contribution from previous collision interval-.

For $n \geq 3$, from Equation (12b) with n decremented by unity and $u = (n-1)q - \varepsilon \to (n-1)q^-$, in the limit as $\varepsilon \to 0$:

$$g_{n-1}((n-1)q^-) = \lim_{\varepsilon \to 0} g_{n-1}((n-1)q - \varepsilon) = \frac{1}{1-\alpha}\int_{(n-2)q}^{(n-1)q} du' g_{n-1}(u'). \tag{13a}$$

Similarly, for $u = (n-1)q + \varepsilon \to (n-1)q^+$ in Equation (12b),

$$g_n((n-1)q^+) = \lim_{\varepsilon \to 0} g_n((n-1)q + \varepsilon) = \frac{1}{1-\alpha}\int_{(n-2)q}^{(n-1)q} du' g_{n-1}(u') \tag{13b}$$

and therefore, comparing to Equation (13a),

$$\Delta g_n((n-1)q) = g_n((n-1)q^+) - g_{n-1}((n-1)q^-) = 0. \tag{13c}$$

Thus, $g(u)$ is a continuous function with the exception of the delta function singularity at $u = 0$ and the finite discontinuity at $u = q$.

2.2.5. General Solution

From $g_n(u), n = 2, 3$, the following pattern emerges:

$$g_n(u) = e^{(u-(n-1)q)/(1-\alpha)} \sum_{k=0}^{n-1} \gamma_{n,k}(u-(n-1)q)^k + \frac{1}{1-\alpha}[1-\Theta(u-q)]\delta_{n,2}, \tag{14}$$

where for $n = 2$

$$\begin{aligned} \gamma_{2,0} &= g_2(q^+) \\ \gamma_{2,1} &= -\tfrac{1}{(1-\alpha)} g_1(0^+), \end{aligned}$$

and $n = 3$

$$\begin{aligned} \gamma_{3,0} &= g_3(2q) \\ \gamma_{3,1} &= -\tfrac{1}{(1-\alpha)} g_2(q^+) \\ \gamma_{3,2} &= \tfrac{1}{2}\tfrac{1}{(1-\alpha)^2} g_1(0^+). \end{aligned}$$

We now confirm the pattern by constructive inductive reasoning.

For $n \geq 4$, assume the form of the solution, Equation (14), is true for $n-1$

$$g_{n-1}(u) = e^{(u-(n-2)q)/(1-\alpha)} \sum_{k=0}^{n-2} \gamma_{n-1,k}(u-(n-2)q)^k. \tag{15a}$$

Differentiating Equation (12b) gives the ODE:

$$\frac{dg_n(u)}{du} = \frac{1}{1-\alpha} g_n(u) - \frac{1}{1-\alpha} g_{n-1}(u-q), \tag{15b}$$

with the following solution over the interval $((n-1)q^+, nq^-)$:

$$\begin{aligned} g_n(u) &= g_n((n-1)q^+) e^{(u-(n-1)q)/(1-\alpha)} \\ &- \tfrac{1}{1-\alpha} \int_{(n-1)q^+}^{u} du' e^{(u-u')/(1-\alpha)} e^{(u'-(n-1)q)/(1-\alpha)} \sum_{k=0}^{n-2} \gamma_{n-1,k}(u'-(n-1)q)^k. \end{aligned} \tag{15c}$$

Performing the integration:

$$g_n(u) = \left[g_n((n-1)q^+) - \frac{1}{1-\alpha} \sum_{k=0}^{n-2} \frac{\gamma_{n-1,k}}{k+1} (u-(n-1)q)^{k+1} \right] e^{(u-(n-1)q)/(1-\alpha)}, \tag{15d}$$

and decrementing index k by unity gives

$$g_n(u) = \left[g_n((n-1)q^+) - \frac{1}{1-\alpha} \sum_{k=1}^{n-1} \frac{\gamma_{n-1,k-1}}{k} (u-(n-1)q)^{k} \right] e^{(u-(n-1)q)/(1-\alpha)}. \tag{15e}$$

Hence, on comparison to Equation (14), for $k = 0$:

$$\gamma_{n,0} = g_n((n-1)q^+), \tag{16a}$$

and for $k = 1, 2, \ldots$

$$\gamma_{n,k} = -\frac{1}{(1-\alpha)k} \gamma_{n-1,k-1} \tag{16b}$$

On solving the recurrence of Equations (16):

$$\gamma_{n,k} = \frac{(-1)^k}{(1-\alpha)^k k!} \gamma_{n-k,0} = \frac{(-1)^k}{(1-\alpha)^k k!} g_{n-k}((n-(k+1))q^+), \tag{16c}$$

we have the conjectured solution of the form of Equation (14) for intervals $n \geq 3$:

$$g_n(u) = e^{(u-(n-1)q)/(1-\alpha)} \sum_{k=0}^{n-1} \frac{(-1)^k}{k!} \left[\frac{u-(n-1)q}{1-\alpha} \right]^k g_{n-k}((n-(k+1))q^+). \tag{16d}$$

Or, from the continuity of $g(u)$ (Equation (13c)),

$$g_n(u) = e^{(u-(n-1)q)/(1-\alpha)} \left[\begin{array}{l} g_{n-1}((n-1)\tilde{q}) \\ + \sum_{k=1}^{n-1} \frac{(-1)^k}{k!} \left[\frac{u-(n-1)q}{1-\alpha} \right]^k g_{n-k}((n-(k+1))\tilde{q}) \end{array} \right], \tag{17a}$$

where

$$(n-(k+1))\tilde{q} = \begin{cases} 0^+, \; k = n-1 \\ 2q^+, \; k = n-2 \\ (n-(k+1))q, \; 1 \leq k \leq n-3. \end{cases} \tag{17b}$$

From Equation (4a), the collision density by collision interval is therefore

$$F_n(u) = e^{(\alpha u-(n-1)q)/(1-\alpha)} \begin{bmatrix} g_{n-1}((n-1)\tilde{q}) \\ +\sum_{k=1}^{n-1} \frac{(-1)^k}{k!} \left[\frac{u-(n-1)q}{1-\alpha}\right]^k g_{n-k}((n-(k+1))\tilde{q}) \end{bmatrix}, \tag{18a}$$

For future use, the jth derivative (using Leibnitz's rule) is:

$$\begin{aligned} &F_n^{(j)}(u) \\ &= e^{(\alpha u-(n-1)q)/(1-\alpha)} \begin{bmatrix} g_{n-1}((n-1)\tilde{q})\delta_{j0} + \sum_{k=1}^{n-1} \frac{(-1)^k}{k!} \left[\frac{1}{1-\alpha}\right]^k g_{n-k}((n-(k+1))\tilde{q}) \\ \bullet \sum_{l=0}^{j} \frac{j!}{l!(j-l)!} \left[\frac{\alpha}{1-\alpha}\right]^{j-l} [u-(n-1)q]^{k-l} \end{bmatrix}. \end{aligned} \tag{18b}$$

We now investigate the singularities of derivatives of $g(u)$.

3. Continuity/Singularities

So far, we have identified the singularities given in Table 1. In this section, all the relevant singularities starting at interval (n) will also be identified. To do so, we require several conjectures concerning the continuity of the collision density.

Table 1. Singularities identified.

Interval	u	Derivative (j)	Type
1	0	0	Infinite *
2	q	0	Finite
2	q	1	Infinite *

* Delta function.

Conjecture 1. *The jth derivatives of g_{n-1} and g_n at $u = (n-1)q$ for $n \geq j+3$ are continuous.*

Symbolically, Conjecture 1 is

$$\Delta g_n^{(j)}((n-1)q) \equiv g_n^{(j)}((n-1)q^+) - g_{n-1}^{(j)}((n-1)q^-) = 0,\ n \geq j+3, \tag{C1.1}$$

which has already been shown for $j = 0$ above (Equation (13c)).

Assuming the conjecture true for $j-1$ gives

$$\Delta g_n^{(j-1)}((n-1)q) = g_n^{(j-1)}((n-1)q^+) - g_{n-1}^{(j-1)}((n-1)q^-) = 0,\ n \geq j+2. \tag{C1.2}$$

We next apply $j-1$ derivatives to the differentiation of Equation (15b) to give

$$g_n^{(j)}(u) = \frac{1}{1-\alpha}\left[g_n^{(j-1)}(u) - g_{n-1}^{(j-1)}(u-q)\right]; \tag{C1.3}$$

moreover, if we reduce n by unity, then

$$g_{n-1}^{(j)}(u) = \frac{1}{1-\alpha}\left[g_{n-1}^{(j-1)}(u) - g_{n-2}^{(j-1)}(u-q)\right]. \tag{C1.4}$$

Evaluating Equations (C1.3) and (C1.4) at $(n-1)q^+$ and $(n-1)q^-$, respectively,

$$g_n^{(j)}((n-1)q^+) = \frac{1}{1-\alpha}\left[g_n^{(j-1)}((n-1)q^+) - g_{n-1}^{(j-1)}((n-2)q^+)\right] \tag{C1.5}$$

$$g_{n-1}^{(j)}\left((n-1)q^-\right) = \frac{1}{1-\alpha}\left[g_{n-1}^{(j-1)}\left((n-1)q^-\right) - g_{n-2}^{(j-1)}\left((n-2)q^-\right)\right] \tag{C1.6}$$

and subtracting

$$\Delta g_n^{(j)}((n-1)q) = \frac{1}{1-\alpha}\left\{\begin{array}{l}\left[g_n^{(j-1)}\left((n-1)q^+\right) - g_{n-1}^{(j-1)}\left((n-2)q^+\right)\right] - \\ -\left[g_{n-1}^{(j-1)}\left((n-1)q^-\right) - g_{n-2}^{(j-1)}\left((n-2)q^-\right)\right]\end{array}\right\}. \tag{C1.7}$$

On re-arrangement,

$$\Delta g_n^{(j)}((n-1)q) = \frac{1}{1-\alpha}\left\{\begin{array}{l}\left[g_n^{(j-1)}\left((n-1)q^+\right) - g_{n-1}^{(j-1)}\left((n-1)q^-\right)\right] - \\ -\left[g_{n-1}^{(j-1)}\left((n-2)q^+\right) - g_{n-2}^{(j-1)}\left((n-2)q^-\right)\right]\end{array}\right\}, \tag{C1.8}$$

which is

$$\Delta g_n^{(j)}((n-1)q) = \frac{1}{1-\alpha}\left\{\Delta g_n^{(j-1)}((n-1)q) - \Delta g_{n-1}^{(j-1)}((n-2)q)\right\}. \tag{C1.9}$$

Since, by assumption, the first term is

$$\Delta g_n^{(j-1)}((n-1)q) = 0,\ n \geq j+2; \tag{C1.10}$$

and with n replaced by $n-1$, the second term is

$$\Delta g_{n-1}^{(j-1)}((n-2)q) = 0,\ n \geq j+3. \tag{C1.11}$$

Thus, Equation (C1.9) vanishes and confirms Conjecture 1, which is therefore true by induction.

Conjecture 2. *The $n-2$ derivative at $u = (n-1)q$ is discontinuous for $g_n, n \geq 2$.*

Symbolically, Conjecture 2 is

$$\Delta g_n^{(n-2)}((n-1)q) \neq 0,\ n \geq 2. \tag{C2.1}$$

We have already shown for $n = 2$:

$$\Delta g_2^{(0)}(q) = g_2(q^+) - g_1(q^-) = -\frac{1}{1-\alpha} \neq 0. \tag{C2.2}$$

Assume conjecture is true for $n-1$:

$$\Delta g_{n-1}^{(n-3)}((n-2))q) \neq 0,\ n \geq 3. \tag{C2.3}$$

Introduce $j = n-2$ into Equation (C1.9):

$$\Delta g_n^{(n-2)}((n-1)q) = \frac{1}{1-\alpha}\left\{\Delta g_n^{(n-3)}((n-1)q) - \Delta g_{n-1}^{(n-3)}((n-2)q)\right\}. \tag{C2.4}$$

However, from Conjecture 1, Equation (C1.1):

$$\Delta g_n^{(j)}((n-1)q) = 0,\ n \geq j+3, \tag{C2.5}$$

which implies for $j = n-3 \geq 0$

$$\Delta g_n^{(n-3)}((n-1)q) = 0; \tag{C2.6}$$

and Equation (C2.4) by assumption becomes

$$\Delta g_n^{(n-2)}((n-1)q) = -\Delta g_{n-1}^{(n-3)}((n-2)q) \neq 0, \tag{C2.7}$$

which is Conjecture 2—again, proved by induction.

Solving the recurrence in Equation (C2.7) gives the discontinuity

$$\Delta g_n^{(n-2)}((n-1)q) = \frac{(-1)^{n-1}}{(1-\alpha)^{n-1}}. \tag{C2.8}$$

Conjecture 3. *The $n-1$ derivative of g_n at $u = (n-1)q$ contains a delta function singularity.*

Symbolically, Conjecture 3 is

$$g_n^{(n-1)}(u) = h_n(u) + \beta_n \delta(u-(n-1)q),\ n \geq 1. \tag{C3.1}$$

We have shown that Conjecture 3 is true for $n = 1$ (and $n = 2$)

$$g_1^{(0)}(u) = h_1(u) + \beta_1 \delta(u) \tag{C3.2}$$

in Equation (8) with

$$\begin{array}{c} \beta_1(u) = 1 \\ h_1 = \frac{1}{1-\alpha} e^{u/(1-\alpha)}. \end{array} \tag{C3.3}$$

Assuming the conjecture is true for $n-1$:

$$g_{n-1}^{(n-2)}(u) = h_{n-1}(u) + \beta_{n-1} \delta(u-(n-2)q),\ n \geq 2. \tag{C3.4}$$

From Equation (C1.2) with $j = n-1$:

$$g_n^{(n-1)}(u) = \frac{1}{1-\alpha}\left[g_n^{(n-2)}(u) - g_{n-1}^{(n-2)}(u-q)\right] \tag{C3.5}$$

and Equation (C3.4) becomes

$$\begin{aligned} g_n^{(n-1)}(u) &= \tfrac{1}{1-\alpha}\left[g_n^{(n-2)}(u) - h_{n-1}(u-q)\right] - \tfrac{\beta_{n-1}}{1-\alpha}\delta(u-q-(n-2)q) \\ &= h_n(u) + \beta_n \delta(u-(n-1)q), \end{aligned} \tag{C3.6}$$

where

$$\begin{array}{c} h_n(u) \equiv \frac{1}{1-\alpha}\left[g_n^{(n-2)}(u) - h_{n-1}(u-q)\right] \\ \beta_n \equiv -\frac{\beta_{n-1}}{1-\alpha}, \end{array} \tag{C3.7}$$

which is Conjecture 3. In addition,

$$\beta_n \equiv \frac{(-1)^{n-1}}{(1-\alpha)^{n-1}}. \tag{C3.8}$$

In summary, from the three conjectures and the analytical solution of Equation (17), one concludes

(a) $g_n(u)$ is continuous at $u = (n-1)q$ for $n \geq 3$ and within the interval

$$(n-1)q \leq u \leq nq;$$

(b) $g_n^{(n-2)}(u)$ has a finite discontinuity at $u = (n-1)q$ for $n \geq 2$ and is otherwise continuous;

(c) $g_n^{(n-1)}(u)$ has a delta function discontinuity at $u = (n-1)q$ for $n \geq 2$ and is otherwise continuous.

Finally, the derivatives of collision density $F(u)$ inherit the continuity properties of $g(u)$, since, by Leibnitz's rule,

$$\begin{aligned} F_n^{(j)}(u) &= \tfrac{d^j}{du^j}\left[e^{-u} g_n(u)\right] \\ &= e^{-u} \sum_{l=0}^{j} (-1)^{l-j} \tfrac{j!}{l!(l-j)!} g_n^{(l)}(u);\ (n-1),\ q \leq u \leq nq. \end{aligned} \tag{19}$$

4. Singularities and Smoothing

Table 2, displaying points of discontinuity of the collision density for scattering against ^{12}C, is based on the above continuity arguments. To the right is increasing lethargy and down the rows increasing derivatives. As is apparent, with increasing lethargy (and disorder), discontinuities become further embedded in the collision density derivatives, making $F(u)$ ever smoother. Embedding of the discontinuities is clearly observed in Figure 3a–c. As noted above, the finite discontinuity at $u = q$, from integration over the delta function source emerges in the collision density itself, as shown in Figure 3a.

Table 2. Embedding of discontinuities in jth derivative of $F(u)$ with collision interval n.

(n − 1)/(n)	**0**	**(1)/(2)**	**(2)/(3)**	**(3)/(4)**	**(4)/(5)**	**(5)/(6)**	...
j/u	0	q	$2q$	$3q$	$4q$	$5q$	...
0	DF	F	C	C	C	C	...
1	-	DF	F	C	C	C	...
2	-	-	DF	F	C	C	...
3	-	-	-	DF	F	C	...
4	-	-	-	-	DF	F	...
5	-	-	-	-	-	DF	...
...	-	-	-	-	-	-	...

C: Continuous; F: Finite discontinuity; DF: Delta function.

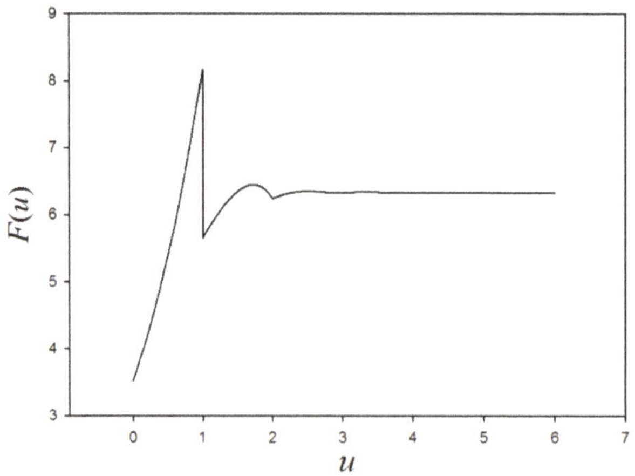

(a)

Figure 3. *Cont.*

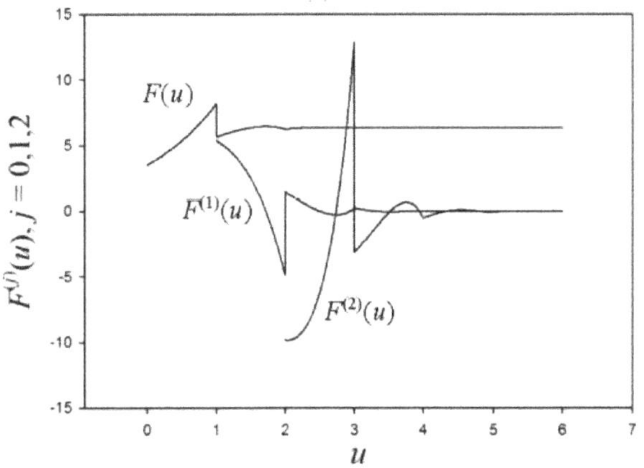

(b)

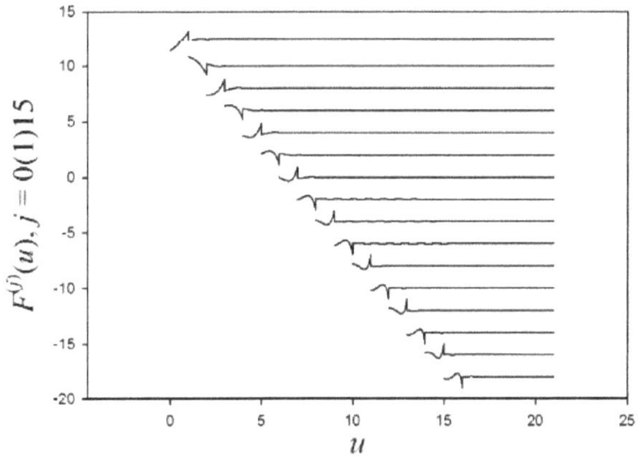

(c)

Figure 3. (**a**) Collision density; (**b**) first two derivative of collision density and the collision density; (**c**) 15 derivatives of collision density.

As neutrons scatter to lower energy (higher lethargy), the memory of the singular source is retained since a delta function (not shown) exists at the beginning of each scattering interval exactly where the previous derivative has a finite discontinuity.

Figure 3b shows several derivatives, as given by Equation (18b). We observe Placzek oscillations in $F(u)$, including the finite discontinuity at $u = q$. The oscillations for increasing u obviously originate from the submerged discontinuities and are indicative of increasing entropy (disorder) and smoothing as the influence of the discontinuities becomes further submerged in the derivatives. Also observed is the constant asymptotic collision density ($1/E$) equilibrium distribution, as shown in Table 3, where

$$F(\infty) = 1/\left[\alpha + \frac{\alpha}{1-\alpha}\ln(\alpha)\right].$$

Figure 3c shows 15 derivatives normalized so that the finite discontinuity in each is the same, and the derivatives are displaced downward for better viewing. The pattern is evident and visualizes how increasing physical smoothing is mathematically linked to the submergence of the initial discontinuities of the source distribution with lethargy.

Table 3. Run-up to asymptotic collision density.

u	$F(u)$
6	6.3383867546
8	6.3383811228
10	6.3383812067
12	6.3383812061
∞	6.3383812061

5. Randomness of Collisions

Another statistical measure of the disorder in the collision density distribution is the randomness of collisions. In the following analysis, we find an expression for the distribution of the collided density in terms of the number of collisions.

As shown in [3], the Laplace transform of Equation (3a) gives the following transform of the collision density:

$$\overline{F}(p) = \frac{1}{1 - Q_s(p)}, \tag{20a}$$

with

$$Q_s(p) = \frac{c}{1-\alpha}\left[\frac{1 - e^{-q(p+1)}}{p+1}\right]. \tag{20b}$$

Here, c is the number of neutrons emitted in a scattering collision relative to the total possible interactions including loss by absorption. In the analysis above, c is unity.

The geometric series representation of Equation (20a) is

$$\overline{F}(p) = \sum_{n=0}^{\infty} Q_s(p)^n = \sum_{n=0}^{\infty} \frac{c^n}{(1-\alpha)^n}\left[\frac{1 - e^{-q(p+1)}}{p+1}\right]^n, \tag{21}$$

where convergence is guaranteed by choice of the complex variable p. Then, using the binomial theorem for the term in brackets, Equation (21) becomes

$$\overline{F}(p) = \sum_{n=0}^{\infty} \frac{c^n}{(1-\alpha)^n} \sum_{l=0}^{n} (-1)^l \frac{n!}{(n-l)!l!}\left[\frac{e^{-ql(p+1)}}{(p+1)^n}\right], \tag{22a}$$

whose analytical inversion becomes

$$F(u) = \delta(u) + \sum_{n=1}^{\infty} c^n \left[e^{-u} \sum_{l=0}^{[u/q]} (-1)^l \frac{n!}{(n-l)!l!}\left(\frac{u - lq}{1-\alpha}\right)^{n-1}\right]. \tag{22b}$$

The upper limit of the second summation $[u/q]$ is the greatest integer contained in u/q. Note that n is now the collision number, not to be confused with the collision interval.

The first term in Equation (22b) is the uncollided collision density at the source and the term in brackets is the nth collided collision density after n collision:

$$F_0(u) = \delta(u) \tag{23a}$$

$$F_n(u) \equiv \frac{e^{-u}}{(n-1)!} \sum_{l=0}^{[u/q]} (-1)^l \frac{n!}{(n-l)!l!}\left(\frac{u - lq}{1-\alpha}\right)^{n-1}; \; n = 1, 2, \ldots, \tag{23b}$$

where the subscript is the number of collisions and c is unity for purely scattering.

For lethargy less than 0.5, Figure 4a shows the variation of the collided density with n, where the hint of a Gaussian distribution is observed. As lethargy increases, as anticipated, the development of the

Gaussian distribution becomes evident in Figure 4b, verifying the randomness of the neutron–nucleus collision. The numerical evaluation of Equation (23b) is highly sensitive to round-off error and requires quadruple precision, which increases the computing time (normally under a minute) on a LENOVO 2.4 GHz YOGA platform by several seconds.

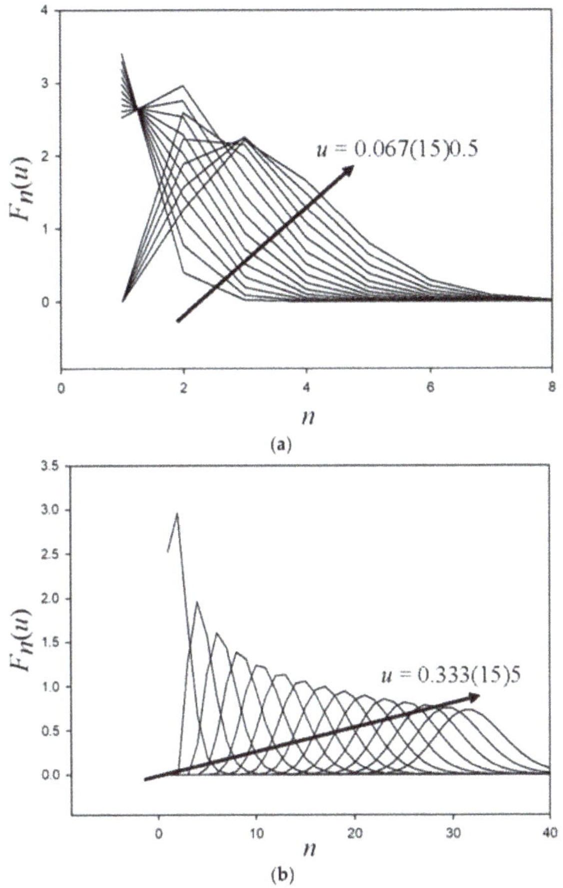

Figure 4. (**a**) Collided collision density to u = 0.5; (**b**) Collided collision density to u = 5.

How do we know that the distribution is indeed Gaussian? This can be shown relatively easily by noting that a normalized Gaussian frequency is

$$f(x) = \frac{1}{\sqrt{2\pi\sigma^2}} e^{(x-\mu)^2/2\sigma^2}. \tag{24a}$$

If each curve of Figure 4b is a Gaussian, then when, at each lethargy u, the distribution is normalized by the area under the curve, which is $F(u)$,

$$\hat{f}(x) = \frac{F_x(u)}{\sum\limits_{n=0}^{\infty} F_n(u)}, \tag{24b}$$

where x is now a continuous analogue of the collision number n. By equating Equations (24a) and (24b) at their maxima μ:

$$\frac{1}{\sqrt{2\pi\sigma^2}} = \hat{f}(\mu(u)), \tag{24c}$$

where the collision number at maximum is

$$\mu(u) = \frac{\sum\limits_{n=0}^{\infty} nF_n(u)}{\sum\limits_{n=0}^{\infty} F_n(u)}, \tag{24d}$$

there results

$$\sigma(u)^2 = \frac{1}{2\pi\hat{f}(\mu(u))^2}. \tag{24e}$$

With the parameters for the Gaussian now known, we can plot the two distributions as shown in Figure 5. They are nearly graphically identical over u = [0.8, 12].

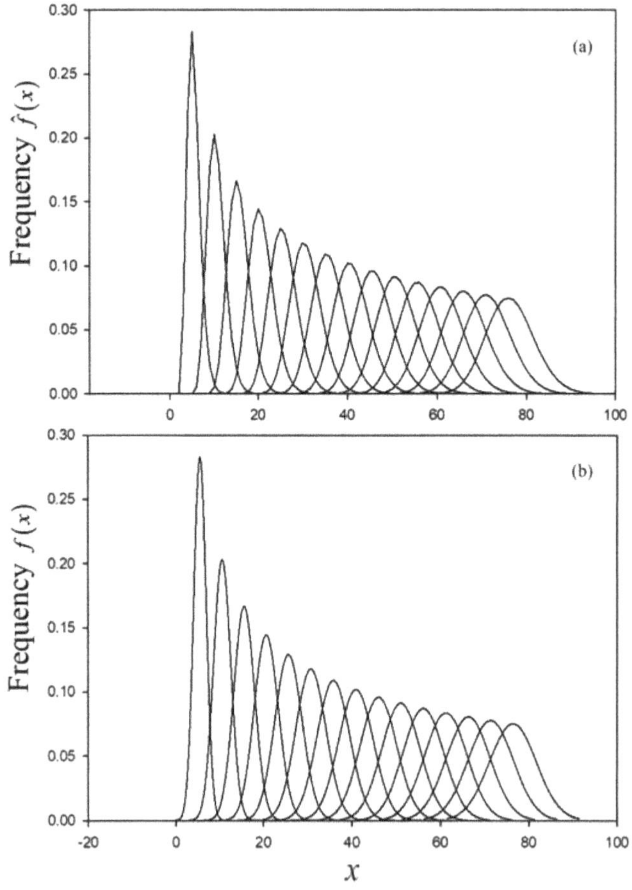

Figure 5. Comparison of calculated frequency of collision (**a**) and the corresponding Gaussian (**b**).

6. Conclusions

Through a rather involved, rigorous mathematical derivation, verification of the obvious was achieved. In particular, the connection between the increased physical smoothness of the collision density distribution with lethargy and consequent singularities from monoenergetic source emission was demonstrated. It was shown that, with increased collisions, the original source singularity becomes submerged in the derivatives of the distribution function, resulting in smoothing of the distribution function. In addition, the tendency of the collision frequency over the number of collisions to become Gaussian with increased lethargy was also demonstrated.

Author Contributions: B.G. conceived of the connection between entropy and neutron scattering after attending the lectures with D.M. given by V.M. on kinetic theory at the university of Bologna. We all discussed the mathematics required to complete the manuscript as well as the figures. B.G. prepared the manuscript and gave the presentation.

Conflicts of Interest: The authors declare no conflict of interest.

References

1. Williams, M.M.R. *Slowing Down and Thermalization of Neutrons*; North-Holland Pub. Co.: Amsterdam, The Netherlands, 1966.
2. Carson, E.M.; Watson, J.R. *Undergraduate Students' Understandings of Entropy and Gibbs Free Energy*; University Chemistry Education–2002 Papers; Royal Society of Chemistry: London, UK, 2002.
3. Ganapol, B.D. *Analytical Benchmarks for Nuclear Engineering Applications*; AEN/NEA: Paris, France, 2008; ISBN 978-92-64-99056-2NEA/DB/DOC.

entropy

MDPI

Article

Energy from Negentropy of Non-Cahotic Systems

Piero Quarati [1,*], **Antonio M. Scarfone** [2,3] and **Giorgio Kaniadakis** [1]

1 Dipartimento di Scienza Applicata e Tecnologia (DISAT-SCUDO)-Politecnico di Torino, 10129 Torino, Italy; giorgio.kaniadakis@polito.it
2 Istituto Sistemi Complessi (ISC–CNR) c/o Politecnico di Torino, 10129 Torino, Italy; antoniomaria.scarfone@cnr.it
3 Istituto Nazionale di Fisica Nucleare (INFN)-Sezione di Torino, 10125 Torino, Italy
* Correspondence: piero.quarati@polito.it; Tel.: +39-011-090-7339

Received: 15 January 2018; Accepted: 7 February 2018; Published: 9 February 2018

Abstract: Negative contribution of entropy (negentropy) of a non-cahotic system, representing the potential of work, is a source of energy that can be transferred to an internal or inserted subsystem. In this case, the system loses order and its entropy increases. The subsystem increases its energy and can perform processes that otherwise would not happen, like, for instance, the nuclear fusion of inserted deuterons in liquid metal matrix, among many others. The role of positive and negative contributions of free energy and entropy are explored with their constraints. The energy available to an inserted subsystem during a transition from a non-equilibrium to the equilibrium chaotic state, when particle interaction (element of the system) is switched off, is evaluated. A few examples are given concerning some non-ideal systems and a possible application to the nuclear reaction screening problem is mentioned.

Keywords: negentropy; non-ideal systems; many-body correlations

1. Introduction

In thermodynamics and in statistical mechanics, the entropy enters, with a central role, into all laws concerning systems in states of equilibrium and non-equilibrium: in reversible and irreversible transformations and phase transitions, in Boltzmann and Gibbs classical and quantum extensive and non-extensive statistics.

Entropy can be expressed as a total, compact, positive expression always increasing in isolated systems during its time evolution towards equilibrium. Often, entropy is composed by a few different terms and, when the system is not in an equilibrium chaotic state, at least one of them represents a negative contribution to its evaluation.

The negative entropy (negentropy) contribution is present because of many different reasons like the quantum Pauli exclusion or boson inclusion principles, the many-body correlations and interactions among the single elements composing the system. Quantum exclusion-inclusion effects as well as correlations and/or interactions produce order to the system decreasing the value of the total entropy [1].

It is interesting to study the negative entropy contribution, different from the positive one, because negentropy represents a stored mobilizable energy in organized systems. This quantity can be spontaneously transferred or exchanged among the elements of the system, or from these elements to an inserted subsystem, or from an organized environment to elements of a system [2,3]. In a non-ideal system, as for instance gas, plasma or liquid and solid metal, composed by interacting or correlated elements in a stationary non-equilibrium (long-life metastable) state, some energy can be transferred to other particles of a subsystem (as for instance deuterons or very light nuclei) implanted into its volume size. The relative energy of the reacting deuteron nuclei can be increased by negentropy of a given number of the system elements, allowing fusion reactions against Coulomb repulsion more easily [4].

Experimental measurements of nuclear fusion rates of deuterons inserted in a liquid or solid metal matrix, impinged by accelerated deuterons of a given kinetic energy much greater than expected, can be explained by means of the role of negentropy of the matrix. Other examples and applications can be found in astrophysical plasmas, where often measured reaction rates are greater than expected [5,6].

The creation of order and energy of high value requires non-equilibrium conditions. If negative entropy of the single elements is responsible for the energy transfer, we expect that small islands of disorder, where negative entropy is spent, and islands of order, where negative entropy is received, will be created. Without any action from outside the system, switching off the interaction among the elements, makes the entropy increase or negative entropy decrease.

Our work is close to the papers by Sato [7,8] and by Yi-Fang Chang [9,10], but differs because we evaluate the energy that can be transferred by means of negentropy to induce and facilitate processes that without this transferred energy would be absent. However, our method is not an alternative to that of Sato and that of Yi-Fang Chang but rather allows the study of particular situations in which the previous methods are less appropriate. Sato has generalized the definition of negentropy that becomes valid under general situations: negentropy represents the potential of work defined by Kullback-Leibler information [11,12]; this is more appropriate than being defined by the difference of entropies.

In this work we calculate the amount of energy that a single element of a system can transfer to other internal elements or clusters of elements spending entropy-lowering (by means of negentropy) when the interaction or the correlations among the elements are switched off or when their intensity is lowered. A few evaluations are reported concerning non-ideal molecular gas, warm dense matter (pseudo white dwarf) and nuclear matter while possible application to the screening of nuclear reactions in plasmas is indicated.

We show the constraints to which negative and positive contributions of entropy and free energy are submitted, explain how negentropy is linked to correlations among the elements of a system and report how energy fluctuation is due to negentropy and therefore to correlations.

In Section 2 we define the different positive and negative contribution of entropy. In Section 3 we report the suggestion by Clausius and Helmholtz concerning the splitting of entropy and free energy in free and bound contribution. In Section 4 we present a different splitting of entropy and free energy while in Section 5 we discuss the corresponding energy transfer mechanism. In Section 6 we give an explicit example to evaluate the energy that can be exported by some elements of the system to an inserted element of a subsystem in a non-ideal system; in Section 7 we report a few applications with numerical examples. Conclusions are given in Section 8.

2. The Different Entropy Contributions

Let us consider a statistical system of elements (for instance laboratory plasma, astrophysical or stellar plasma, solid and liquid metal, non-ideal gas, warm dense matter; the system can also be an environment or a medium) that contains a subsystem (implanted nuclei, accelerated ions through the system, special cluster of different particles). The system is composed of N elements much larger than the number of elements of the subsystem. The subsystem interacts with a finite number M of elements of the system with $M \ll N$. For instance, in a time Δt during successive elastic collisions or after a mean free path, the subsystem enters into an interaction with M elements of the system. We assume that the system is not in a global thermodynamic equilibrium; it is not in a full disorder chaotic state but has a certain degree of order. To be more precise, the system is in a stationary state with a long lifetime (metastable state), its entropy is the sum of a positive and a negative contribution because correlations and/or interactions are present, which are responsible, for isolated systems, for a lower value of entropy. The negative terms are expression of the order and represent the correlation entropy contribution. Without any action from outside of the system, by switching off the interaction, we impose the entropy to increase or the negative entropy to reduce.

Of course, the evolution in time of entropy is always positive for isolated systems; this is the same for the entropy of an ensemble in equilibrium. For instance, a classical ideal gas is always the sum of positive contributions.

However, if we introduce particle correlations, for instance and for simplicity, for quantum effects like the Pauli exclusion or the inclusion-exclusion principle, that give a sort of order to the gas, the related contributions to entropy are negative and the entropy decreases compared to its value for classical, uncorrelated ideal gas.

Therefore, a system with correlations or interactions among the N elements, with a given potential energy, has negative contribution in the entropy expression:

$$S = S^{\mathrm{ideal}} - S^{\mathrm{corr}} , \tag{1}$$

where S^{ideal} is the maximum value of entropy that is the entropy of the equilibrium state and S^{corr} is the contribute due to the correlations, S^{corr} was also called by Gibbs [13] "capacity of entropy or negentropy", by Brillouin [14] "input or information", by Obukhov [15] "deficiency of entropy" and by Sato [7,8] "potential of work or negentropy".

If the interaction is a weak short-range, two-body interaction, the equilibrium distribution function is still Maxwellian and the entropy is the sum of positive terms only.

A state of thermal equilibrium is chaotic at zero value of controlling parameters (an example of a controlling parameter of interest here is the interaction two-body potential $U(r)$ as discussed in Section 7) and entropy is a measure of chaoticity or deviation from equilibrium of a state.

Non-equilibrium state does not remain constant after isolation because of irreversible processes that are responsible for transitions to a state of equilibrium on the same energy shell. A stationary state of non-equilibrium of a system is a time invariant state of an open system. The entropy of an equilibrium state is greater than in state of non-equilibrium with non-zero value of control parameters. Moreover, we can compare states of a different order (or chaoticity) only if they belong to the same energy shell.

3. Clausius and Helmholtz Suggestion

Following Clausius and Helmholtz, it seems useful to divide the energy of a system in two parts E^b and E^f, where b is for bound and f for free [16–18]:

$$E = T\,S + F = E^b + E^f , \tag{2}$$

with $E^b = T\,S$ the part of energy bound in thermal motion of molecules and $E^f = F$ the part of energy free to perform work and coincides with the thermodynamical free energy.

In a system at equilibrium, with a given volume V, temperature T and number of elements N, the value of entropy S is maximum and the value of free energy F is minimum. The more work that can be produced, the greater the value of the entropy.

Like energy, entropy can also be divided into a bound and a free part

$$S = S^b + S^f , \tag{3}$$

where S^b is the entropy bound in the microscopic motion of molecules or the N elements of the system, while S^f is the entropy available for information processing or to produce work outside the system. We argue that S^f can also be exchanged among the elements or group of elements of the system or with elements of an inserted subsystem.

Because of the splitting of S, it would be more correct, in place of the above definitions $E^b = T\,S$ and $E^f = F$, to write:

$$\widetilde{E}^b = T\,S^b \qquad \text{and} \qquad \widetilde{E}^f = T\,S^f + F\,. \tag{4}$$

Then, the energy available to produce work becomes $\widetilde{E}^f$. In addition and for completeness, it would be better also to split the free energy F in:

$$F = F^b + F^f\,, \tag{5}$$

where the quantities F^b and F^f can be understood as it follows: at equilibrium F is minimum ($F^{eq} = F^b$ which is a negative quantity) and $F^f = 0$. Moving toward non-equilibrium state, taking T, V and N constants, the positive quantity F^f increases and the total F increases too.

By grouping all the terms with index b together and all the terms with index f together we obtain:

$$\overline{E}^b = T\,S^b + F^b\,, \qquad \text{and} \qquad \overline{E}^f = T\,S^f + F^f\,, \tag{6}$$

with

$$E = E^b + E^f = \widetilde{E}^b + \widetilde{E}^f = \overline{E}^b + \overline{E}^f\,. \tag{7}$$

Therefore, the energy bound to microscopic motion is $\overline{E}^b$ and the energy available for work is $\overline{E}^f$. Clearly, at equilibrium total energy is $\overline{E}^b$ and $\overline{E}^f = 0$.

If during the transition from equilibrium to non-equilibrium, or vice versa, temperature, volume and number of elements do not change, then S^b and F^b do not change, as well as $\overline{E}^b$.

4. A Different Subdivision of Free Energy and Entropy

In the following, we consider a system of independent elements and a system with elements correlated or interacting through a potential and therefore with an entropy value lower than that of a system with independent elements. We focus our attention on the subdivision of entropy and free energy into the contributions related to equilibrium, where contributions to produce work are absent if V, N and T remain constant, and into the contributions due to the presence of correlations or interactions among the elements of the system. These are the contributions that can produce work and can be spent internally and locally between elements of the system or can be spent outside the system or to an inserted subsystem. In this way, the system will lose order and become more chaotic.

We may divide the total entropy in two parts: one positive and one negative. Accordingly and differently from Decomposition (3), we consider:

$$S = S^+ - S^-\,, \tag{8}$$

with $S^- > 0$.

The contribution S^- is due to correlations or interactions and brings order to the system; it can also be thought of as an environment whose entropy at global thermodynamic equilibrium is given by S^+. Negentropy S^- is available to transfer energy to other subjects and to do work.

Let us divide also the free energy F into two parts:

$$F = F^+ - F^-\,, \tag{9}$$

with $F^- > 0$; the part F^+ can do work.

Since, the energy of the system is $E = F + T\,S$, we may write:

$$E = (F^+ - F^-) + T\,(S^+ - S^-)\,. \tag{10}$$

As a consequence, not all the free energy F contributes to make work and the entire quantity of T, S does not represent work.

Comparing Equations (8) and (10) with Equations (3) and (5), we can write the following relations

$$S^b = S^+ \,, \qquad \text{and} \qquad S^f = -S^- \,, \tag{11}$$

as well as

$$F^b = -F^- \,, \qquad \text{and} \qquad F^f = F^+ \,, \tag{12}$$

so that

$$\overline{E}^b = T\,S^+ - F^- \,, \qquad \text{instead of} \qquad E^b = T\,S \,, \tag{13}$$

and

$$\overline{E}^f = -T\,S^- + F^+ \,, \qquad \text{instead of} \qquad E^f = F \,. \tag{14}$$

5. Transfer of Energy

A system in global thermodynamic equilibrium with entropy $S = S^+$ does not have energy to spend outside. On the other hand, a system, with entropy $S = S^+ - S^-$ can transfer energy to an inserted subsystem that, entering in interaction during a mean free path with M elements, can receive an average energy from negentropy up to the quantity $M\,T\,S^-/N$. The system loses order, and the subsystem gains energy.

If the M elements of the system that interact with the subsystem transfer their available energy to the subsystem at constant temperature T, the variation of energy is:

$$\Delta E = T\,\Delta S + \Delta F \,. \tag{15}$$

For simplicity of notation we define entropy $\mathcal{S}$, energy $\mathcal{E}$ and free energy $\mathcal{F}$ per element, that is $\mathcal{S} = S/N$, $\mathcal{E} = E/N$ and $\mathcal{F} = F/N$.

The total energy variation of the system, after transfer of some amount of energy to the subsystem with a transition from the initial state 1 to the final state 2, is:

$$\begin{aligned} \Delta\mathcal{E} = \mathcal{E}_2 - \mathcal{E}_1 &= T\left(\mathcal{S}_2^+ - \mathcal{S}_1^+\right) - T\left(\mathcal{S}_2^- - \mathcal{S}_1^-\right) + \left(\mathcal{F}_2^+ - \mathcal{F}_1^+\right) - \left(\mathcal{F}_2^- - \mathcal{F}_1^-\right) \\ &= T\left(\Delta\mathcal{S}^+ - \Delta\mathcal{S}^-\right) + \Delta\mathcal{F}^+ - \Delta\mathcal{F}^- \\ &= \left(-T\,\Delta\mathcal{S}^- + \Delta\mathcal{F}^+\right) + \left(T\,\Delta\mathcal{S}^+ - \Delta\mathcal{F}^-\right) \\ &= \Delta\mathcal{E}^f + \Delta\mathcal{E}^b \,. \end{aligned} \tag{16}$$

Toward the equilibrium, the free energy F must diminish and the entropy S must increase, therefore $\Delta\mathcal{F} < 0$, $\Delta\mathcal{S} > 0$, that is $\Delta\mathcal{F}^+ < \Delta\mathcal{F}^-$ and $\Delta\mathcal{S}^+ > \Delta\mathcal{S}^-$, or:

$$\mathcal{F}_2^+ - \mathcal{F}_2^- - \;<\; \mathcal{F}_1^+ - \mathcal{F}_1^- \,, \tag{17}$$

$$\mathcal{S}_2^+ - \mathcal{S}_2^- \;>\; \mathcal{S}_1^+ - \mathcal{S}_1^- \,. \tag{18}$$

If the system loses energy ($\Delta\mathcal{E} < 0$, i.e., $\Delta\mathcal{E}^b + \Delta\mathcal{E}^f < 0$), the work to the subsystem is made by $\mathcal{F}^+$ and/or $\mathcal{S}^-$ and

$$\Delta\mathcal{F}^+ - \Delta\mathcal{F}^- < T\left(\Delta\mathcal{S}^- - \Delta\mathcal{S}^+\right) \,. \tag{19}$$

The total exchanged energy E_{ex} between the system and the subsystem, at constant temperature, is:

$$E_{\text{ex}} = M\left(-T\,\mathcal{S}_1^- + \mathcal{F}_1^+\right) \,, \tag{20}$$

so that the residual total energy E_{res} of the system becomes:

$$\begin{aligned} E_{res} &= E_1 - E_{ex} \\ &= N\left[T\left(\mathcal{S}_1^+ - \mathcal{S}_1^-\right) + \mathcal{F}_1^+ - \mathcal{F}_1^-\right] - M\left(-T\,\mathcal{S}_1^- + \mathcal{F}_1^+\right) \\ &= -(N-M)\,T\,\mathcal{S}_1^- + (N-M)\,\mathcal{F}_1^+ + N\left(T\,\mathcal{S}_1^+ - \mathcal{F}_1^-\right). \end{aligned} \tag{21}$$

In the transfer of energy from the initial state to the final state, all the above relations should be satisfied because of the increase of total entropy S and the decrease of F.

The important point is that the expression to evaluate the transferred energy on the work that can be performed is not $E = F = F^+ - F^-$. It is instead given by the difference of the positive free energy F^+ and the negentropy contribute $T\,S^-$. This means that a part of entropy also contributes to perform work and a part of free energy contributes to produce heat or non-information. F^+ and S^- are different from zero when the elements of the ensemble are submitted to correlations or interactions. These two quantities go to zero when the system goes to equilibrium. In this case, the remaining quantities different from zero are F^- and $S^+ \equiv S^{eq}$, while S^- can be identified with negentropy or lowering of entropy $S^- = S^{eq} - \Delta S$.

In conclusion, let us describe two examples to evaluate the energy that can be exported or imported by the system during a process.

We recall that sometimes [9,10], the variation of entropy of a system is divided into internal S^{int} and external S^{ext} entropy (the internal entropy is related to irreversible processes):

$$\Delta S = \Delta S^{ext} + \Delta S^{int}\,, \tag{22}$$

$\Delta S^{int} \geq 0$ is required by the second law of thermodynamics.

- Case of exporting entropy

We study first the conditions required to prepare a system with an entropy lower than its greatest value when at equilibrium. Since the order increases if the entropy decreases, the system can have lower entropy not only exporting entropy but also importing negentropy. Of course, the system that exports negentropy increases its entropy.

We want $\Delta S < 0$ to increase the order of the system, so that:

$$-\Delta S^{ext} > \Delta S^{int} \geq 0\,. \tag{23}$$

This expression represents the thermodynamic condition for self-organization. We must export more entropy than the entropy variation for the irreversible internal transformations. In fact, from relation:

$$\Delta S = (S_2^{ext} - S_1^{ext}) + \Delta S^{int} < 0\,, \tag{24}$$

and using $\Delta S^{int} \geq 0$, we get the condition $S_1^{ext} > S_2^{ext}$, that is, the external entropy should decrease.

We repeat the relations of the above transition by using our definition of entropy (8), that is:

$$\Delta S = \Delta S^+ - \Delta S^- < 0\,. \tag{25}$$

By identifying

$$\Delta S^+ \equiv \Delta S^{int} \geq 0\,, \tag{26}$$

and

$$\Delta S^- \equiv -\Delta S^{ext}\,, \tag{27}$$

the condition (25) gives

$$0 \le \Delta S^- < \Delta S^+ \,. \tag{28}$$

Therefore, negentropy increases, the order increases and the positive contribution must increase in a quantity to keep valid the condition $\Delta S < 0$.

- Case of importing entropy

The opposite case is when the system increases its entropy $\Delta S > 0$. Since the order decreases if the entropy increases, the system can have higher entropy not only importing entropy but also exporting negentropy.

We want

$$\Delta S^{\text{ext}} > -\Delta S^{\text{int}} \,, \tag{29}$$

that is, internal entropy increases, therefore external entropy may decrease quite a lot despite total entropy increasing. However, the disorder shall increase.

In this case, by using our definition, we have:

$$\Delta S^+ > \Delta S^- \,, \tag{30}$$

that is assured if $\Delta S^- < 0$ since $\Delta S^+ \equiv \Delta S^{\text{int}} \ge 0$. Therefore, positive contribution increases more than negative contribution is exported. Entropy increases because negentropy is exported.

6. An Explicit Example to a Non-Ideal System

We define free energy F and entropy S for a non-ideal system of elements interacting by the two-body potential $U(r)$ following [19]. The two quantities F^- and F^+ reads:

$$F^- = k\,T\,\ln A(T)\,, \tag{31}$$

$$F^+ = k\,T\,N\,n\,B(T)\,, \tag{32}$$

with $n = N/V$, where

$$A(T) = \frac{V^N}{N!}\left(\frac{2\,\pi\,m\,k\,T}{h^2}\right)^{3\,N/2}\,, \tag{33}$$

$$B(T) = 2\,\pi\int_0^\infty\left(1 - \exp\left(-\frac{U(r)}{k\,T}\right)\right)r^2\,dr\,, \tag{34}$$

and from the thermodynamical relation

$$S = -\left(\frac{\partial F}{\partial T}\right)_{V,N}\,, \tag{35}$$

and using the Decompositions (8) and (9), we have

$$S^+ = \left(\frac{\partial F^-}{\partial T}\right)_{V,N}\,, \qquad S^- = \left(\frac{\partial F^+}{\partial T}\right)_{V,N}\,. \tag{36}$$

Therefore, we get

$$T\,S^+ = k\,T\,\ln A(T) + \frac{3}{2}\,N\,k\,T\,, \tag{37}$$

$$T\,S^- = k\,T\,N\,n\,B(T) - N\,n\,C(T)\,, \tag{38}$$

where

$$C(T) = 2\pi \int_0^{\infty} U(r) \exp\left(-\frac{U(r)}{kT}\right) r^2 dr\,, \tag{39}$$

so that, the energy becomes

$$E = F + TS = N\left(\frac{3}{2} kT + n\,C(T)\right)\,. \tag{40}$$

We consider the two following cases:

1. Transition from nonequilibrium to equilibrium by switching off the interaction: from the state $(T, V, N, U(r) \neq 0)$ to $(\hat{T}, V, N, U(r) = 0)$. The energy difference is

   $$\Delta E = N\left(\frac{3}{2} k\Delta T + n\,\Delta C(T)\right)\,, \tag{41}$$

   with $\Delta T = \hat{T} - T$.

   The second term $\Delta C(T)$ derives from terms containing the interaction $U(r)$.

   Let us consider a system in interaction whose potential can be modelled by an average energy value $\langle U(r)\rangle$ with an interaction radius equal to R_c. Then, Equations (37) and (38) still hold where now

   $$B(T) = \frac{2}{3}\pi R_c^3 \left(1 - \exp\left(-\frac{\langle U(r)\rangle}{kT}\right)\right)\,, \tag{42}$$

   $$C(T) = \frac{2}{3}\pi \langle U(r)\rangle R_c^3 \exp\left(-\frac{\langle U(r)\rangle}{kT}\right)\,. \tag{43}$$

   When the interaction is switched off, the same amount of energy calculated above can transfer to one element of the system itself (or a cluster of elements) by at most $M \ll N$ elements during a time Δt between two successive elastic collisions while the element performs a mean path length. The accelerated element (or cluster) can later give back the acquired energy to other elements. The energy of an element, during the non-equilibrium to equilibrium transition with switching off the interaction or correlation $U(r)$, fluctuates of the order of TS^-/N and the energy per particle gained before is $\frac{3}{2}k\Delta T + n\,\Delta C(T)$. The bulk properties of the system do not change, however, locally, a modification of the features of the site can be observed.
2. Transition from equilibrium state $(T, V, N, U(r) = 0)$ to the equilibrium state $(T, \hat{V}, N, U(r) = 0)$:

   $$\Delta F \equiv \Delta F^- = NkT \ln\left(\frac{\hat{V}}{V}\right) = -p\Delta V\,, \tag{44}$$

   where $\Delta V = \hat{V} - V$ and $p\Delta V$ is the work performed outside the system. This relation implies $\Delta S = \Delta S^+$, so that

   $$Q = -NkT \ln\left(\frac{\widetilde{V}}{V}\right)\,, \tag{45}$$

   is the heat received from the outside to maintain T constant and $\Delta E = 0$.

7. Some Possible Applications

We report a few applications to several non-ideal systems described by simple models just to have an indication of the order of magnitude we might expect from the negentropy effect described in the previous sections.

- Molecular gas

We consider a non-ideal gas with two-body interaction modelled by $\langle U(r)\rangle = 2\times 10^{-2}\,\mathrm{eV}$ with $R_c = 5\times 10^{-8}\,\mathrm{cm}$. At temperature $T = 300\,\mathrm{K}$, corresponding to the thermal energy $kT = 2.6\times 10^{-2}\,\mathrm{eV}$ per donor, and a particle density $n = 10^{20}\,\mathrm{cm}^{-3}$, the negentropy transfer of energy is $T\mathcal{S}^- = 4\times 10^{-6}\,\mathrm{eV}$. Otherwise, changing the average potential value in $\langle U(r)\rangle = 10^{-1}\,\mathrm{eV}$ and the particle density in $n = 10^{22}\,\mathrm{cm}^{-3}$, we obtain $T\mathcal{S}^- = 5\times 10^{-2}\,\mathrm{eV}$.

- Warm dense matter (a pseudo white dwarf) and nuclear matter

We consider a non-ideal neutral gas, composed by a background of α-particles (He^4 nuclei) and a degenerate Fermi gas of electrons. The system looks like a pseudo white dwarf made of a warm, dense matter. Let us focus our attention to the inert α-particle gas with the following parameters: $n = 10^{30}\,\mathrm{cm}^{-3}$; $T = 10^7\,\mathrm{K}$, corresponding to the thermal energy $kT = 860\,\mathrm{eV}$, and $R_c = 10^{-10}\,\mathrm{cm}$ (the range of average correlated $\alpha-\alpha$ screened potential).

We evaluate the energy that a single element can transfer as a function of $\langle U(r)\rangle/kT$ (the value of $\langle U(r)\rangle$ depends on the assumption of the screening model) given by (38). If we take $\langle U(r)\rangle/kT = 10^{-2}$, each element can transfer 18 eV of energy that reduces to 9 eV if $\langle U(r)\rangle/kT = 5\times 10^{-3}$.

During a mean free path λ inside the system, an α-particle or an ion or deuteron inserted into the system interacts with $M \ll N$ elements and can transfer to it the energy given by

$$\begin{aligned}\frac{M}{N}TS^- &= n\pi R_c^2\lambda T\mathcal{S}^- \\ &= \frac{2}{3}\pi^2\frac{R_c^5}{\sigma}nkT\left[1-\left(1+\frac{\langle U(r)\rangle}{kT}\right)\exp\left(-\frac{\langle U(r)\rangle}{kT}\right)\right].\end{aligned} \tag{46}$$

We assume a mean free path $\lambda = 10^{-9}\,\mathrm{cm}$ and recalling that $\sigma = 1/(n\lambda)$, the elastic cross section is $\sigma = 10^{-21}\,\mathrm{cm}^2$. The value of M is about 30 particles so that the total energy transferable is 550 eV if $\langle U(r)\rangle/kT = 10^{-2}$ and reduces to 280 eV if $\langle U(r)\rangle/kT = 5\times 10^{-3}$.

This amount of energy does not change the bulk property of the system, nevertheless it can improve the probability of an $\alpha-\alpha$ or an α-nucleus fusion reaction inside the pseudo white dwarf. A description and result analogous to those of a pseudo white dwarf can be made for a system of nuclear matter where the kinetic energy of an element (nucleon) could be increased by means of M elements donors of negentropy.

- Liquid metals

A great number of liquid metals have a negentropy of about 2.5 eV [20,21]. An inserted subsystem can receive a few hundred eV when it interacts with 100 elements during a mean free path.

For instance, in liquid metals with implanted deuterons, the fusion reaction rate with deuterons accelerated through the metal increases because the term MTS^- increases the kinetic energy of the colliding nuclei [4]. This permits an easier passage through the Coulomb barrier by the tunnel effect. This effect is in addition to electron screening, stopping power and other microscopic effects.

- Nuclear reaction in plasmas

Another important application concerns the screening of nuclear reactions in plasmas slightly far from equilibrium. It is well known that the rate of nuclear fusions in a plasma grows with the plasma screening effect. In the weak screening regime with plasma parameter $\Gamma < 1$, Salpeter's approach works. A colliding pair of equal nuclei with relative kinetic energy receive an extra kinetic energy given by $(Ze)^2/\lambda_D$, where the Debye-Hückel length is $\lambda_D^2 = kT/n\,(Ze)^2$, and the Coulomb tunneling becomes easier.

In addition to this increment, we can introduce another one due to the negative entropy effect of plasma to the colliding pair. The expression of negentropy can be derived from internal and excess free energy given in [22,23]. Beyond the Debye-Hückel, other terms representing negative contributions to entropy permit the surrounding particles, while two nuclei are tunneling toward one another, to transfer energy making close approach easier against the repulsive Coulomb potential.

The same plasma negentropy effect can be introduced in measurements of correlations enhanced collision rates using cryogenic non-neutral plasmas [24].

- Fluctuations

As a final example that deserves further attention, let us now consider a part of the system itself instead of a subsystem of a different nature from the system or medium.

We divide the system of N elements in W cells of an elementary volume $\tau = h/\pi$, where h is the Planck constant and π is the phase space momentum volume. The cell A may receive in time Δt the energy $Z T \mathcal{S}^-$, being Z the number of elements interacting with the cell A ($Z \ll N$). In a successive time Δt, the cell A can lose the energy received because other cells of the system enter into an interaction with it. Therefore, the energy of each element fluctuates based on the quantity $T\mathcal{S}^-$. The time Δt is given by the uncertainty principle: $\Delta t \approx h/T\mathcal{S}^-$. For instance, a system at $kT = 10\,\text{eV}$, with $\mathcal{S}^- = 3.5 \times 10^{-4}\,\text{eV/K}$, we obtain $\Delta t = 10^{-16}\,\text{s}$, that is the time between two elastic collisions of the cell. In this time, any element of the cell fluctuates in energy of the quantity $T\mathcal{S}^- \approx 40\,\text{eV}$. We conclude that the energy fluctuation of a system in a given time is induced by the presence of negentropy, which occurs because of the existence of correlations among the elements of the system, as it can be expected (all details on the evaluation of this effect will be given elsewhere).

A list of systems with important negative contributions to entropy is reported in works by Yi-Fang Chang [9,10]. Chang reported the decrease of entropy in macroscopic thermodynamics quantities in many different isolated systems submitted to gravitational interactions, attractive electromagnetic interactions, and strong nuclear interaction (weak nuclear being excluded). From microscopic nuclear, atomic, molecular, and nano-physical theories, one can obtain the decrease of entropy in macroscopic thermodynamics.

8. Conclusions

Since the early works by Szilard [25], Schrödinger [26] and Brillouin [27], many researchers have learned how useful it is to apply the concept of negentropy in many complex problems in physics, biology and information science. Later, Sato [7,8] has generalized the definition of negentropy and his last proposal was linked to the Kullback-Leibler approach to statistical mechanics [11,12]. This definition allows the use of negentropy not only in isolated systems but also in a wider class of systems. Yi-Fang Chang [9,10] has recently remarked that negentropy enters into the description of many different physical situations.

Our contribution in this paper is that by separating the positive from the negative contribution of entropy, we may suggest that a subsystem inserted into a non chaotic system can benefit from a certain amount of energy transferred by a group of elements of the system, those that within a certain amount of time enter into an interaction with the subsystem itself.

In a few cases we have made approximate calculations of the energy that each element of the system can transfer. The number of elements in the subsystem is much smaller than those of the system and interact with a small part of elements before participating in processes, like nuclear fusion and Coulomb tunnelling that, without the transferred energy, would be absent. The system in this transfer of energy through the negentropy does not change its bulk properties. We have evaluated in a few cases the amount of energy that can be transferred per element. Further applications to turbulent motion and screening problems of thermonuclear reactions are in progress.

Author Contributions: All authors performed the theoretical calculations, discussed the results, prepared the manuscript and commented on the manuscript at all stages. All authors have read and approved the final manuscript.

Conflicts of Interest: The authors declare no conflict of interest.

References

1. Quarati, P.; Lissia, M.; Scarfone, A.M. Negentropy in the many-body quantum systems. *Entropy* **2016**, *18*, 63.
2. Ho, M.-W. What is (Schrödinger's) negentropy? *Mod. Trends Biothermokin* **1994**, *3*, 50–61.
3. Mahulikar, S.P.; Herwing, H. Exact thermodynamics principles for dynamic order existence and evolution in chaos. *Chaos Solitons Fractals* **2009**, *41*, 1939–1948.
4. Coraddu, M.; Lissia, M.; Quarati, P.; Scarfone, A.M. The role of correlation entropy in nuclear fusion in liquid lithium, indium and mercury. *J. Phys. G Nucl. Part. Phys.* **2014**, *41*, 125105–125112.
5. Quarati, P.; Scarfone, A.M. Modified Debye-Hučkel electron shielding and penetration factor. *APJ* **2007**, *666*, 1303–1310.
6. Dappen, W.; Mussack, K. Dynamic screning in solar and stellar nuclear reactions. *Contrib. Plasma Phys.* **2012**, *52*, 149–152.
7. Sato, M. Proposal of an extension of negentropy by Kulback-Leibler information (Definition and exergy). *Bull. JSME* **1985**, *28*, 2960–2967.
8. Sato, M. Proposal of an extension of negentropy by Kulback-Leibler information (Proportional relation between negentropy and work). *Bull. JSME* **1986**, *29*, 837–844.
9. Chang, Y.-F. Entropy decrease in isolated system and its quantitative calculations in thermodynamics of microstructure. *Int. J. Mod. Theor. Phys.* **2015**, *4*, 1–15.
10. Chang, Y.-F. Entropy, fluctuation magnified and internal interactions. *Entropy* **2005**, *7*, 190–198.
11. Kullback, S.; Leibler, R.A. On Information and sufficiency. *Ann. Math. Stat.* **1951**, *22*, 79.
12. Kullback, S. *Information Theory and Statistics*; John Wiley: New York, NY, USA, 1968.
13. Gibbs, J.W. A method of geometrical representation of the thermodynamic properties of substances by means of surfaces. *Trans. Conn. Acad. Arts Sci.* **1873**, *2*, 382–404.
14. Brillouin, L. The negentropy principle of information. *J. Appl. Phys.* **1953**, *24*, 1152–1163.
15. Obukhov, A. Structure of the temperature field in turbulent flows. *Izv. Akad. Nauk (Geogr. Geophys. Ser.)* **1949**, *13*, 58–69.
16. Clausius, R. *Die Mechanische Warmtheorie*; Vieweg: Braunschweig, Germany, 1865. (In German)
17. Helmholtz, H. *Wissenschaftliche Abhandlungen, I–III*; Teubner: Leipzig, Germany, 1882. (In German)
18. Ebeling, W.; Sokolov, I.M. *Statistical Thermodynamics and Stochastic Theory of Nonequilibrium Systems*; World Scientific: Singapore, 2005.
19. Walecka, J.D. *Introduction to Statistical Mechanics*; World Scientific: Singapore, 2011.
20. Wallace, D.C. *Statistical Physics of Crystals and Liquids*; World Scientific: Singapore, 2002.
21. Quarati, P.; Scarfone, A.M. Non-extensive thermostatistics approach to metal melting entropy. *Physica A* **2013**, *392*, 6512–6522.
22. Tanaka, S.; Mitake, S.; Yan, X.-Z.; Ichimaru, S. Theory of interparticle correlations in dense, high-temperature plasmas. III. Thermodynamic functions. *Phys. Rev. A* **1985**, *32*, 1779.
23. Ichimaru, S. Nuclear fusion in dense plasmas. *Rev. Mod. Phys.* **1993**, *65*, 255.
24. Anderegg, F.; Dubin, D.H.; Affolter, M.; Driscoll, C.F. Measurements of correlations enhanced collision rates in the mildly correlated regime ($\Gamma \sim 1$). *Phys. Plasmas* **2017**, *24*, 09218.
25. Szilard, L. Über die entropieverminderung in einem thermodynamischen system bei eingriffen intelligenter wesen. *Z. Phys.* **1929**, *53*, 840.
26. Schrödinger, E. *What Is Life?*; Cambridge University Press: Cambridge, UK, 1945.
27. Brillouin, L. *Science and Information Theory*; Academic Press: New York, NY, USA, 1962.

MDPI
St. Alban-Anlage 66
4052 Basel
Switzerland
Tel. +41 61 683 77 34
Fax +41 61 302 89 18
www.mdpi.com

Entropy Editorial Office
E-mail: entropy@mdpi.com
www.mdpi.com/journal/entropy

www.ingramcontent.com/pod-product-compliance
Lightning Source LLC
Chambersburg PA
CBHW051850210326
41597CB00033B/5847
9783038974697